OPTICAL IMAGING OF BRAIN FUNCTION AND METABOLISM

ADVANCES IN EXPERIMENTAL MEDICINE AND BIOLOGY

Recent Volumes in this Series

Volume 325
RECOVERY FROM BRAIN DAMAGE: Reflections and Directions
Edited by F. D. Rose and D. A. Johnson

Volume 326
THE USE OF RESEALED ERYTHROCYTES AS CARRIERS AND BIOREACTORS
Edited by Mauro Magnani and John R. DeLoach

Volume 327
GENETICALLY ENGINEERED VACCINES
Edited by Joseph E. Ciardi, Jerry R. McGhee, and Jerry M. Keith

Volume 328
ENZYMOLOGY AND MOLECULAR BIOLOGY OF CARBONYL METABOLISM 4
Edited by Henry Weiner, David W. Crabb, and T. Geoffrey Flynn

Volume 329
DENDRITIC CELLS IN FUNDAMENTAL AND CLINICAL IMMUNOLOGY
Edited by Eduard W. A. Kamperdijk, Paul Nieuwenhuis, and Elisabeth C. M. Hoefsmit

Volume 330
THE UNDERLYING MOLECULAR, CELLULAR, AND IMMUNOLOGICAL FACTORS IN CANCER AND AGING
Edited by Stringner Sue Yang and Huber R. Warner

Volume 331
FRONTIERS IN CEREBRAL VASCULAR BIOLOGY: Transport and Its Regulation
Edited by Lester R. Drewes and A. Lorris Betz

Volume 332
MECHANISM OF MYOFILAMENT SLIDING IN MUSCLE CONTRACTION
Edited by Haruo Sugi and Gerald H. Pollack

Volume 333
OPTICAL IMAGING OF BRAIN FUNCTION AND METABOLISM
Edited by Ulrich Dirnagl, Arno Villringer, and Karl M. Einhäupl

---

OPTICAL IMAGING OF BRAIN FUNCTION AND METABOLISM

Edited by

Ulrich Dirnagl

University of Munich
Munich, Germany and
Humboldt University
Berlin, Germany

Arno Villringer

University of Munich
Munich, Germany

and

Karl M. Einhäupl

Humboldt University
Berlin, Germany

PLENUM PRESS • NEW YORK AND LONDON

Library of Congress Cataloging-in-Publication Data

Optical imaging of brain function and metabolism / edited by Ulrich Dirnagl, Arno Villringer, and Karl M. Einhäupl.
p. cm. -- (Advances in experimental medicine and biology ; v. 333)
"Proceedings of a Symposium on Optical Imaging of Brain Function and Metabolism, held October 21-22, 1991, in Garmisch-Partenkirchen, Germany"--T.p. verso.
Includes bibliographical references and index.
ISBN 0-306-44528-X
1. Brain--Imaging--Congresses. 2. Infrared imaging--Congresses. 3. Brain--Blood-vessels--Imaging--Congresses. 4. Brain--Metabolism--Congresses. I. Dirnagl, Ulrich. II. Villringer, Arno. III. Einhäupl, Karl. IV. Symposium on Optical Imaging of Brain Function and Metabolism (1991 : Garmisch-Partenkirchen, Germany) V. Series.
[DNLM: 1. Brain--physiology--congresses. 2. Diagnostic Imaging--methods--congresses. W1 AD559 v.333 1993 / WL 300 O62 1991]
RC386.6.D52O64 1993
591.1'88--dc20
DNLM/DLC
for Library of Congress 93-24703
CIP

Proceedings of a symposium on Optical Imaging of Brain Function and Metabolism, held October 21–22, 1991, in Garmisch-Partenkirchen, Germany

ISBN 0-306-44528-X

A Division of Plenum Publishing Corporation
233 Spring Street, New York, N.Y. 10013

Printed in the United States of America

PREFACE

"....At last the doctor will be freed from the tedious interpretation of screens and photographs. Instead, he will examine and scan through his patient directly. Wearing optical-shutter spectacles and aiming a pulsed laser torch, he will be able to peer at the beating heart, study the movement of a joint or the flexing of a muscle, press on suspect areas to see how the organs beneath respond, check that pills have been correctly swallowed or that an implant is savely in place, and so on. A patient wearing white cotton or nylon clothes that scatter but hardly absorb light, may not even have to undress....".

David Jones, Nature (1990) 348:290

Optical imaging of the brain is a rapidly growing field of heterogenous techniques that has attracted considerable interest recently due to a number of theoretical advantages in comparison with other brain imaging modalities: it uses non-ionizing radiation, offers high spatial and temporal resolution, and supplies new types of metabolic and functional information. From a practical standpoint it is important that bedside examinations seem feasible and that the implementations will be considerably less expensive compared with competing techniques. In October 1991, a symposium was held at the Eibsee near Garmisch, Germany to bring together the leading scientists in this new field. It was the purpose of this meeting to discuss new results, correlate these to reference techniques, compare different approaches, and examine the impact and perspective of optical imaging for basic and clinical neuroscience. In addition to the presentations of the meeting several authors were invited to contribute to this volume in order to cover the full spectrum of optical brain imaging techniques.

U. Dirnagl
A. Villringer
K.M. Einhäupl

ACKNOWLEDGEMENTS

Sponsored by the

VOLKSWAGEN STIFTUNG

Additional support was provided by:

Bayer, Biorad, FIDIA, Hamamatsu, IBM, Kontron, Molecular Dynamics, Sanofi, Somanetics, Zeiss

CONTENTS

NEAR INFRARED SPECTROSCOPY OF THE BRAIN

IMAGING OF BRAIN FUNCTION USING INTRINSIC SIGNALS

THERMAL IMAGING OF THE BRAIN

OPTICAL MEASUREMENT OF ION CONCENTRATIONS IN BRAIN CELLS AND TISSUES

OPTICAL ACCESS TO THE BRAIN MICROCIRCULATION

EMERGING OPTICAL TECHNIQUES

NON-OPTICAL REFERENCE TECHNIQUES

NMR AND TIME-RESOLVED OPTICAL STUDIES OF BRAIN IMAGING

B. Chance

University of Pennsylvania
Department of Biochemistry and Biophysics
Johnson Foundation
Philadelphia, USA

My assignment in this symposium, as I see it, is to bridge the gap between non-invasive "whole body" technologies of NMR and time resolved spectroscopy, to give some perspective of what they do best, and where the future lies for the two technologies. The assignment has been made much more pleasant and less difficult because of the highly significant contributions of a number of my colleagues, particularly on NMR and time resolved spectroscopy, and my report here will emphasize those aspects that link the two technologies which might have been overlooked by my colleagues. Since the symposium focuses on oxygen in tissues, ^{31}P NMR is of greatest interest and its ability to detect tissue oxygen relies upon the general equation for oxidative phosphorylation (Chance, et al., 1985; Chance, et al., 1986):

$$3ADP + 3P_i + 1/2O_2 + NADH + H^+ \xrightarrow{E} 3\ ATP + NAD^+ + H_2O \quad (1)$$

in which oxygen is not just one of, but the key participant because its reduction to water gives the driving force for ATP synthesis. In brain, heart, and skeletal muscle (Chance, et al., 1985; Chance, et al., 1986), and more recently in transvected mice, by assuming Michaelis-Menten kinetics for oxidative phosphorylation (Chance and Williams, 1955) the effect of oxygen on the phosphocreatine (PCr)/inorganic phosphate (P_i) ratio is indicated by Eq. 2 Cr-kinase equilibrates a large pool of PCr with a smaller Creatine pool of ATP and permits the determination of ADP itself from the known equilibrium constant of the creatine kinase reaction. However, our interest is in detection of intracellular oxygen. The validity of Michaelis-Menten kinetics for the respiratory chain enables us to show the response of PCr/P_i value to oxygen through the following equation:

Optical Imaging of Brain Function and Metabolism, Edited by
U. Dirnagl *et al.*, Plenum Press, New York, 1993

$$\frac{V}{V_m} = \frac{1}{1 + \frac{PCr}{P_i K'} + \frac{K''_m}{KO_2}} \qquad (2)$$

The conditions where tissue oxygen falls to a level of its affinity constant (10^{-8} - 10^{-6} M depending upon metabolic activity) PCr/P_i will fall together with oxygen in a compensatory relationship. As oxygen falls to its K_m, so PCr/P_i will fall to compensate for the oxygen deficiency and maintain a homeostasis of V/V_m as required by Claude Bernard's (1878) ideas on tissue steady states. This relationship has proved most useful in distinguishing the response of mitochondrial ATP synthesis from the optical signals obtained from spectroscopy of hemoglobin. No change of PCr/P_i can be observed until tissue pO_2's causing large deoxygenations of hemoglobin are observed for tissues that are normally perfused and have normally large oxygen gradients between the capillary bed and the mitochondria. In pathological cases, however, this oxygen gradient can increase to the extent that it appears that hemoglobin deoxygenates parallel to the decrease of mitochondrial ATP synthesis.

Optical Spectroscopy

Optical spectroscopy measures concentration of the principle absorber; one absorber per wavelength. In the case of oxy- and deoxyhemoglobin two wave-lengths are required, and at least four if hemoglobin and cytochrome are to be measured. A general application of Beer's law seems appropriate to such systems with the important caveat that the *in vitro* extinction coefficients are transferable to the *in vivo* system only if the pathlength appropriate to the particular wavelength and absorbances are measured (Sevick and Chance, 1991). The pathlengths may be the average of many, particularly if continuous light is used. The integral over all pathlengths is probably the most appropriate parameter to be inserted into the Beers Law equation

$$- \text{Log } I = eCL.$$

Time-Resolved Spectroscopy

When time resolved spectroscopy is used, the pathlength is a function of time after the input light pulse and the absorbance may be measured at short, medium or long pathlengths depending upon the desired optical field, i.e., shallow for short times, and deep for long times. Frequency domain techniques can be used at low frequencies for long pathlengths, and high frequencies for short pathlengths. The phase shift or amplitude modulation parameters may be appropriately calculated. The diffusion equation by time and frequency domain measurements has been shown by Patterson and others (Patterson, et al., 1989; 1990) to be adequately applicable for optical pathlengths that involve a large number of scattering events. The great advantages of the applicability of this equation is that it indicates particular times in TRS or frequencies in PMS at which the absorption factor, μ_a, and the scattering factor, μ_s', or their ratio be separately detected. For example, in the latter part of the photon decay curves for μ_a or the earlier part of the photon kinetics for μ_s. With phase and amplitude modulation, dual wavelength techniques can be used to obtain ratios of μ_a's from which concentration can be calculated, or at appropriate frequencies μ_s'/μ_a. An extensive analysis of the optimum frequency where these determinations has been provided by Jacques (1991).

Continuous light tissue spectrophotometers involve errors in the calculation of concentration due to the unknown pathlength and changes thereof. Delpy has proposed the differential pathlength factor (DPF) to be applied generally to such spectrophotometers to be a multiplicative factor on the input/output separation of the optical fibers (Delpy et al. 1988). The question is, the constancy of these factors and their transferability from model systems to humans or from human subject to human subject. With respect to the values themselves, the DPF itself is a function of absorption and scattering of light along the particular set of optical path ways when using the continuous light system. As described above, it is clear that the frequency resolved methods select out a particular set of optical paths, while continuous light does not. It is recommended that the pathlengths be continuously monitored to ensure the validity and the applicability of Beers Law. DPF seems relatively constant for neonates but the population is necessarily limited (see Benaron to the contrary). In the case of adults, the brain tissue seems to be much more variable in its scattering and absorption parameters to make transferability and usefulness of DPF somewhat restricted. On the other hand, more recent data on the human mammary gland suggests a much tighter distribution of absorption values whilst, characteristically, the scattering varies.

We may conclude that continuous light spectroscopy is appropriate for relative changes in systems where the concentration of absorbers varies narrowly and the optical path itself is short. The pulse oximeter operates under these assumptions, namely, that the physiologically feasible range of arterial saturations is narrow and the optical pathlength through the finger, the ear, the nose, etc. is relatively short. However the variation of the scattering factor is ill determined. The application of these algorithms to the measurement of tissue oxygen levels, for example, the use of pulse oximetry to measure the brain tissue saturation in neonates is particularly risky because of the fallibility of clinical decisions which may be based upon these measurements.

The identification of critical oxygen saturations for the human brain are necessarily fraught with uncertainty when using continuous light measurements. Time and frequency resolved technology is strongly recommended to be used continuously to monitor pathlength (a review of the derivation of the DPF equation has been presented by Sevick and Chance (1991) and it is shown to depend upon μ_a.

Saturation of hemoglobin in the vascular bed of a tissue can be determined simply and directly by time and frequency modulation systems at two wave-lengths (Sevick, et al., 1991). Under these conditions, the scattering factor cancels and for TRS the simple equation is used where the ratio of the rates of decay of photons, $\mu_a l_1/\mu_a l_2$ are taken at longer times where maximal penetration into the tissue is obtained and the effects of superficial absorber/ scatterer such as skin, bone, arachnoid space, etc. are minimal. A similar expression has been derived for phase modulation. and TRS and PMS give appropriate calibration curves with respect to both absorber and scatterer, verifying the applicability of the diffusion equation to photon kinetics under the particular conditions. Optimal conditions of time and frequency for such determinations have been suggested by Sevick and Chance (1991) and Jacques (1991). It appears that a robust and sensitive method for determination of hemoglobin concentration in tissue is available. The extension of the method to precise determination of cytochrome concentrations has been suggested by Cope and colleagues (Delpy, et al. (1988), where time resolved data has been used as DPF. Not withstanding the caveats mentioned above, significantly improved determinations of cytochrom a/a_3 have been obtained.

Imaging

Imaging of hidden absorbers in highly scattering tissue is a topic of research in many laboratories over the world. Generally, the approach has been to employ ballistic photons in relatively small and thin objects using various opto-electronic technologies for accepting only photons which have traveled the shortest path. The resolution would naturally approach that available for any unscattered optical system and indeed resolutions of hundreds of microns are readily possible depending upon the size of the optical field and the number of elements in the detector.

Early time gating of the photon kinetics may identify those few photons which have undergone a minimum number of scattering events so that their pathlength approximates the input/output separation, particularly in transmission geometries. Here the resolution of an absorbing object can be high and indeed approach that obtainable in transparent media. The advantages of the coherent light spectroscopy for image formation is obvious, and dual wave-length detection of concentrations follows closely that of continuous light spectroscopy except in that absorption and scattering are not resolved. The terms coined to indicate that the pathlength in "coherent light imaging" is signifi-cantly longer than the input/output separation is the use of the "snake imaging" by Alfano and colleagues (Yoo and Alfano, 1990). Even the early arrival photons may have undergone a number of scattering events in a transit between input and output. The more appropriate term might be minimally scattered; it implies that changes of absorption and scattering upon which imaging depends must in themselves shorten the pathlength for increased absorption, or the pathlength for increased scattering. No doubt, however, that a combination of these methods will be employed to produce an image of a hidden absorber/scatterer of optimal resolution but only in thin tissues.

The overlap of the two approaches to imaging becomes obvious when it is considered that the most sharply defined optical field for time resolved spectroscopy occurs in the early part of the photon kinetics; the earliest part can be called the "coherent imaging region" and the latter part the "diffusive wave region".

The resolution with TRS and PMS appears to be remarkably good and considering that the optical field may be rather large compared to the size of the object, typically a 1 cm^2 object can be detected at 2-3 cm within a highly scattering medium with a FWHM of the edges of 4 mm. Numerous studies in such systems using gated ICCD systems are underway. However the absorbers have been black and scatterers have not yet been studied extensively. Major research efforts are underway in some laboratories on the quantitation of the differential absorption/ scattering gradients that can be detectable under these conditions, a topic which we have coined "diffusometry". Diffusometry is a term for the quantitation of photon diffusion in the heterogeneous scattering materials, analogous to the topic of relaxometry in nuclear magnetic resonance, the study of the factors that cause differential relaxation between a host material and a hidden object. Preliminary results in Lakowicz's laboratory indicate that the high signal to noise ratio of the phase and amplitude modulation systems may lend them better to the detection of incremental changes of absorption and scattering of objects hidden in a background scatterer, as in the case of time resolved spectroscopy. However, detailed studies in progress are as yet incomplete. They may of course be of highly significant clinical importance in the detection of stroke volumes in the brain and tumor volumes in the mammary.

Motional properties

It is implicit in the algorithm by which a fixed object can be detected within a highly scattering medium by scanning the periphery of the medium with fixed input/output positions, motion of the absorber in the scattering medium would be detected. This aspect has been highly developed by Yodh, et al. (1990) and others in the detection of motional properties of molecules in highly scattering colloids. Such modulations of the photon migration pattern have been detected both by continuous and pulsed, and phase modulated light. In fact, in this case, the continuous light system has only minor disadvantages of sensitivity and major advantages of signal-to-noise ratio because of the high light fluxes which are available from continuous light sources.

Cognition

We have found that the frontal projections of a mentation process from the visual cortex to the temporal region would exhibit a "motional signal" because of the intensity/time variation of the position and intensity of the blood concentration changes in localized tissue volumes that are momentarily active. To test this hypothesis, a series of human subjects have been studied to which a visual stimulus in the form of analogies have been presented serially over an interval of ten minutes. The presentation over the ten minute interval has been followed by an equivalent interval of "rest" under which no presentation of material is made to the subject. The blood concentration was made by a simple dual wavelength continuous light spectrophotometer detecting hemoglobin saturations changes by the difference of absorbance between 760 and 850 nm and for blood concentration changes by the sum of the signals at the two wavelengths. The former mode was employed, no significant changes of hemoglobin oxygenation were observed, or indeed, were expected. Similar stimuli have been shown by PET scanning to evoke blood concentration changes in the left temporal cortex.

In order to determine whether or not there was a "motional" or time varying portion of the photon migration signal, the output of the spectrophotometer was connected to two channels of a fast Fourier transform oscilloscope (Nicolet). Figure 1 displays the individuals result for presentation of the abstractions and the subsequent control interval in the frequency range 0-5 Hz and the voltage range of 10's of microvolts. The rest interval has sharp peaks and the stimulated interval has additional peaks. In order to eliminate the rest peaks and visualize only the additional peaks caused by mentation, the two Fouriers are subtracted and displayed. Additional frequency components in the 0.5-2.5 Hz region are observed and are verified by 28 studies of 9 subjects over a several weeks interval. The conclusions are: No two subjects have identical frequencies; insufficient data were obtained to determine whether the same subject over time will exhibit the same frequencies. As a part of these tests, several intervals have been taken where no stimulation was presented for the two ten minute intervals. In addition to these tests, additional controls where different stimulatory algorithms have been used, i.e., noun/verb, or nonsense words, as suggested by Raichle and others and the success of presentation of the abstractions shown here gives by far most distinct signals.

These studies find parallels with the studies by PET, MEG, and EEG correspond and the optical studies presented in this meeting. Fourier transform analysis is readily applied to EEG and to MEG (S. Williams, person. commun.), the optical method on the exposed monkey and cat cortex does not seem yet to have been analyzed in this fashion. Grinvald indicates preliminary data which show higher frequencies with considerably broader frequency distributions in the cat cortex (personal communication).

Two explanations are afforded for the nature of the measurement here: 1) that a single site has a characteristic repetition frequency of blood concentration changes

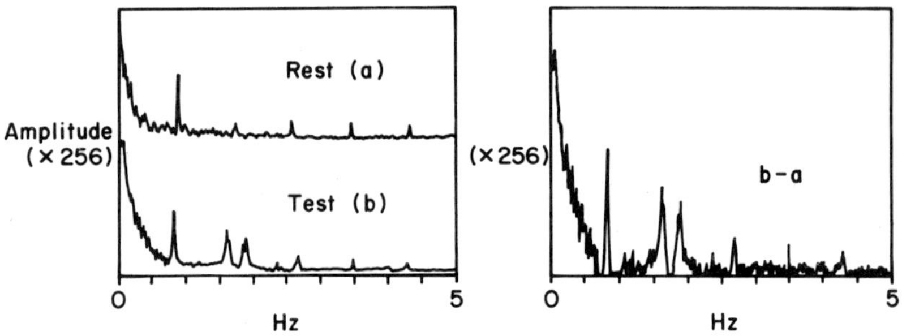

FIGURE 1. Cognitive test as described in text. Fourier transform of blood concentration signals on human head. a (no stimulation); b (visual stimulation by analogies; (b-a) difference. [Subject U, 07/31/91, 81 BPM, n=16]

during mentation, and none or very little, in the absence of it. 2) The other explanation is that several sites are involved and that the repetition frequency of activation of these sites is constant. These two explanations could be resolved by a high degree of localization of the response that might distinguish single and multiple sites of activation. The different optical fields might include more or less sites of activation. Resolution of a single vs multiple sites explanations may be difficult. This simple and economical spectrophotometric method for brain mentation may have significant applications in evaluating mentation capability in an objective way in those whose may have been afflicted by neuronal degeneration.

SUMMARY

The classical optical method of the biochemist used in so many applications to determine stoichiometry kinetics, quantitative analysis of biochemical systems which was subsequently applied to tissue studies by G.A. Millikan and others seems now to have a new lease on life provided by time and frequency resolved studies which appear to afford the essential quantification lacking in the earlier methods. The fact that localization seems possible as well may bring the optical technology into the arena of new non-invasive technology capable of high sensitivity, high resolution and rapid data acquisition.

ACKNOWLEDGMENTS. Supported in part by National Institutes of Health Grants NS RR 03365; and NS 27346

REFERENCES

Baldeck, P.L., Raccah, F. and Alfano, R.R.(1987) Observation of Self-focusing in Optical Fibers with Picosecond Pulses. Optics Lett. 12:588-589

Bernard, C., Les Phenomenes de la Vie, Vols 1,2, Paris, J.B. Balliere et fils, 1878.

Chance, B. and Williams, G. Respiratory Enzymes in Oxidative Phosphorylation. I. Kinetics of Oxygen Utilization. J. Biol. Chem. 217:383-393. 1955.

Chance, B., Leigh, Jr. J.S., Clark, B.J., Maris, J., Kent, J. and Smith, D. (1985) Control of Oxidative Metabolism and Oxygen Delivery in Human Skeletal Muscle: A Steady-state Analysis of the Work/Energy Cost Transfer Function. Proc. Natl. Acad. Sci. USA 82:8384-8388.

Chance, B., Leigh, Jr. J.S., Kent, J., McCully, Nioka, S., Clark, B.J., Maris, J.M. and Graham, T. (1986) Multiple Controls of Oxidative Metabolism of Living Tissues as Studied by ^{31}P MRS. Proc. Natl. Acad. Sci. USA. 83:9458-9462.

Delpy, D.T., Cope, M., van der Zee, P., Arridge, S., Wray, S., and Wyatt, J. (1988) Estimation of Optical Pathlength through Tissue from Direct Time of Flight Measurement. Phys. Med. Biol. 33:1433-1442.

Jacques., S.L. (1991) Principles of Phase-Resolved Optical Measurements. in SPIE Conference: Future Trends in Biomedical Applications of Lasers 1525:143-153

Patterson, M.S., Chance, B. and Wilson, B.C. (1989) Time Resolved Reflectance and Transmittance for the Noninvasive Measurement of Tissue Optical Properties. J. Appl. Optics 28:2331-2336.

Patterson, M.S., Moulton, J.D., Wilson, B.C. and Chance, B. (1990) Applications of Time-Resolved Light Scattering Measurements to Photodynamic Therapy Dosimetry. In Photodynamic Therapy: Mechanisms II Proc. Soc. Photo Optical Instrum. Engr. 1203:62-75.

Sevick, E.M. and Chance, B. (1991) Quantitation of Time-and Frequency-Resolved Optical Spectra for the Determination of Tissue Oxygenation. Anal. Biochem. 195: 330-351.

Sevick, E., Weng, J., Maris, M. and Chance,. B. (1991) Analysis of Absorption, Acattering, and Hemoglobin Saturation using Phase-modulation Spectroscopy In: Proceedings of Time-Resolved Spectroscopy and Imaging of Tissues, SPIE (B. Chance, ed) SPIE, Bellingham, WA, 1431:274-275.

Wray, S., Cope, M., Delpy, D.T., Wyatt, J.S. and Reynolds, E.O.R. (1988) Characterization of the Near Infrared Absorption Spectra of Cytochrome aa3 and Haemoglobin for the Non-invasive Monitoring of Cerebral Oxygenation Biochim. Biophys. Acta 933:184-192.

Yodh, A.G., Kaplan, P.D. and Pine, D.J. (1990) Pulsed Diffusing-Wave Spectroscopy: Pathlength Specific Observation of Speckle Fluctuation Spectra from Dense Colloids. in Ultrafast Phenomena VII (Harris, C.B., Ippen, E.P., Mourou, G.A. and Zewail, A.H., eds.) Springer-Verlag, Berlin. Vol. 53, pp. 169-171.

Yoo, K.M. and Alfano, R.R. (1990) Time-Resulved Coherent and Incoherent Components of Forward Light Scattering in Random Media. Optics Lett. 15:320-322.

WAVELENGTH DEPENDENCE OF THE DIFFERENTIAL PATHLENGTH FACTOR AND THE LOG SLOPE IN TIME-RESOLVED TISSUE SPECTROSCOPY

M. Essenpreis*, M. Cope*, C.E. Elwell*, S.R. Arridge+,
P. van der Zee*, and D.T. Delpy*

*Department of Medical Physics and Bioengineering
+Department of Computer Science
University College London
London, UK

INTRODUCTION

The monitoring of tissue oxygenation by the technique of near infrared spectroscopy (NIRS) was first described by Jöbsis in 1977 (Jöbsis, 1977). The technique relies upon the relative transparency of tissue to near infrared (NIR) light to enable measurements of changes in optical attenuation across many centimetres of tissue. Early NIRS measurements could only derive qualitative changes in tissue and blood oxygenation from the observed variations in tissue attenuation (Brazy et al., 1985). However, data on the optical pathlength of light in tissue, measured by time resolved techniques employing picosecond laser pulses, have now permitted a quantitative analysis of attenuation measurements to be made (Delpy et al., 1988; Wyatt et al., 1990a). By incorporating information on the optical pathlength into a modified Beer-Lambert law it is possible to quantify *changes* in chromophore concentration from the measured *changes* in tissue attenuation. The optical pathlength needed in this calculation, the Differential Pathlength (DP) is defined as the local gradient in a plot of the attenuation measured in a scattering medium versus the absorption coefficient of the medium (Cope et al., 1991a; Cope, 1991b). It has been shown in previous studies (Delpy et al., 1988) that the DP can be approximated by measuring the mean distance that a picosecond light pulse travels across the tissue. Furthermore, a dimensionless multiplying factor, the Differential Pathlength Factor (DPF), can be obtained when the DP is divided by the geometric distance between light source and detector on the tissue surface. This factor has been shown, both theoretically and experimentally, to be approximately constant for any tissue once the optode spacing is larger than about 25 mm (van der Zee et al., 1990; van der Zee et al., in press), enabling clinical NIRS measurements to be made with varying optode geometries. Two instruments are now commercially available which make use of this property of the DPF to calculate variations in the concentrations of haemoglobin and cytochrome aa_3 from spectral changes measured at the bedside (NIRO 1000 and NIRO 500, Hamamatsu Photonics KK, Japan). Using these instruments, methods for absolute quantitation of cerebral blood flow and blood volume have been developed, and measurement of these parameters made in newborn infants (Edwards et al., 1988; Wyatt et al., 1990b) and adult humans (Elwell et al., in press).

It is known from theoretical and experimental studies that both the absorption and the scattering coefficient effect the Temporal Point Spread Function (TPSF) of an object, the latter being defined as the temporal response of the light intensity exiting the object following an input impulse function (Delpy et al., 1988). Time resolved measurements on tissue at a single wavelength have demonstrated these effects, the DPF for individual tissues being different (van der Zee et al., in press), and varying slightly with changes in attenuation (Delpy et al., 1989). It is the purpose of this paper to investigate the wavelength dependence of the TPSF of various tissues since the addition of this information into the modified Beer Lambert law would further improve the accuracy of the calculated chromophore concentrations. The TPSF is a complex function of both the fundamental optical properties of the tissue and of the measurement geometry, but in this study, the variation with wavelength of two parameters characterizing the TPSF are analyzed: The Differential Pathlength Factor and the logarithmic slope of the intensity decay with time (K_{ls}). Experimental measurements of these parameters have been made over a wavelength range between 740 nm and 840 nm, on the normoxic adult head, calf and forearm, and across the head of a rat at varying levels of blood oxygenation and blood content.

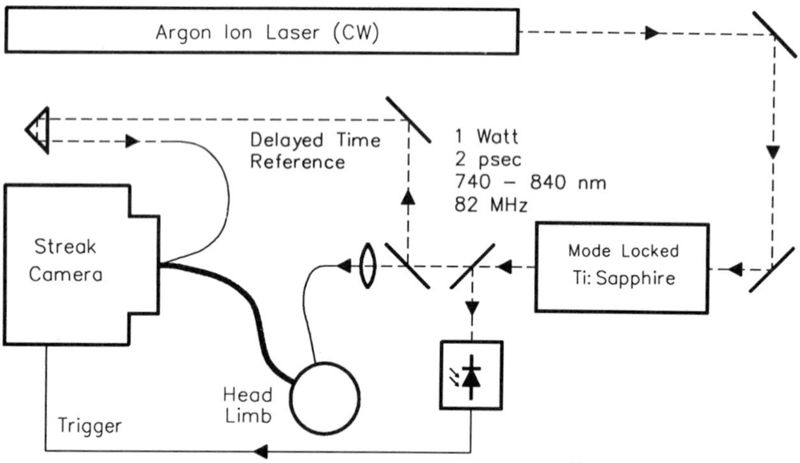

FIGURE 1. Experimental system for the measurement of the wavelength dependence of the TPSF in tissue.

EXPERIMENTAL TECHNIQUE

The experimental system used to measure the TPSF is shown in Figure 1. The light source was a mode-locked Ti:Sapphire laser (Tsunami, Spectra Physics Ltd) pumped by a 12 W Argon ion laser (Model 2040, Spectra Physics Ltd). The mirror set of the Ti:Sapphire laser allowed for tunability between 740 nm and 840 nm. The pulse width of a single pulse from the laser was about 2 ps (full width half maximum) and the repetition rate was 82 MHz.

To allow flexible application of the light to the tissue, the laser output was coupled into a single low dispersion fibre (Corning SDF, 125 μm diameter, 0.6 m length). Part of the laser output was used both as a time reference and as a trigger for the detector, a synchroscan streak camera (C1587, Hamamatsu Photonics Ltd). The streak camera was operated at its slowest sweep speed in order to sample a large part of the TPSF in one streak image. In this mode, the temporal calibration of the camera was τ=4.95 ps per channel and the streak width 2.53 ns. The overall temporal resolution of the system was about 10 ps. A one metre long fibre bundle with a circular cross section (1.9 mm diameter) was used to collect light emerging from the tissue. At the camera end, the fibres were aligned in a linear array across the streak camera input slit. The individual fibres in the bundle (Corning SDF) were all the same length (± 1 mm) to minimise transit time distribution. Incorporated into the row of fibres was a single additional fibre into which the time reference light pulse was coupled. Since the light pulse passing through this reference fibre was acquired on the same streak image together with the light from the sample, it could be used to accurately time the photons emerging from the tissue and to correct for the colour delay of the streak camera.

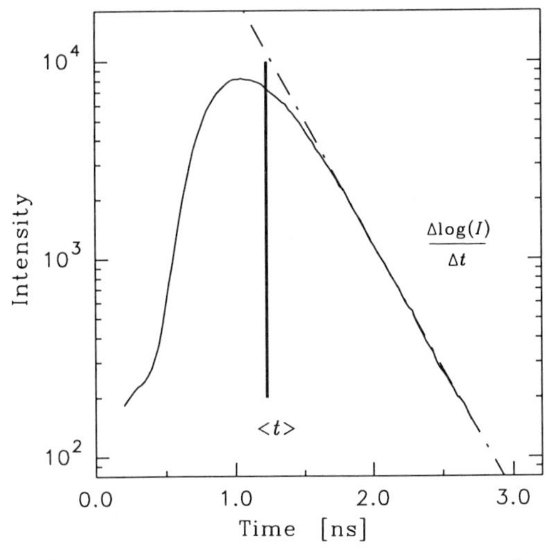

FIGURE 2. A typical Temporal Point Spread Function showing the mean time <t> and the logarithmic slope of the intensity decay (measured on the head of an adult).

Characterization of Temporal Point Spread Functions

Light is multiply scattered in tissue, so the photons injected in a picosecond laser pulse take different paths of varying length in travelling from the source to the detector fibre. Thus, photons which travel over a different total distance will arrive at different times at the detecting fibre bundle. This gives rise to an intensity distribution with time which is spread out broadly and shifted in time with respect to the picosecond input pulse. An example of such an intensity distribution is shown in Figure 2. The data shown was collected at 760 nm on the head of an adult using an optode spacing of

40 mm. Indicated on this figure are the mean time <t> of the intensity distribution and $\log(I)/t$, the gradient of the linear part of the intensity decay following the initial intensity maximum. <t> was calculated by integrating over the TPSF:

$$<t> = \frac{\int_0^\infty t\, I(t)\, dt}{\int_0^\infty I(t)\, dt} \tag{1}$$

where $I(t)$ is the intensity of the light recorded by the streak camera as a function of time t. K_{ls} was calculated using linear regression within a window defined by percentages of the intensity maximum. Regression started at 10 % and ended at 2 % of the maximum intensity for each TPSF. K_{ls} (in units of length^{-1}) can then be expressed as:

$$K_{ls} = -\frac{\Delta \log(I(t))}{\Delta l} = -\frac{n_t}{c} \cdot \frac{\Delta \log(I(t))}{\Delta t} \tag{2}$$

with $\Delta l = \Delta t \times c/n_t$, c being the speed of light in vacuum and n_t the refractive index of the tissue. For all the data described here, a value for n_t of 1.40 was assumed (Bolin et al., 1989). The DPF is calculated from the measured TPSF by multiplying the mean time with the speed of light in tissue and dividing by the physical length (chord length) d between source and detector fibres:

$$DPF = \frac{DP}{d} \approx \frac{1}{d} \cdot \frac{c}{n_t} <t> \tag{3}$$

Experimental Procedure for Studies on Adult Humans

As in studies reported previously (van der Zee et al., in press), the TPSF has been measured on the head, forearm and calf of a group of male and female adult volunteers. The age of the subjects was between 17 and 55 years (median 26 years), and they had no known muscular or circulatory disorders. Measurements were made on the limbs of 6 subjects and on the head of 7 subjects. A wavelength scan consisted of thirteen separate measurements between 740 nm and 840 nm at intervals of approximately 5 nm between 750 nm and 770 nm, and 10 to 15 nm below 750 nm and above 770 nm. More frequent sampling was applied around 760 nm to accurately measure the more rapid spectral change of the TPSF predicted to occur around the 760 nm absorption peak of deoxy-haemoglobin (Wray et al., 1988). A full set of measurements on one subject was typically completed within 30 minutes.

For measurements on the leg or arm, the subjects were seated comfortably and the limb was placed in a U-shaped rest into which holes were drilled which enabled both the transmitting and receiving fibres to be placed in contact with the skin from below. The fibres were either positioned over the muscles on the medial aspect of the forearm away from any palpable bone, or on the rear surface of the calf muscle at the broadest part of the leg. The optode spacing was approximately 40 mm for all the studies. Care was taken that the blood supply to the tissues was not obstructed whilst the limb lay in the apparatus by supporting the wrist or heel on an adjustable rest. Interference from background light was reduced by wrapping the limb with a black cloth.

The procedure for measurements made on the head was similar to that reported in previous studies (van der Zee et al., in press). The receiving fibre bundle was positioned over the temple, with the transmitting fibre on the upper forehead approximately 40 mm away. Figure 3 shows the TPSF at several wavelengths measured on the head of one subject. The change in both the position and the shape of the TPSF with wavelength can be clearly seen.

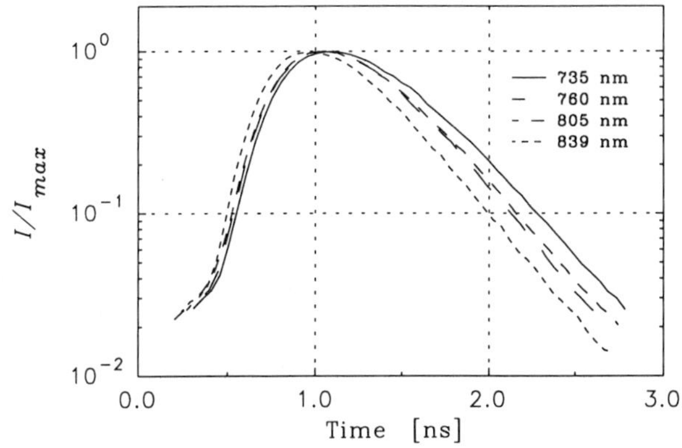

FIGURE 3. TPSFs measured at selected wavelengths (735 nm, 760 nm, 805 nm and 839 nm) on the head of one subject.

Experimental Procedure and Preparation for Study on Adult Rat Brain

TPSF measurements were also made on the brain of an adult Wistar rat. Following anaesthesia (Urethane 36 % w/v intraperitoneal, 5 ml kg^{-1}), the temporoparietal muscles were reflected, the skull exposed and cleared of residual surface tissue. A tracheal cannula was inserted to allow easy ventilation and manipulation of inspired gas concentrations, and a femoral artery was cannulated for blood sampling. The head of the animal was immobilised in a stereotactic frame. The transmitting fibre and the receiving fibre bundle were fixed in diametrically opposed positions across the head, the ends of the fibres being in contact with the skull via an optical coupling gel (Math Associates Inc, Westbury, NY, USA). Measurements of the TPSF were made with the animal breathing 47 % and 30 % oxygen (balance N_2). Finally, with the animal breathing 100% oxygen, an exchange transfusion with saline was performed until the animal died (residual haematocrit < 5 %). A TPSF scan was then performed immediately post mortem. The rat head diameter (and hence optode spacing) was measured at the end of the study.

RESULTS

The DPF and K_{ls} were calculated for each set of TPSFs and then linearly interpolated to 5 nm intervals. For each tissue studied, the values for the DPF and K_{ls} at each wavelength were then averaged over all the subjects, and the corresponding standard deviations (SD) were calculated. Figure 4 summarises these results, showing the DPF and K_{ls} as a function of wavelength for the head (Figure 4(a)), the forearm (Figure 4(b)), the male calf (Figure 4(c)) and the female calf (Figure 4(d)).

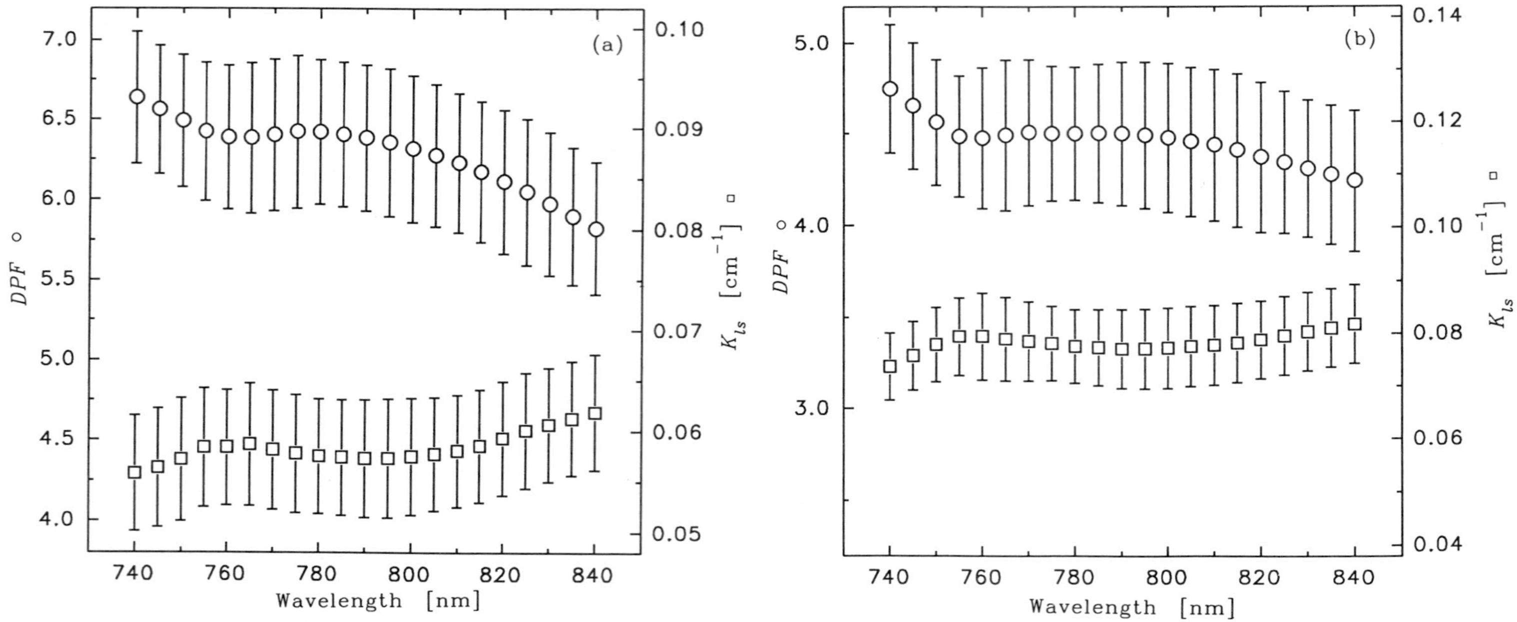

FIGURE 4 a/b: Changes with wavelength of the DPF and K_{ls} in the adult head (a) and the adult forearm (b). Data are the average for all subjects ± one standard deviation.

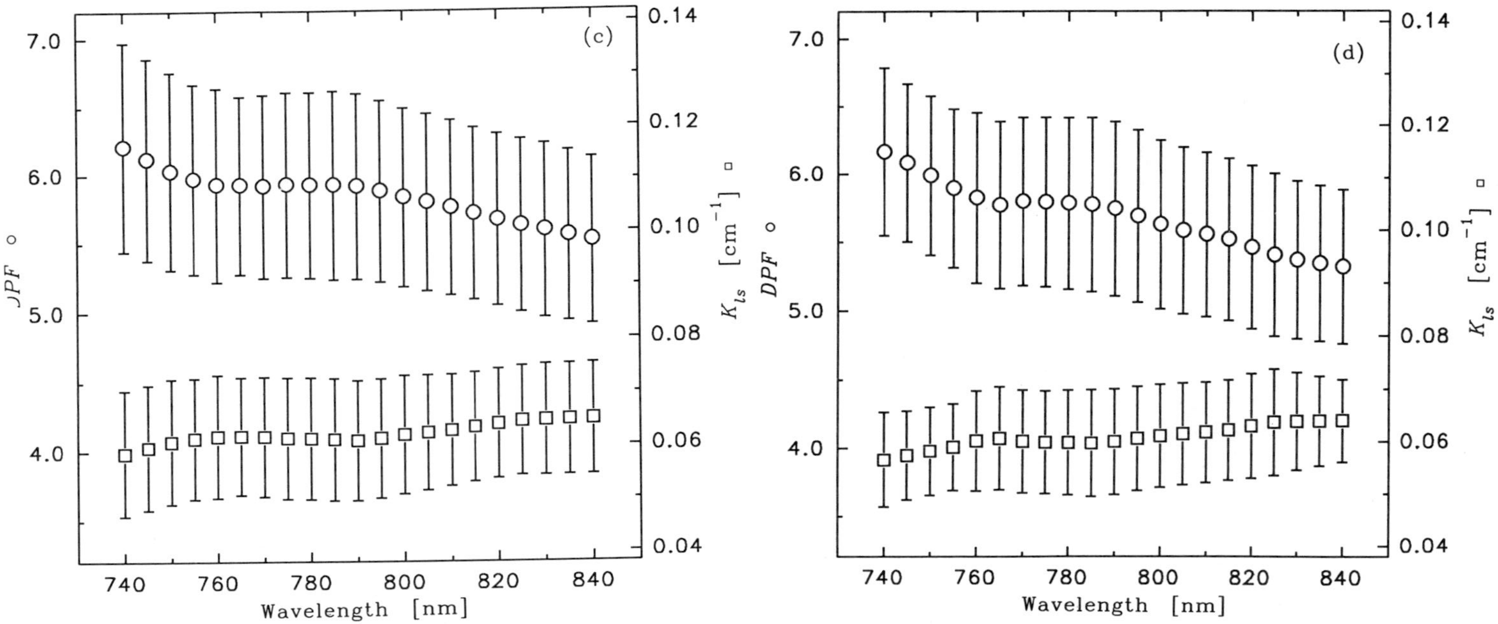

FIGURE 4 c/d: Changes with wavelength of the DPF and K_{ls} in the adult male calf (c) and the adult female calf (d). Data are the average for all subjects ± one standard deviation.

The data for the DPF measured on the rat brain are shown in Figure 5(a) and the corresponding data for K_{ls} in Figure 5(b). For the sake of clarity, error bars for the inherent SD of the measurement of the DPF have been omitted. In a previous study (van der Zee et al., in press) these have been shown by repeated measurement at one site to be less than 5%.

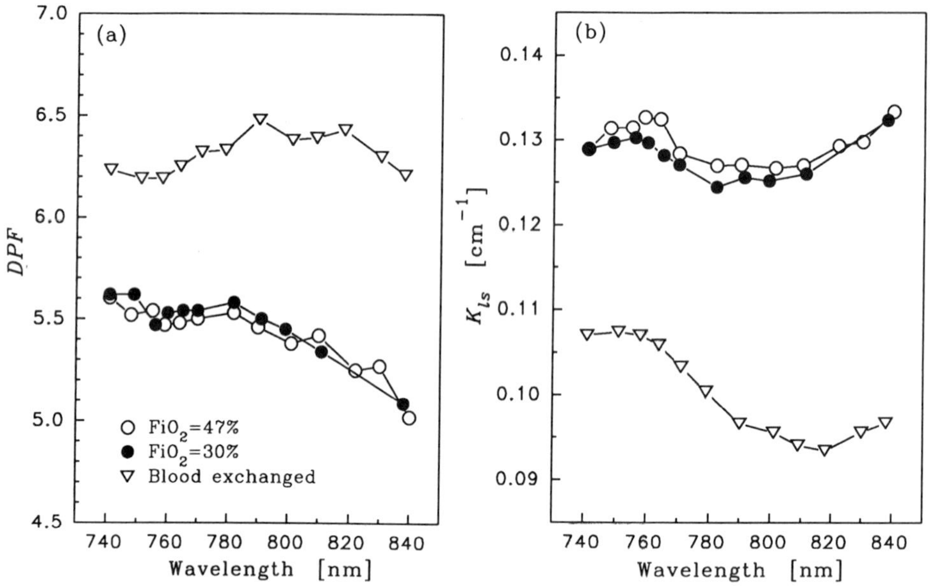

FIGURE 5. Changes with wavelength of the DPF (a) and K_{ls} (b) in the brain of an adult rat at two levels of FiO_2, and following saline exchange transfusion.

DISCUSSION

A number of features common to the DPF and K_{ls} spectra of all the tissues studied can be observed.

Firstly, a local minimum in the DPF can be seen around 760 nm. This coincides with a known peak in the absorption spectrum of deoxyhaemoglobin (Hb). Since all of the studies on humans were made under normoxic conditions, the amount of Hb would be expected to be small, and this is reflected in the magnitude of the observed peak. The general reduction of the DPF due to the increased tissue absorption is in agreement with experimental and theoretical predictions (Delpy et al., 1989; Cope et al., 1991a). The general fall in the DPF beyond 800 nm can be partly explained by the same argument, since the absorption of oxyhaemoglobin (HbO_2) increases beyond 800 nm. However, the DPF spectrum also shows a decline over the whole of the 740 - 840 nm range. This is a consequence of the effects of tissue scattering. Light scattering by tissue is very anisotropic in its angular distribution (van der Zee and Delpy, 1988), a feature which is a characteristic of large particle or Mie scattering. The wavelength dependence of Mie scattering results in a fall in both the reduced scattering coefficient (van der Zee et al., 1991) and the scattering efficiency (Bohren and Huffman, 1983), and both of these would result in a reduction of DPF with wavelength (Cope et al., 1991a). The data for DPF in each of the human tissues are consistent with these explanations, the DPF measured in the arm and calf (where myoglobin is present as an additional absorber of light) being smaller than those for the head.

Secondly, the K_{ls} spectra are approximately mirror images of their corresponding DPF spectrum. The value of K_{ls} has previously been interpreted as representing the absorption coefficient of the tissues (Chance et al., 1988a; Chance et al., 1988b), although this is in fact only true in extremely limited circumstances (Jacques and Flock, 1991; Patterson et al., 1990). In these results, the K_{ls} values for calf and forearm are higher than those for the head, although the differences are smaller than expected. In addition, the absolute magnitudes are not in agreement with published data for the absorption coefficient of muscle (Cheong et al., 1990) and brain (Sterenborg et al., 1990). The reasons for this discrepancy are still the subject of study, although it is thought most likely to be due to the effects of measurement geometry and finite boundaries (Madsen et al., 1991).

Thirdly, both the DPF and K_{ls} data in Figure 4 exhibit a large standard deviation. This is not due to error in the measurement technique, which has been shown previously to be smaller than 5%, but instead represents a true inter subject variation. The absolute magnitudes of the mean DPF obtained in this study are also different from those reported previously, but measured on a different group of volunteers (van der Zee et al., in press). The most striking difference between these two studies are in the data from the adult calf. In the previous study, a significant difference in DPF was observed between male and female subjects, and this was attributed to male female differences in fat/muscle ratio. In this study, the difference although present, is not significant. One can speculate that since the major difference appears to be in the DPF of the male calf (at a wavelength of 760 nm the DPF is 5.93 in this study compared to 3.98 in the previous study), the subjects employed in this present study were significantly less fit! The overall inter subject variability probably arises from variation both in the anatomy of the tissues (Maughan et al., 1984), and in its physiological status, especially the tissue blood volume and oxygenation (Sakai et al., 1985). These variations set a limit to the accuracy with which spectroscopic data can presently be quantitated. Recent developments of techniques for measuring optical pathlength at the bedside should overcome this limitation (Lakowicz and Berndt, 1990; Sevick and Chance, 1991).

The general features of the DPF and K_{ls} spectra measured on the rat brain (Figure 5) are similar to those of the human head. A small, but still observable local minimum in DPF occurs around 760 nm, and DPF decreases with wavelength beyond 800 nm. Again, the K_{ls} spectrum appears to be approximately a mirror image of the DPF spectrum. There was no significant change in both spectra as the FiO_2 was lowered from 47% to 30%, but this is only to be expected, since PaO_2 only dropped from 12.4 to 12.1 kPa, a value which would still result in an arterial haemoglobin saturation of 97 %. A change in the tissue absorption coefficient is therefore not expected and the very similar DPF and K_{ls} spectra (Figure 5) reflect this. Note however that the magnitude of K_{ls} is far higher than that measured on the human brain. This marked overestimate is a consequence of the effects of the much closer tissue boundary in the small rat head. The effects of the exchange transfusion are dramatic, the DPF increasing by approximately 16% and K_{ls} falling by 21%. This change, due to an overall decrease in tissue attenuation is similar to that reported previously (Delpy et al., 1989), and again illustrates a possible source of error in the calculation of NIRS data.

CONCLUSIONS

We have shown experimentally the spectral dependence of the Differential Pathlength Factor and the log slope in the NIR. In general, the experimental results are in agreement with theoretical predictions. The DPF increases with wavelength and shows a local minimum at 760 nm, coinciding with the absorption peak of Hb. K_{ls} was found to show inverse behaviour to the DPF spectrum, increasing with wavelength and revealing a local maximum at 760 nm. However, the absolute magnitude of K_{ls} was not equal to the absorption coefficient of the tissues known from previous studies.

The importance of the DPF in quantitating NIRS data has previously been established. The inter subject variation in the absolute magnitude of the DPF shown in this and in previous studies sets a limit on the accuracy with which chromophore concentrations can be calculated. This variation in the DPF is typically about $\pm$ 10%, and its source of error could only be eliminated if the DPF were measured on each patient. In addition, the spectral variation in DPF shown in this study demonstrates that further improvements in the accuracy of chromophore estimation can be obtained if a wavelength dependent DPF is used in the modified Beer Lambert law. The most accurate estimation of tissue chromophore concentration will only come from NIRS systems that not only measure changes in attenuation at several wavelengths, but also continuously monitor optical pathlength at each wavelength.

ACKNOWLEDGEMENTS -The work described in this paper was supported by grants from the Wellcome Trust, SERC, MRC, the Wolfson Foundation and Hamamatsu Photonics KK. We also thank Spectra Physics UK for the loan of the Tsunami laser system. M. Essenpreis is grateful for the financial support of the Gottlieb Daimler- und Karl Benz-Stiftung.

REFERENCES

Bohren C.F. and Huffman D.R. (1983) Absorption and Scattering of Light by Small Particles. J. Wiley (New York).

Bolin F.P., Preuss L.E., Taylor R.C., Ference R. (1989) Refractive index of some mammalian tissues using a fibre optic cladding method. Applied Optics, 28:2297-2302.

Brazy J.E., Lewis D.V., Mitnick M.H., Jöbsis F.F. (1985) Noninvasive monitoring of cerebral oxygenation in preterm infants: preliminary observations. Pediatrics, 75:217-225.

Chance B., Leigh J.S., Miyake H., Smith D.S., Nioka S., Greenfeld R., Finander M., Kaufmann K., Levy W., Young M., Cohne P., Yoshioka H., Boretsky R. (1988a) Comparison of time resolved and unresolved measurements of deoxyhaemoglobin in brain. Proc. Nat. Acad. Sci., 85:4971-4975.

Chance B., Nioka S., Kent J., McCully K., Fountain M., Greenfeld R., Holtom G. (1988b) Time resolved spectroscopy of haemoglobin and myoglobin in resting and ischaemic muscle. Anal. Biochem., 174:698-707.

Cheong W.-F., Prahl, S.C., Welch A.J. (1990) A review of the optical properties of biological tissues. IEEE J. Quant. Electron., 26:2166-2185.

Cope M., van der Zee P., Essenpreis M., Arridge S.R., Delpy D.T. (1991a) Data analysis methods for near infrared spectroscopy of tissue: problems in determining the relative cytochrome aa_3 concentration. Proc. SPIE, 1431:251-262.

Cope M. (1991b) The development of a near infrared spectroscopy system and its application for non invasive monitoring of cerebral blood and tissue oxygenation in the newborn infant. PhD Thesis, University of London.

Delpy D.T., Cope M., van der Zee P., Arridge S.R., Wray S., Wyatt J.S. (1988) Estimation of optical pathlength through tissue from direct time of flight measurement. Phys. Med. & Biol., 33:1433-1442.

Delpy D.T., Arridge S.R., Cope M., Edwards A.D., Reynolds E.O.R., Richardson C.E., Wray S., Wyatt J.S., van der Zee P. (1989) Quantitation of pathlength in optical spectroscopy. Adv. Exp. Med. & Biol., 248:41-46.

Edwards A.D., Wyatt J.S., Richardson C.E., Delpy D.T., Cope M., Reynolds E.O.R. (1988) Cotside measurement of cerebral blood flow in ill newborn infants by near infrared spectroscopy. Lancet ii, 770-771.

Elwell C.E., Cope M., Edwards A.D., Wyatt J.S., Reynolds E.O.R., Delpy D.T. (in press) Measurement of cerebral blood flow in adult humans using near infrared spectroscopy - methodology and possible errors. Adv. Exp. Med. & Biol.

Jacques S.L. and Flock S.T. (1991) Effect of surface boundary on time resolved reflectance: measurement with a prototype endoscopic catheter. Proc. SPIE, 1431:12-20.

Jöbsis F.F. (1977) Non invasive, infrared monitoring of cerebral and myocardial oxygen sufficiency and circulatory parameters. Science, 198:1264-1267.

Lakowicz J.R. and Berndt K. (1990) Frequency domain measurements of photon migration in tissues. Chem. Phys. Lett., 166:246-252.

Madsen S.J., Patterson M.S., Wilson B.C., Park Y.D., Moulton J.D., Jacques S.L., Hefetz Y. (1991) Time resolved diffuse reflectance and transmittance studies in tissue simulating phantoms: a comparison between theory and experiment. Proc. SPIE, 1431:42-51.

Maughan R.J., Watson J.S., Weir J. (1984) The relative proportions of fat, muscle and bone in the normal human forearm as determined by computer assisted tomography. Clin. Sci., 66:683-689.

Patterson M.S., Moulton J.D., Wilson B.C., Chance B. (1990) Applications of time resolved light scattering measurements to photodynamic therapy dosimetry. Proc. SPIE, 1203:62-75.

Sakai F., Nakazawa K., Tazaki Y., Ishii K., Hidetada H., Igarashi H., Kanda T. (1985) Regional cerebral blood volume and haematocrit measured in normal human volunteers by single-photon emission computed tomography. J. Cereb. Blood Flow and Metabolism, 5:207-213.

Sevick E.M. and Chance B. (1991) Photon migration in a model of the head measured using time and frequency domain techniques: potentials of spectroscopy and imaging. Proc. SPIE, 1431:84-96.

Sterenborg H.J.C.M., van Gemert M.J.C., Kamphorst W., Wolbers J.G., Hogervorst W. (1989) The spectral dependence of the optical properties of human brain. Lasers in Med. Sci., 4:221-227.

Wray S., Cope M., Delpy D.T., Wyatt J.S., Reynolds E.O.R. (1988) Characterisation of the near infrared absorption spectra of cytochrome aa_3 and haemoglobin for the non invasive monitoring of cerebral oxygenation. Biochim. Biophys. Acta., 933:184-192.

Wyatt J.S., Cope M., Delpy D.T., van der Zee P., Arridge S.R., Edwards A.D., Reynolds E.O.R. (1990a) Measurement of optical pathlength for cerebral near infrared spectroscopy in newborn infants. Dev. Neuroscience, 12:140-144.

Wyatt J.S., Cope M., Delpy D.T., Richardson C.E., Edwards A.D., Wray S.C., Reynolds E.O.R. (1990b) Quantitation of cerebral blood volume in newborn infants by near infrared spectroscopy. J. Appl. Physiol., 68(3):1086-1091.

van der Zee P. and Delpy D.T. (1988) Computed point spread functions for light in tissue using a measured volume scattering function. Adv. Exp. Med. & Biol., 222:191-197.

van der Zee P., Arridge S.R., Cope M., Delpy D.T. (1990) The effect of optode positioning on optical pathlength in near infrared spectroscopy of brain. Adv. Exp. Med. & Biol., 277:79-84.

van der Zee P., Essenpreis M., Delpy D.T., Cope M. (1991) Accurate determination of the optical properties of biological tissues using a Monte Carlo inversion technique. *Proc. ICO Meeting on Atmospheric, Volume and Surface Scattering and Propagation*, Florence, Italy, August 27-30, 1991.

van der Zee P., Cope M., Arridge S.R., Essenpreis M., Potter L.A., Edwards A.D., Wyatt J.S., McCormick D.C., Roth S.C., Reynolds E.O.R., Delpy D.T. (in press) Experimentally measured optical pathlengths for the adult head, calf and forearm and the head of the newborn infant as a function of interoptode spacing. Adv. Exp. Med. & Biol.

TOWARDS HUMAN BRAIN NEAR INFRARED IMAGING: TIME RESOLVED AND UNRESOLVED SPECTROSCOPY DURING HYPOXIC HYPOXIA

M. Ferrari[*], R.A. De Blasi[+], F. Safoue[+], Q. Wei[^], and G. Zaccanti[^]

[*]Dipartimento di Scienze e Tecnologie Biomediche, Università dell'Aquila, Italy and Laboratorio di Biologia Cellulare, Istituto Superiore di Sanità,[+]Istituto di Anestesiologia e Rianimazione, I Università di Roma, Roma Italy;[^]Dipartimento di Fisica, Universita' di Firenze, Firenze, Italy

INTRODUCTION

Near infrared (IR) spectroscopy (NIRS) on animals and humans originated from the work of Jobsis (1977). Early cerebral and muscle NIRS measurements using multi-wavelength photometers were only qualitative. Fast scanning spectrophotometry has been utilized for a better spectral analysis and quantification purposes (Ferrari et al., 1989). Only pulse time and phase modulation spectroscopies can calculate pathlength in brain and muscle. Optical imaging in the near IR using these time-resolved methods is currently pursued by different groups (Chance, 1991 for a review). There are however significant technical and clinical problems related to the imaging technique yet to be solved before it can become truly a routine clinical tool.

When transmitting picosecond laser pulses and detecting photons on the opposite side of the object, the contrast can be strongly enhanced by detecting only the photons with the shortest traveling time. Short time resolution would restrict the scattering path and could provide further amelioration in image definition. Recently, Hitachi research laboratories (Shinohara et al., 1991, and this volume) obtained images of rat brain oxygenation during middle cerebral artery ischemia using dual wavelength picosecond near IR laser technique combined with conventional back projection image reconstruction. The aim of this paper was to investigate human brain oxygenation changes during hypoxic hypoxia using two wavelength picosecond near IR laser spectroscopy. Pathlengths were mapped on the human forehead. By combining spectral information obtained by a fast scanning spectrometer with pathlength data it was possible to quantify oxy- and deoxy-hemoglobin (Hb) concentration changes during hypoxic hypoxia.

Optical Imaging of Brain Function and Metabolism, Edited by
U. Dirnagl *et al.*, Plenum Press, New York, 1993

METHODS

All measurements were obtained on adult volunteers. Each subject was submitted to graded steps of hypoxic hypoxia (FiO_2 21, 10, 7, 6%). Arterial Hb saturation was measured by pulse oximetry. The brain was illuminated using an optic fiber (70 cm length, 0.3 cm active diameter) by ultrashort light pulses (duration 4 ps, repetition rate 76 MHz, wavelengths 760 and 800 nm) from a Coherent 700 dye laser (Fig. 1).

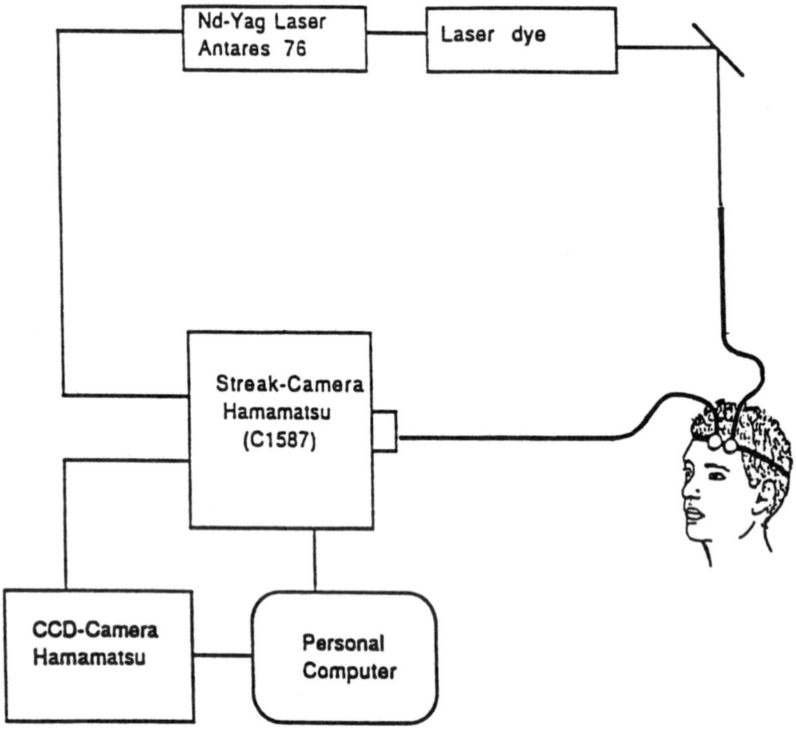

FIGURE 1. Experimental set-up used in human brain time-resolved spectroscopy.

The average power was 50-100 mW at the tip of the fiber. The light emerging from tissue was collected with an optic fiber (70 cm length, 0.5 cm active diameter) positioned at 3-5 cm distance. Light was measured over a 5 sec period by a syncroscan streak camera (Hamamatsu Photonics). The resolution of this equipment is a few picoseconds and the temporal window 1500 psec. Measurements were collected every 30 sec. Using this equipment the time of flight and temporal dispersion of light pulses through a multiple scattering medium can be studied. Optical density changes were calculated from the integral of the measured received pulses. The transit times (t) of the IR photons can be converted into distance travelled through the medium (d) using the formula $d=ct/n$ ($n=1.4$, tissue refractive index). The mean path length was calculated from the weighted average of time in the pulse shape. The differential pathlength factor (DPF) was calculated dividing the mean pathlength by the interfiber distance. The slope of the logarithmic curve was calculated fitting the signal over a time interval of about 300 psec starting from the time of half peak amplitude. The slope changes are associated with the variations of the absorption coefficient σ (cm^{-1}) of the medium. This time interval was chosen taking into account the linearity of this portion.

Spectral measurements were made using a fast scanning spectrophotometer (mod. 6500, NIRSystems, Silver Spring, MD). The spectral analysis procedure has been recently described (Ferrari et al., 1989). Two optic fibers (200 cm length and 0.5 cm active diameter) were applied 3-4 cm apart by a black rubber support so that a stable

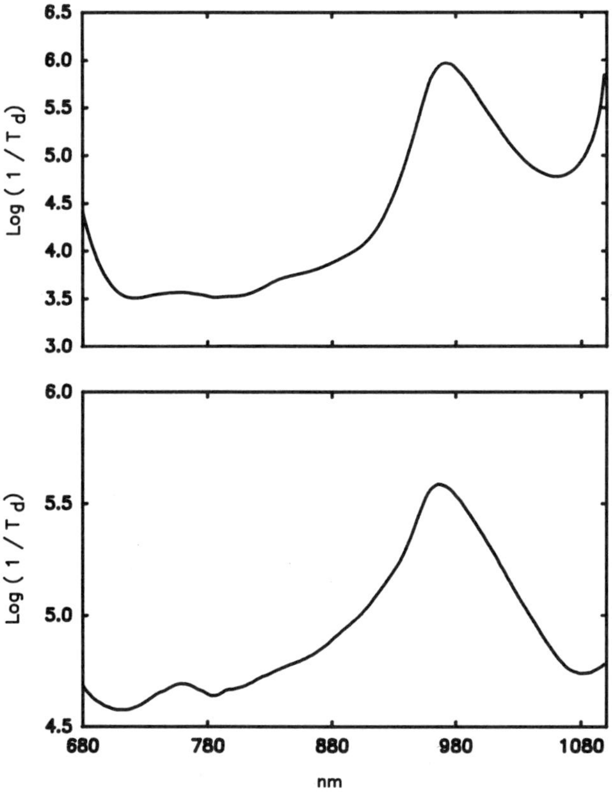

FIGURE 2. Typical near IR spectra of an human forehead (lower panel) and an apple (upper panel). The absorption band centered around 960 nm is attributable to water overtones. Brain presents also a band centered at around 755 nm attributable to deoxy Hb. T_d, diffuse transmittance.

fiber geometry was achieved. NIRSystems software was utilized to automatically collect a scan every 5 sec. Spectra were analyzed according to a modified Lambert-Beer law in order to obtain quantification of Hb changes during the experimental procedures. Difference spectra (ΔA) of the brain tissue were calculated relative to the pre-hypoxic period. These were converted into brain absorption coefficient ($\Delta\mu_a$) using $\Delta\mu_a = \Delta A/(Bd)$ where d was the physical separation of the optodes on skin surface and B was the DPF. Changes in brain absorption coefficient were assumed to result only from changes in the concentration of oxy- and deoxy-Hb (Wray et al., 1988). The results were expressed as micromoles per liter of tissue ($\mu M/L$).

The tissue absorption coefficient spectra ($\Delta\mu_a$) were split into Δ[Hb] and Δ[HbO_2] using multilinear regression analysis (Cope et al., 1988) of the Hb- and HbO_2-spectra (Wray et al., 1988). The regression analysis was performed over the wavelength region 750 to 900 nm with data points at 2 nm intervals.

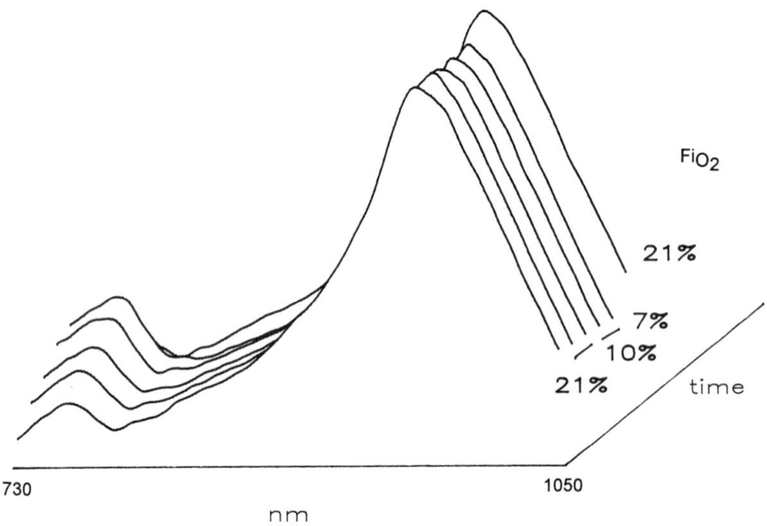

FIGURE 3. Successive near IR spectra from forehead of a representative subject during hypoxic hypoxia. Hypoxia caused an increase of the peak at 755 nm more consistent at the most severe tested hypoxia level (FiO_2 7%). The last spectrum was collected 1 min after recovery in air.

RESULTS

Near IR spectra of human brain are very similar to previously published cat and dog spectra. Fig. 2 shows a spectrum measured on the human forehead compared to the spectrum of an apple. Both spectra, obtained with comparable optic fiber geometry, showed the water peak. The 755 nm deoxy-Hb peak is the only notable spectral feature which differentiates brain and apple spectra. When a volunteer is subjected to hypoxic hypoxia as shown in Fig. 3, the 755 nm peak rises constantly with increasing hypoxic hypoxia. Fig.4 shows typical pulse shapes measured on the forehead with different distances of the optic fibers. Small changes of the DPF were found when the DPF was mapped on the forehead (Fig. 5).

Fig.6 shows typical pulses measured at 760 nm before and during 10 min hypoxic hypoxia. Consistent changes of the pulse amplitude and slope were observed only at the most severe tested hypoxic hypoxia level. Fig.7 shows the time course of σ measured from a light pulse at 760 nm and 800 nm during graded hypoxic hypoxia and recovery. Using 760 nm light, σ increased gradually during hypoxia. No changes were observed using 800 nm light suggesting that no consistent blood volume changes occur during moderate hypoxia. σ measured at 760 nm rapidly recovered with room air ventilation.

By combining spectral information with pathlength data it was possible to quantify oxy and deoxy Hb-concentration changes during hypoxic hypoxia. Fig. 8 shows the time course of Hb-saturation, Hb-content and cytochrome oxidase (cyt a-a_3) redox changes during hypoxic hypoxia. Hb-content and cytochrome redox state did not change during hypoxia. Hb-content transiently raised only when FiO_2 returned to 21%.

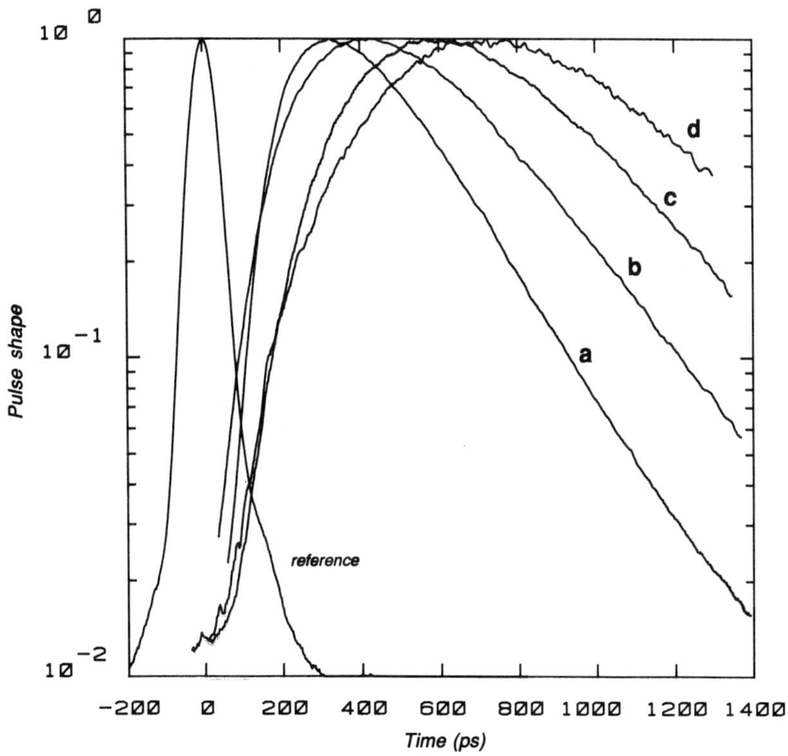

FIGURE 4. Pulse shape measured with the optic fibers at different distances (d) on the forehead. The maximum of each pulse was normalized to 1. Curves a-d correspond to d=2.0, 2.5, 3.0, and 4.0 cm respectively. The reference pulse was obtained with fibers aligned in air. For any curve the time t=o corresponds to the arrival time for the unscattered transmitted beam. Differential pathlength factor for curves a-c was 5.9, 5.9 and 6.4 respectively.

CONCLUSIONS

NIRS has been developed experimentally and clinically to non-invasively monitor brain and muscle Hb/myoglobin-oxygenation and cytochrome a-a_3 redox state. Relevant human brain clinical applications are summarized in Table 1. Optical imaging in the near IR using time-resolved methods is currently pursued by different groups. Using these methods, at the Hitachi research laboratories (Shinohara et al., 1991, and this volume) images of rat brain oxygenation were obtained. To the best of our knowledge, the present results constitute the first report of human brain oxygenation monitoring by optic fiber time resolved spectroscopy on volunteers during graded hypoxic hypoxia. These results and the reported recent findings on time gate imaging of phantoms (Chance, 1991) suggest that human brain IR imaging might be performed in the near future using new tools such as ultrafast lasers, tunable over a wide spectral range, and ultrafast electronics.

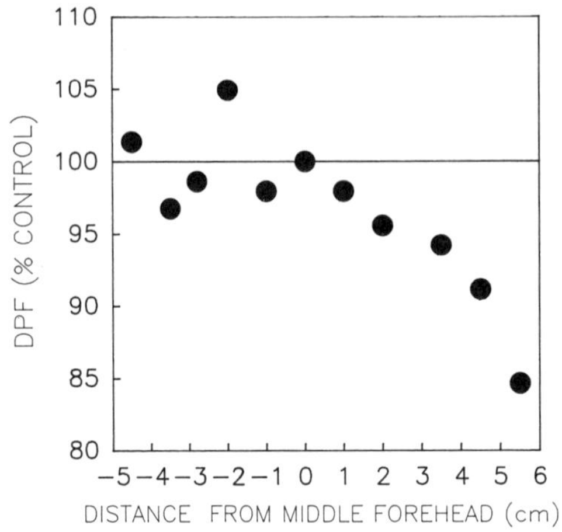

FIGURE 5. Differential pathlength factor (DPF) changes in the human forehead. The optic fibers, positioned at 3 cm distance, were moved along a plane perpendicular to the body sagittal axis. The reference plane was positioned at the level of the forehead center.

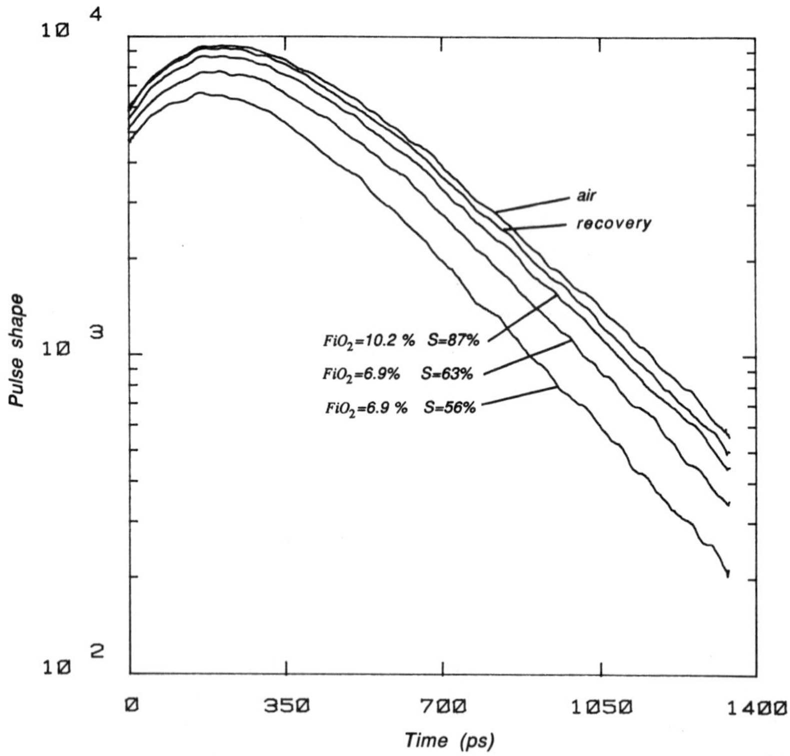

Figure 6. Pulse shape (arbitrary units) measured at 760 nm before and during 10 min hypoxic hypoxia. Consistent changes of pulse amplitude and slope were observed at the most severe tested hypoxic hypoxia level. Pulse slope recovered when the volunteer returned to air ventilation.

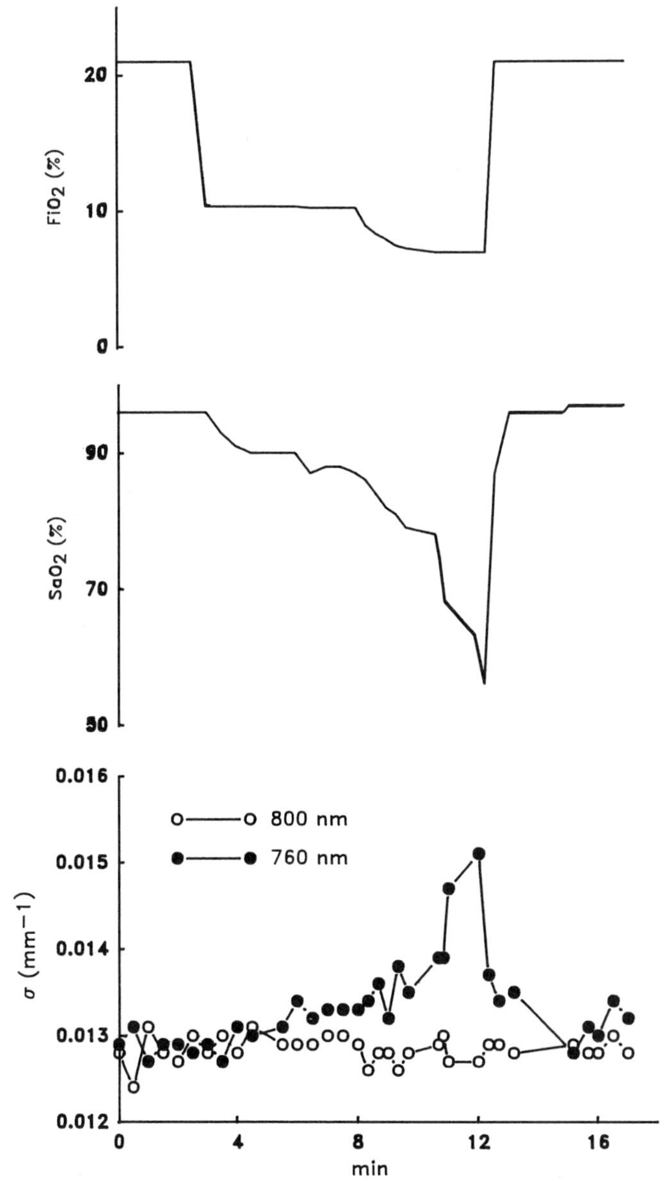

FIGURE 7. Time course of σ measured from the slope of the light pulse at 760 nm and 800 nm during graded hypoxic hypoxia and recovery. Using 760 nm light, σ increased gradually during hypoxia. No changes were observed using 800 nm light suggesting that no consistent blood volume changes occurred during moderate hypoxia. σ measured at 760 nm rapidly recovered with room air ventilation. FiO_2 (inspired oxygen fraction); SaO_2 (Hb saturation measured by pulse oximetry on the nose).

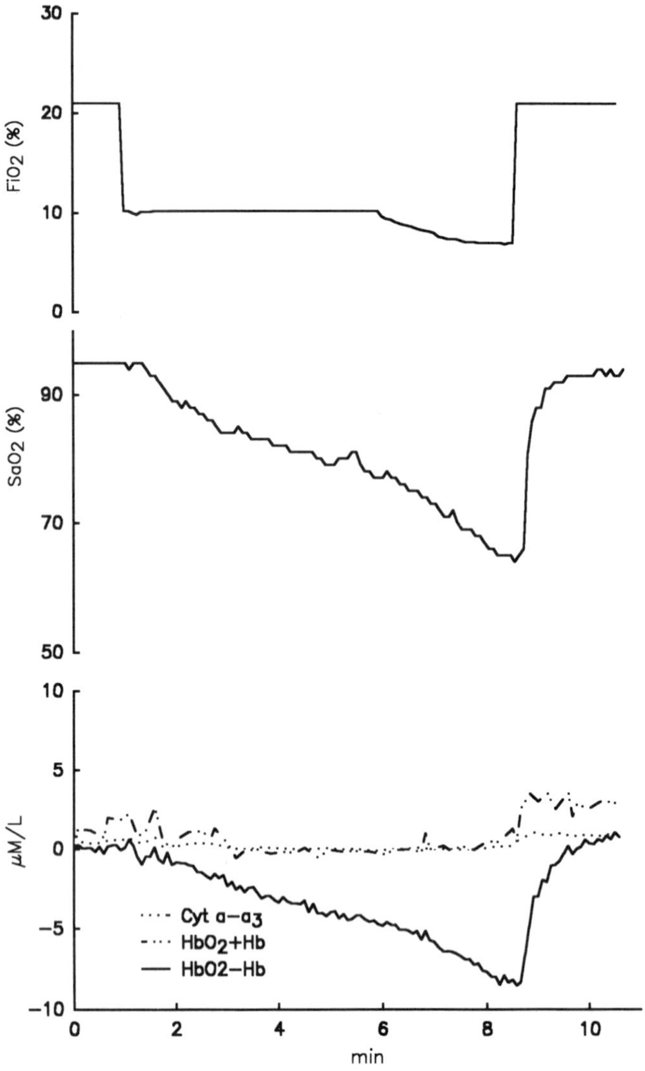

FIGURE 8. Time course of Hb saturation, Hb content and cytochrome oxidase (cyt a-a_3) redox changes during hypoxic hypoxia. Hb volume and cytochrome redox state did not change during hypoxia. Hb volume raised when FiO_2 returned to 21%. Symbols as Fig. 7.

ACKNOWLEDGEMENTS

We wish to thank Prof. S. Califano, Director of the European Laboratory for Non-linear spectroscopy, University of Florence. This research was supported in part by CNR contribution 91.00253.04. This research has been performed in the framework of the "Centro di ricerca interuniversitario: studio dei meccanismi molecolari coinvolti nel danno tissutale da ipossia e iperossia e di molecole che modificano tali lesioni".

Table 1. Main clinical applications of NIRS

FIELD OF APPLICATION	INVESTIGATED PARAMETER	AUTHORS
<u>ADULT</u>		
HYPOXIC HYPOXIA IN HEALTHY VOLUNTEERS	HbV, HbS, blood flow, transit time	Elwell et al.,1992, Hampson et al.,1990 McCormick et al.,1991)
GENERAL ANESTHESIA	HbV, HbS	(Fox et al., 1985)
VASCULAR PATIENTS	HbV, HbS	(Ferrari et al., 1987)
CARDIAC SURGERY	HbS during cardio-pulmonary bypass	(Tamura,1991)
	HbS during surgery	(Koorn and Silvay,1991)
CARDIAC ARREST	Hyperoxia-reperfusion	(Smith,1990)
CENTRIFUGAL ACCELERATION	HbV, HbS	(Glaister and Jöbsis,1988)
<u>CHILDREN</u>		
CARDIAC SURGERY	HbV, HbS during Cardiopulmonary bypass	(Steven et al.,1990)
<u>NEWBORN</u>		
INTENSIVE CARE	HbS monitoring during spontaneous hypoxia	(Brazy et al.,1985)
	HbV, HbS	(Wyatt et al.,1986)
	Blood flow, volume	(Edwards et al.,1988, 1990)
	Blood volume changes during hypercapnia	(Wyatt et al.,1991)
	Crying influence on HbV	(Brazy,1988)
	HbV changes during hypertension	(Brazy et al.,1986)
	Hypoxia, bradicardia effects on HbS	(Livera et al.,1991)
CARDIAC SURGERY	HbS during extracorporeal circulation	(Liem et al.,1992)
<u>FETUS</u>		
LABOUR	Oxytocin effects on HbV, HbS	(Peebles et al.,1991)

Hemoglobin content, HbV; Hemoglobin saturation, HbS

REFERENCES

Brazy J.E., Lewis D.V., Mitnick M.H., Jobsis F.F. (1985) Non invasive monitoring of cerebral oxygenation in preterm infants: preliminary observations. Pediatrics 75:217-225.

Brazy J.E., Lewis D.V. (1986) Changes in cerebral blood volume and cytochrome aa_3 during hypertensive peaks in preterm infants. J Pediatr 108:983-987.

Brazy J.E. (1988). Effects of crying on cerebral blood volume and cytochrome aa_3. J Pediatr 112:457-461.

Chance B. (1991) In: Time Resolved Spectroscopy and Imaging of Tissue, B. Chance, Editor, Proc. SPIE 1431.

Cope M., Delpy D.T., Reynolds E.O.R., Wray S., Wyatt J., van der Zee P. (1988) Methods of quantitating cerebral near infrared spectroscopy data. Adv Exp Med Biol. 222:183-189.

Edwards A.D., Richardson C., Cope M., Wyatt J.S., Delpy D. T., Reynolds E.O.R. (1988). Cotside measurement of cerebral blood flow in ill newborn infants by near infrared spectroscopy. Lancet 2:770-771.

Edwards A.D., Wyatt J.S., Richardson C.E., Potter A., Cope M., Delpy D.T., Reynolds E.O.R. (1990) Effects of indomethacin on cerebral haemodynamics and oxygen delivery investigated by near infrared spectroscopy in very preterm infants. Lancet 2:1491-1495.

Elwell C.E., Cope M., Edwards A.D., Wyatt J.S., Reynolds E.O.R., Delpy D.T. (1992) Measurement of cerebral blood flow in adult humans using near infrared spectroscopy - methodology and possible errors. Adv Exp Med Biol (in press).

Ferrari M., Zanette E., Sideri G,. Giannini I., Fieschi C., Carpi, A. (1987) Effects of carotid compression, as assessed by near infrared spectroscopy, upon cerebral blood volume and hemoglobin O_2 saturation. J Royal Soc Med 80:83-87.

Ferrari M., Wilson D.A., Hanley D.F., Hartman J.F., Traystman R.J., Rogers M.C. (1989) Non invasive determination of venous hemoglobin saturation in the dog by derivative near infrared spectroscopy. Am J Physiol 256:H1493-H1499.

Fox E., Jobsis F.F., Mitnick M.H. (1985) Monitoring cerebral oxygen sufficiency in anesthesia and surgery. Adv Exp Med Biol 191:849-854.

Glaister D.H., Jobsis F.F. (1988) A near-infrared spectro-photometric method for studying brain O_2 in man during $+G_z$ acceleration. J Aerospace Med 59:199-207.

Hampson N.B., Camporesi E.M., Stolp B.W., Moon R.E., Shook J.E., Grebel J.A., Piantadosi C.A. (1990) Cerebral oxygen availability by NIR spectroscopy during transient hypoxia in humans. J Appl Physiol 69:907-913.

Jobsis F.F. (1977) Non invasive infrared monitoring of cerebral and myocardial oxygen sufficiency and circulatory parameters. Science 198:1264-1267.

Koorn R., Silvay G. (1991) Evaluation of the Somanetics^R Invos^R 3100 cerebral oximeter during cardiac surgery. 13th Ann Meet Soc Cardiovascular Anesthesiol p.176.

Liem K.D., Hopman J.C.W., Kollée L.A.A., Oeseburg B. (1992) Assessment of cerebral oxygenation and hemodynamics by near infrared spectrophotometry during induction of ECMO. Adv Exp Med Biol (in press).

Livera L.N., Spencer S.A., Thorniley M.S, Wickramasinghe Y.A., Rolfe P.(1991) Effects of hypoxemia and bradycardia on neonatal cerebral haemodynamics. Arch Dis Child 66:376-380.

McCormick P.W., Stewart M., Goetting M.G., Balakrishnan G. (1991) Regional cerebrovascular oxygen saturation measured by optical spectroscopy in humans. Stroke 22:596-602.

Peebles D.M., Edwards A.D., Wyatt J.S., Bishop A.P., Cope M., Delpy D.T., Reynolds E.O.R.(1991) Effect of oxytocin on fetal brain oxygenation during labour. Lancet 338:254-255.

Shinohara Y., Haida M., Kawaguchi F., Ito Y., Takeuchi H. (1991) Hemoglobin oxygen-saturation image of rat brain using near infrared light. J Cerebral Blood Flow Metab 11:S459.

Smith D.S., Levy W., Maris M., Chance B. (1990) Reperfusion hyperoxia in brain after circulatory arrest in humans. Anesthesiology. 73:12-19.

Steven J.M., Kurth C.D., Nicholson S.C., Phoon C., Chance B. (1990) Continuous non invasive assessment of brain oxygenation and blood volume during cardiopulmonary bypass in children undergoing ADS closure. Anesthesiology 73:A457.

Tamura M.(1991) Non-invasive monitoring of brain oxygen metabolism during cardiopulmonary bypass by near infrared spectrophotometry. Jap Circ J 55:330-335.

Wray S., Cope M., Delpy D.T., Wyatt J.S., Reynolds E.O.R. (1988) Characterization of the near infrared absorption spectra of cytochrome a,a_3 and haemoglobin for the non invasive monitoring of cerebral oxygenation. Biochem Biophys Acta 933:184-192.

Wyatt J.S., Cope M., Delpy D.T., Wray S., Reynolds E.O.R. (1986) Quantification of cerebral oxygenation and hemodynamics in sick newborn infants by near infrared spectrophotometry, Lancet. 2:1063-1066.

Wyatt J.S., Edwards A.D., Cope M., Delpy D.T., McCormick D.C., Potter A., Reynolds E.O.R. (1991) Response of cerebral blood volume to changes in arterial carbon dioxide tension in preterm and term infants. Ped Res 29:553-557.

van der Zee P., Cope M., Arridge S.R., Essenpreis M., Potter L.A., Edwards A.D., Wyatt J.S., McCormick D.C., Roth S.C., Reynolds E.O.R., Delpy D.T. (1992) Experimentally measured optical pathlengths for the adult head, calf and forearm and the head of the newborn infant as a function of inter optode spacing. Adv Exp Med Biol (in press).

MEASUREMENT OF HUMAN HYPOTHERMIC CEREBRAL OXYGEN METABOLISM BY TRANSMISSION SPECTROSCOPY

P.W. McCormick, M.C. Stewart, G.D. Lewis, and J.M. Zabramski

Department of Neurosurgery, Henry Ford Hospital
Detroit, MI, USA
Division of Neurosurgery, Barrow Neurologic Institute
Phoenix, AZ, USA

INTRODUCTION

Near infrared light (650 - 1100 nm) is capable of penetrating the human scalp and skull (Wan et al. 1981). Hemoglobin and oxyhemoglobin circulating in the cerebrovascular compartment have unique infrared absorption spectra which influence wavelength specific attenuation in this wavelength band (Tamura et al. 1985). The human brain contains 600 to 1000 mg of hemoglobin per 100 grams of tissue; thus, the brain is an excellent organ for *in vivo* optical spectroscopic (INVOS) measurements of hemoglobin oxygen saturation. Direct and continuous spectroscopic evaluation of cerebral oxygen saturation is possible (McCormick et al. 1991a/b).

Mixed oxyhemoglobin/hemoglobin transmission spectra can be recorded from the exposed and unexposed brain (Jobsis et al. 1977, Wray et al. 1988). These spectra are reducible to a qualitative measure of cerebral oxyhemoglobin, and such paradigms track qualitative changes in aggregate cerebral oxyhemoglobin and hemoglobin content of the cerebral vasculature over time. Quantitative measurement of cerebral oxyhemoglobin content from attenuation data requires that photon tissue path length be measured, but *in situ* measurement of photon path length for clinical monitoring is impractical at this time.

Another quantitative measure is hemoglobin oxygen saturation. This value can be derived from cerebral transmission spectra because the variable, path length, drops from the calculation under the assumption that this value is wavelength independent over the narrow band of 700 - 850 nm (see below). Hemoglobin oxygen saturation is a clinically meaningful parameter, and in this case represents an aggregate measure of the arterial, venous, and capillary oxygen content of the sampled brain. This aggregate measure is referred to as "cerebrovascular oxygen saturation" (McCormick et al. 1991a/b).

No report to date has shown clinical data tracking changes in cerebrovascular saturation during events of reduced or arrested cerebral oxygen delivery, nor has data been published suggesting a unique monitoring role for this technique.

Optical Imaging of Brain Function and Metabolism, Edited by
U. Dirnagl *et al.*, Plenum Press, New York, 1993

This report presents data demonstrating cerebrovascular saturation changes resulting from lowered cerebral oxygen delivery during human hypothermic circulatory arrest. Primate data demonstrating cerebral hypothermic oxygen metabolism are presented to aid in understanding human cerebrovascular saturation changes. In addition, human cerebrovascular hemoglobin saturation measurements seem to function as a unique monitoring strategy for preventing coincidental cerebral hypoxic injury.

BACKGROUND

Instrumentation

This report is based on prototype instrumentation. Infrared light generated by an incandescent source is focused through a band pass filter mounted on a chopper wheel. Sequential filters divide the beam into five wavelength bands: 672, 726, 750, 803, and 840 nm. An internal microcomputer controls the timing of the delivery of this light to the fiber optic transmission cables.

These cables deliver light directly to the subject; the tips protrude slightly from the surface of the fixture to ensure tight coupling with the scalp. A single fixture houses the incident light cable and one or more fiber optic receiving bundles to collect transmitted light. The receiving light guides are separated from the source by variable distances (5.0 - 35.0 mm) to allow variation of the mean photon path length through the tissue. Often three fiber optic bundles are used such that the protrusions form a tripod to accommodate the convexity of the cranium in coupling the fiber optics orthagonal to the surface.

The light returning from the subject passes through short (2.5 cm) light pipes directly to photo diodes, which are on miniature printed circuit boards along with pre-amplifiers and drivers. The data are fed directly to a 12 bit analog-to-digital converter over coaxial cables. The microcomputer controls the time interval in which the light of a particular wavelength enters the tissue and over which analog data are accepted into memory. Signal integrity is maintained by rejecting any signal with a signal-to-noise ratio less than 30:1.

The analysis of absorption data is facilitated by normalizing to a reference wavelength. A convenient reference wavelength is the isosbestic point of oxyhemoglobin and hemoglobin (803 nm). This reduces the effects of other sources of attenuation such as optical boundaries. The transmission intensity and incident intensity at each wavelength are processed through an algorithm based on the Beer-Lambert law:

$$\ln\frac{I_{(\omega)}}{I_{(\omega_o)}} = \sum_{j=1}^{N} \alpha_{(\omega,j)} C_{(j)} s \tag{1}$$

In this expression, $I_{(\omega)}$ represents intensity of transmitted light at wavelength ω, $I_{(\omega o)}$ represents the intensity of the incident light at wavelength ω_o, α represents the molar extinction coefficient of oxyhemoglobin or hemoglobin, C represents the concentration of this molecule in the tissue, and s represents the photon path length in the tissue.

If absorption at a second wavelength, ω', is subtracted from absorption at ω, the following expression is derived:

$$-\ln\frac{I_{(\omega)}}{I_{(\omega_o)}}+\ln\frac{I_{(\omega')}}{I_{(\omega'_o)}}=\sum_{j=1}^{N}(\alpha_{(\omega,j)}-\alpha_{(\omega',j)})C_{(j)}s$$

(2)

This expression is solved by making enough (N+1) measurements to solve for $C_{(1)}s$ (oxyhemoglobin) and $C_{(2)}s$ (hemoglobin). These values do not represent actual chromophore content but are proportional to it. Although a function of hemoglobin concentration, the variable s is considered invariant with wavelength over the 600 - 1000 nm range. Thus, the effects of unknown photon path length and variable total hemoglobin can be removed by expressing the solutions of equation 2 as the ratio:

$$\frac{C_{(2)}s}{C_{(1)}s}=\frac{C_{(2)}}{C_{(1)}}=Hr$$

(3)

The variable Hr represents the ratio of hemoglobin to oxyhemoglobin, which is converted to % hemoglobin saturation.

$$\frac{100}{(1+Hr)}=100\times\frac{[HbO_2]}{([Hb]+[HbO_2])}=\%\textit{hemoglobin saturation}$$

(4)

CLINICAL EVALUATION OF INSTRUMENTATION

Detecting intracranial attenuation

The ability of the above described prototype spectroscopy unit to detect intracranial cerebrovascular attenuation was established in a clinical protocol approved by the Henry Ford Hospital Human Rights in Experimentation Committee (Detroit, MI).

Five adults undergoing routine carotid endarterectomy had a small bolus (1.0 mg in 1.0 cc normal saline) of the infrared attenuating tracer indocyanine green (ICG) (Becton Dickinson Co, Baltimore, MD) placed into the exposed internal carotid artery (ICA) via cannulation with a 22 gauge needle. The external carotid artery was temporarily occluded to prevent reflux of tracer into this vessel.

Diffuse transmission spectroscopy was carried out over the ipsilateral frontal scalp 4.0 cm anterior to the coronal suture and 3.0 cm lateral to the sagittal suture.

Two receivers tuned to 803 nm were used to simultaneously collect infrared light signal 1.0 cm and 2.7 cm from the infrared light source.

After the ICG bolus was given into the ICA, a series of transmission spectroscopic measurements, 29 per second, were made for 1.0 minute. The transit of infrared absorber through the intracranial vascular compartment was detected as the change in optical density at 803 nm with time. The measurement was then repeated, allowing the ICG tracer to flow only into the external carotid artery.

Bolus infusion in the ICA did not significantly alter the optical density recorded in the channel with a 1.0 cm source - receiver separation distance. This was expected because the superficial, extracranial tissue sampled by this source - receiver configuration is not vascularized by the internal carotid artery. Large changes in optical density were, however, recorded in the channel with a 2.7 cm source - receiver separation distance. A significant portion of the attenuating events experienced by photons reaching this receiver are predicted to take place intracranially where there is tracer induced attenuation. Change in optical density in the two channels with an ICA bolus of tracer was reproducible in the five subjects.

When the tracer was allowed to flow exclusively into the external carotid artery, both receivers generated activity curves similar to those in the 2.7 cm source - receiver separation channel during ICA infusion. Here the tracer is in extracranial vascular beds, and therefore in the field of both receivers.

The signal-to-noise ratio for detection of ICG in the cerebrovasculature in the 2.7 cm source -receiver separation channel was more than 100:1. Signal was defined as the maximum increase in optical density associated with ICG infusion, and noise was defined as the peak-to-peak baseline optical density oscillation. The apparent noise was due primarily to systolic and diastolic fluctuations in blood concentration.

Correlation of cerebrovascular saturation with reference value

Under this protocol, nine critically ill patients with presumed abnormal cerebral oxygen delivery and an abnormal or fluctuating neurological examination were studied. In each patient, a jugular bulb catheter was placed with its position confirmed by radiography.

Each patient underwent a series of spectroscopic measurements of cerebrovascular saturation over the frontal lobe of the brain. Samples of arterial and cerebral mixed venous blood were simultaneously drawn (1.0 cc over 45 sec) during spectroscopy. The arterial blood was representative of the blood present in the cerebral arteries, and cerebral mixed venous blood was representative of the blood in the veins and distal capillaries of the brain. Blood samples were immediately placed on ice and analyzed for hemoglobin oxygen saturation on an IL 282 CO-Oximeter (Instrument Laboratories Inc., Lexington MA).

A total of 68 measurements were made on 9 patients. Because the instantaneous size of the arterial and venous portions of the cerebrovascular compartment was an unknown, available published data on regional arterial and venous cerebral blood volume was used to estimate cerebrovascular hemoglobin oxygen saturation in the brain (Mchedlishvili et al. 1986):

$$\text{Estimated saturation} = (1-x)(SaO_2) + (x)(SvO_2) \quad (5)$$

Here x represents the percentage of regional cerebral blood volume which is arterial; SaO_2 represents systemic arterial oxygen saturation; and SvO_2 represents mixed cerebral venous oxygen saturation.

Correlation of the spectroscopic and estimated cerebrovascular oxygen saturation was performed. In the physiologic range of cerebral venous blood volume weighting (0.70 - 0.90), significant correlation was noted (Mayevski et al. 1988). For example, if venous weighting were taken to be 0.77 (equation 5), the positive linear correlation would be as noted ($n=68$, $r=0.74$, $s=3.5$) (McCormick et al. 1991c).

METHODOLOGY

Primate investigations

Other than by optical spectroscopy, there is no medically permissible independent measurement of cerebral hemoglobin oxygen saturation in humans during deep hypothermic cardiac arrest. Accordingly, a primate analogue was used. Three adult baboons (25 Kg) were enrolled in a protocol approved by the Animal Rights Investigation Committee of Barrow Neurologic Institute (Phoenix, AZ USA). In this protocol, the animals were anesthetized with intravenous pentobarbital and xylazine (Rompum, Haver, Shawnee, KS) and intubated. A femoral arterial line and retrograde jugular bulb catheter were placed. Femoral - femoral bypass was used to allow extracorporeal oxygenation and cooling of the blood and the animals were fully heparinized. The core body and brain surface temperatures were measured as the blood was slowly cooled. At approximately 25°C, the heart rhythm became ventricular fibrillation and this was converted to asystole with intravenous potassium.

The animal's brain temperature was then lowered to less than 3°C by progressively cooling the blood. To avoid thermal injury to plasma proteins and red blood cells, below 15°C whole blood was harvested from the animal and stored. Blood volume was maintained with a saline-dextran replacement fluid. Cerebral oxygen extraction was intermittently monitored by simultaneously collecting arterial and cerebral mixed venous blood and recording the difference (arterial less venous) in the measured oxygen content of each. The animals were rewarmed and identical measurements were made. Approximately 25 measurements were made during cooling and 25 during rewarming.

The data were analyzed to determine whether a critical temperature existed at which measured cerebral oxygen extraction was zero.

Human measurements

Spectroscopic monitoring of cerebrovascular saturation was carried out in seven patients undergoing hypothermic circulatory arrest for the treatment of complex cerebrovascular malformations at Henry Ford Hospital. The details of the perioperative management of these patients and the use of extracorporeal femoral circulation for cooling and ultimately arresting these patients has been previously described (Spetzler et al. 1988).

The average age of the seven patients was 46 years (range 17 - 64). Three patients presented with neurologic deficits secondary to mass effect, and two patients presented with acute subarachnoid hemorrhage and depressed levels of consciousness. The remaining patients were neurologically intact.

Prior to surgery, the tissue infrared spectrometer described above was positioned for monitoring over the middle frontal gyrus on the nonoperative side. Optical coupling to the scalp was protected from ambient light with an adhesive shield.

Cerebrovascular saturation measurements were stored every 5.5 seconds during patient cooling, circulatory arrest and rewarming. Simultaneous recordings of patient mean arterial blood pressure and core body temperature at five minute intervals were made. The patients were cooled to 18°C (esophageal temperature) over an average of 43 $\pm$ 22 min. At an average temperature of 28 $\pm$ 3°C, spontaneous cardiac asystole occurred.

RESULTS

Primate measurements

Cerebral oxygen extraction progressively dropped as brain temperature fell below 30°C. At a brain temperature of 15 - 18°C, it was approximately 30% of baseline (figure 1). Cerebral oxygen extraction was demonstrable until brain temperature was 5°C.

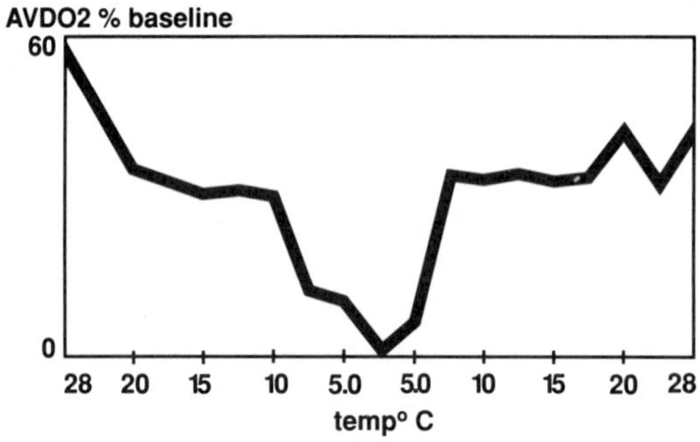

FIGURE 1. The cerebral arterial-venous oxygen content difference is shown as a function of brain temperature. Note that oxygen extraction does not fall below 10% of baseline until brain temperature reaches 5.0°C. These data were recorded in a 25 kg male baboon.

Human measurements

Spectroscopic measurements made during passive cooling, active cooling, circulatory arrest, and reperfusion were divided into three minute epochs (33 measurements). The three minute epochs at the end of passive cooling, at the end of active cooling, at the end of circulatory arrest, and immediately following reperfusion were considered representative of these major intraoperative phases and were statistically compared by analysis of variance (figure 2).

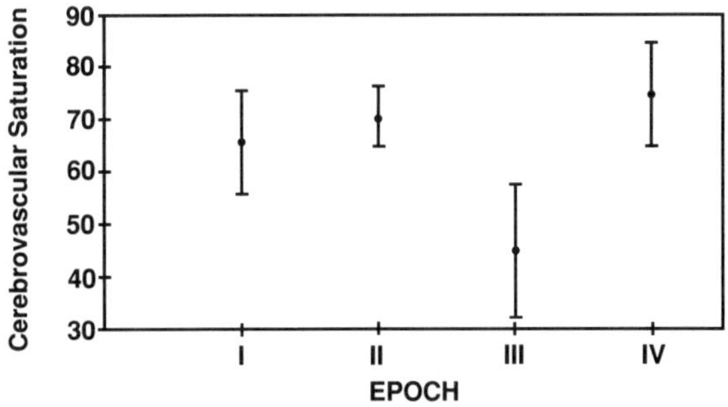

FIGURE 2. Cerebrovascular saturation is shown at four different epochs during human hypothermic circulatory arrest: Epoch I is at the end of passive cooling; Epoch II is at the end of active cooling to 18°C; Epoch III is at the end of circulatory arrest; Epoch IV is following recirculation of blood. Epoch III is significantly different ($p<0.01$) from the other conditions.

No significant difference between cerebrovascular saturation measured during passive cooling, active cooling, or reperfusion was found (figure 2). There was a significant difference between each of the above conditions and the final three minutes of circulatory arrest ($p<0.01$). This difference demonstrated desaturation during circulatory arrest and resaturation during reperfusion.

The duration of circulatory arrest varied (10 - 65 min) and had a negative linear correlation with cerebrovascular saturation ($n=7$, $r=-0.87$, $s=7$). The smallest drop recorded in saturation was 15%. At the opposite extreme, in one patient, cerebrovascular saturation fell to 30% after 45 minutes of circulatory arrest, and in another patient, saturation fell to 34% after 65 minutes of arrest.

The two patients with prolonged circulatory arrest periods (45 and 65 min) had poor outcomes. One patient had significant neurologic injury going into surgery and the outcome may not have been related to the duration of circulatory arrest. The other patient, however, had little neurologic injury prior to surgery. This patient at postmortem had histopathologic changes consistent with diffuse cerebral ischemia.

Of the remaining five patients, four had excellent outcomes and returned to premorbid levels of activity. One patient had a good outcome with a new onset of mild left hemiparesis.

DISCUSSION

Measurement Methodology

It is firmly established that near infrared light penetrates human tissue to a depth of several centimeters (Chance et al. 1988, Wan et al. 1981). This light is attenuated during transmission through tissue by scatter and absorption events. Absorption is due primarily to the few molecules in human tissue which absorb near infrared light; these include hemoglobin, oxyhemoglobin, and cytochrome aa3 (Jobsis et al. 1977, Wray et al. 1988). Because the cerebral content of hemoglobin is approximately 600 mg per 100 gms of tissue and is several orders of magnitude higher than cytochrome aa3, hemoglobin accounts for the majority of absorption events.

Recently, the concept of diffuse transmission spectroscopy has been adapted to generating cerebral transmission spectra. This involves the use of a single infrared point source and one or more diode receivers placed one centimeter or more from the source (McCormick et al. 1991a, Smith et al. 1990). In this arrangement, light propagates along a high probability course down into the tissue of interest and then back to the surface. This arrangement has the unique advantage of depth resolution which allows cerebral hemoglobin to be measured independently of scalp and skull hemoglobin.

In deep tissue spectroscopy, all hemoglobin (arterial, capillary, and venous) contributes to the measured saturation; in the case of cerebral hemoglobin this aggregate measurement is referred to as "cerebrovascular" saturation.

In patients undergoing hypothermic circulatory arrest, the major cortical arteries and veins empty. Under these conditions, hemoglobin left in the microcirculation becomes the predominant source of infrared attenuation. The aggregate "microcirculatory saturation" due to this residual hemoglobin can be followed during circulatory arrest.

Cerebral hypothermic protection

Hypothermia is known to reduce neuronal injury associated with hypoxia (O'Conner et al. 1986). The mechanism is related to a reduction in metabolism and therefore the oxygen requirements of neurons.

In dogs at 20°C, the cerebral metabolic rate of oxygen was reduced to 18% $\pm$ 2% of baseline demonstrating that although cellular oxygen demand is reduced, it is not zero (Tanaka et al. 1988). The half-life of cerebral high energy phosphates in rabbits during hypothermia is increased about three-fold (Stocker et al. 1986). These investigators were unable to identify a temperature below which cerebral oxygen metabolism stopped; the lowest temperature tested was 21°C.

The primate work described here demonstrates that cellular oxygen demand persists during hypothermic circulatory arrest to temperatures well below those used clinically. At cerebral temperatures in the clinical range (15 - 18°C), oxygen extraction was about 30% of baseline (anesthetized) values. The oxygenated hemoglobin left behind in the microcirculation of the brain should release oxygen to move down its partial pressure gradient to the mitochondria. The mitochondrial pO_2 is normally 3 to 8 torr, giving a gradient of >80 torr between the capillary and the mitochondria. As the partial pressure of residual cranial blood drops, the oxyhemoglobin desaturates. Desaturation at 15 - 18°C is less than at 37°C because the affinity of hemoglobin for oxygen increases as temperature drops.

At 20°C, a hemoglobin saturation of 30% is associated with a partial pressure of oxygen of 7 - 10 torr. At this pO_2 there is no longer a capillary-to-mitochondria gradient and hemoglobin desaturation stops. This may explain why a drop below 30% saturation was not detected in these patients. In the patient who had a 45 min. arrest, a saturation of 30% was reached; in the patient who had a 65 min. arrest, a saturation of 34% was reached.

Residual microcirculatory oxyhemoglobin during circulatory arrest will theoretically act as an oxygen reserve to maintain cellular metabolism. This situation is fundamentally analogous to critical tissue pO_2 (or venous pO_2). Here cerebral oxygen metabolism is availability dependent. As oxyhemoglobin desaturation proceeds, the pO_2 gradient diminishes until it is insufficient to meet metabolic demand; lactic acid production and hypoxic cellular injury result.

The point at which hypoxic cellular injury begins is uncertain, but it logically would occur following the "critical" point where anaerobic metabolism begins. This suggests that keeping residual hemoglobin saturation well above 30% at 18°C may prolong neuronal protection.

The limit of the oxyhemoglobin reserve is determined by the amount of residual oxyhemoglobin, the metabolic rate of oxygen (positively correlated with temperature) and the affinity of hemoglobin for oxygen (negatively correlated with temperature). Therefore, the residual oxyhemoglobin utilization hypothesis predicts that in each patient an arrest temperature exists where cerebral oxygen metabolism and cerebral oxygen availability are optimal for best cerebral protection during ischemic hypoxia.

The existence of an optimal temperature is demonstrable. The neurologic tolerance of baboons to hypothermic circulatory arrest is worse at 10°C than 15°C (Zimmerman et al. 1990). An explanation for this observation is the concept of an available oxygen reserve.

The ability to measure microvascular saturation makes it possible to monitor cerebral oxygen delivery and extraction, thus providing objective information which may affect intervention. For example, when saturation values approach 30%, transient recirculation of blood will increase cerebral oxygen availability and perhaps improve the safety of clinical hypothermic cardiac arrest.

REFERENCES

Chance B., Leigh J., Miyake H., Smith D., Nioka S., Greenfield R., Finlander M., Kaufman K., Levy W., Young M., Cohen P., Yoshioka H. (1988) Comparison of time resolved and unresolved measurements of deoxygenation in brain. Proc Natl Acad Sci USA 85:4971-4975.

Jobsis F. (1977) Noninvasive, infrared monitoring of cerebral and myocardial oxygen sufficiency and circulatory parameters. Science 198:1264-1267.

Mayevsky A., Nikoda S., Chance B. (1988) Fiber optic surface fluorometry/ reflectometry and 31-P-NMR for monitoring the intracellular energy state in vivo. Adv Exp Med Biol 222:365-374.

McCormick P., Stewart M., Goetting M., Dujovny M., Lewis G., Ausman J.(1991a) Noninvasive cerebral optical spectroscopy for monitoring cerebral oxygen delivery and hemodynamics. Crit Care Med 19:89-97.

McCormick P., Stewart M., Lewis G. (1991b) Noninvasive measurement of regional cerebrovascular oxygen saturation in humans using optical spectroscopy. Proceedings SPIE 1431:294-302.

McCormick P., Stewart M., Goetting M., Balakrishnan G. (1991c) Regional cerebrovascular oxygen saturation measured by optical spectroscopy in humans. Stroke 22:596-602.

Mchedlishvili G. (1986) Arterial Behavior and Blood Circulation in the Brain, Plenum Press, New York, 55-60.

O'Conner J., Wilding T., Farmer P., Sher J., Ergin M., Griepp R. (1986) The protective effect of profound hypothermia on the canine nervous system during one hour of circulatory arrest. Ann Thorac Surg 41:255-259.

Smith D., Levy W., Maris M., Chance B. (1990) Reperfusion hyperoxia in brain after circulatory arrest in humans. Anesthesiology 73:12-19.
Spetzler R., Hadley M., Rigamonti D., Carter L., Raudzens P., Shedd S., Wilkinson E. (1988) Aneurysms of the basilar artery treated with circulatory arrest, hypothermia, and barbiturate cerebral protection. J Neurosurg 68:868-879.
Stocker F., Herschkowitz N., Bossi E., Stoller M., Cross T., Aue W., Seelig J. (1986) Cerebral metabolic studies in situ by 31P-nuclear magnetic resonance after hypothermic circulatory arrest. Pediatric Research 20:867-871.
Tamura M., Hazeki O., Nioka S., et al. (1988) The simultaneous measurements of tissue oxygen concentration and energy state by near infrared and nuclear magnetic resonance spectroscopy. Adv Exp Med Biol 222:359-363.
Tanaka J., Shiki K., Asou T., Yasui H., Tokunaga K. (1988) Cerebral autoregulation during deep hypothermic nonpulstile cardiopulmonary bypass with selective cerebral perfusion in dogs. J Thoracic and Cardiovascular Surgery 95:124-132.
Wan S., Parrish J., Anderson R., Madden M. (1981) Transmittance of nonionizing radiation in human tissues. Photochemistry and Photobiology 34:679-681.
Wray S., Cope M., Delpy D., Wyatt J., Reynolds E. (1988) Characterization of the near infrared absorption spectra of cytochrome aa3 and hemoglobin for the noninvasive monitoring of cerebral oxygenation. Biochemica et Biophysica Acta 933:184-192.
Zimmerman R., Hargraves R., Zabramski J., Spetzler R. (1990) Cardiac standstill and hypothermia: an evaluation of time and temperature. Abstract American Association of Neurologic Surgeon: #1030

OPTICAL CT IMAGING OF HEMOGLOBIN OXYGEN-SATURATION USING DUAL-WAVELENGTH TIME GATE TECHNIQUE

Y. Shinohara, S. Takagi, N. Shinohara, F. Kawaguchi*, Y. Itoh*, Y. Yamashita*, and A. Maki*

Department of Neurology, Tokai University School of Medicine, Isehara, Japan
*Central Research Laboratory, Hitachi Ltd, Kokubunji, Japan

INTRODUCTION

Several attempts have been made to measure hemoglobin oxygen-saturation level in various tissues in vivo by using near-infrared transmission or reflection spectroscopy (Jobsis, 1977, Chance et al. 1988, Delpy et al. 1988). This is based on the fact that oxyhemoglobin and deoxyhemoglobin have different absorption spectra in the near infrared region, where light absorption by other tissues is relatively weak. However, although tissues are quite transparent to near-infrared light, strong light scattering occurs, making it difficult to measure accurately the absorption due to hemoglobin contained in the tissue. Because of this multiple light scattering, oxygen-saturation imaging of tissues is very difficult, especially using the conventional filtered back-projection CT algorithm mainly employed in X-ray CT. We have already reported our method for in vivo visualization of the pattern of hemoglobin oxygen-saturation in the rat brain (Shinohara et al. 1991). The purpose of this presentation is to describe the results obtained with dual-wavelength time gate technique using different time gate widths for imaging of hemoglobin oxygen-saturation.

MATERIALS AND EQUIPMENT

Nineteen 10-week-old Wistar rats were used under light anesthesia with chloralose and urethane. The method and equipment have already been reported by us (Shinohara et al. 1991). Briefly, two kinds of ultra fast semiconductor pulse lasers having wavelengths of 770 nm and 812 nm were used as near-infrared light sources. The duration of each emitted light pulse was 60 ps, and the repetition frequency was 100 Mhz. In other words, the tissue was pulse-irradiated every 10 ns, and pulsing was repeated 7×10^8 times. As a specific absorption band for deoxyhemoglobin the wavelength of 770 nm was selected, while 812 nm was selected as a wavelength near the isosbestic point of oxy- and dexoy- hemoglobin. A light beam emitted by a

Optical Imaging of Brain Function and Metabolism, Edited by
U. Dirnagl *et al.*, Plenum Press, New York, 1993

semiconductor laser was passed through the intact skull, after depilation, into the rat brain by means of an optical fiber having an outer diameter of 1 mm. Light transmitted through the brain was detected by 7 optical fibers held in place on the opposite side of the skull by a circular aluminium holder, which was set on the skull. The light detected by the 7 optical fibers was passed to a streak camera to obtain 7 time-reolved spectral data simultaneously. The streak camera used for this study was a Hamamatsu C 1587 System.

RESULTS AND COMMENTS

Fig. 1 shows a schematic view of the input light pulse and the output signal from the streak camera corresponding to the light detected at one optical fiber after transmission through rat brain. These data were transferred to a computer and a time-gating process was conducted. The width of the time gate was altered between 60 and 300 ps, and the total light intensity included in the time window was calculated. Although the obtained output signals were far smaller in the case of 60 ps than those in the case of, for example, 300 ps, the former excludes most of the scattering.

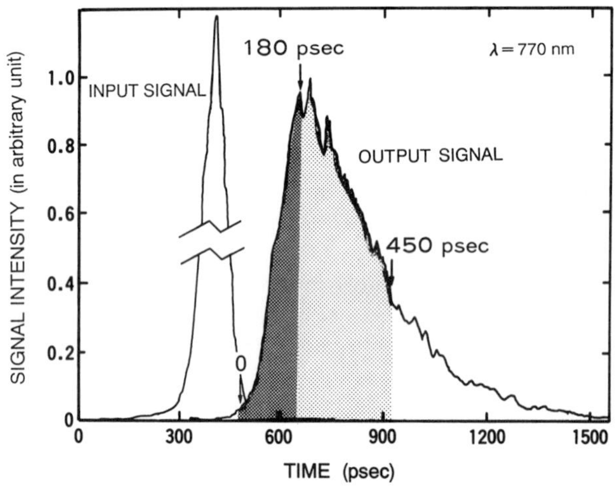

FIGURE 1. Schematic view of the input signal and the output signal obtained by our method. The black area indicates the portion of the output signal that is utilized when the time gate is set at 180 ps.

The zero point of the time gate for each spectrum was arbitrarily set at the time when the integrated intensity of the output signal reached 1% of its total value. This seems to be reasonable, because the zero point set in this way occurred about 100 ps after the peak of the input signal in most cases. Since light travels about 2.3 cm in 100 ps, the zero point corresponds to light transmitted directly from the input optical fiber to the detecting optical fiber. These measurements were done for one wavelength first, for example 770 nm, then for the other wavelength, 812 nm. Each set of measurements was repeated 12 times from 12 different directions around the skull to obtain a complete set of data for a CT image.

Twelve times 7 projection data for each wavelength were used to obtain an image of oxygen-saturation distribution using a conventional back-projection CT image reconstruction method. In this reconstruction process we assumed the absorption level out of the field of detection to be zero.

Fig. 2 shows a hemoglobin oxygen-saturation image of the brain of a lightly anesthetized rat. In this example the width of the time gate was set at 300 ps. The posterior fossa seems to have a high oxygen saturation level compared to the cerebrum in this condition, although this result is qualitative.

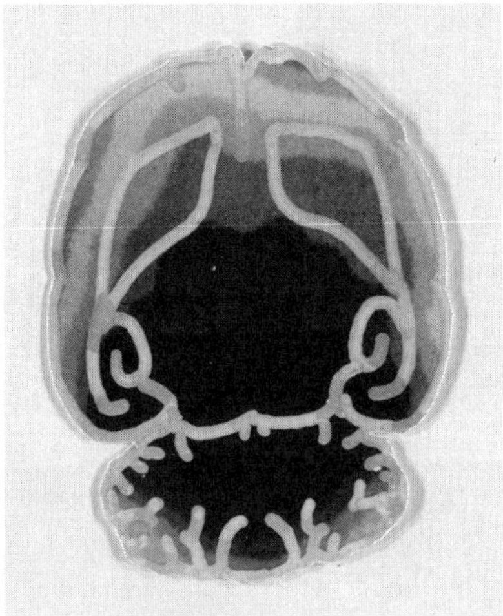

FIGURE 2. A hemoglobin oxygen-saturation image of the brain of a lightly anesthetized rat (time gate at 300 ps). Black indicates high hemoglobin oxygen-saturation level and white indicates low level.

Fig. 3 shows hemoglobin oxygen-saturation images of the brain of a lightly anesthetized rat with right middle cerebral artery occlusion, obained with different time gate settings. As can be seen in the upper figure, the shorter time gate (60 ps) seems to give the best result.

SUMMARY

Our results may be summarized as follows:
1. In-vivo hemoglobin oxygen-saturation images of rat brain can be obtained using near-infrared light.
2. The transmitted light has multiple scattered components, which degrade spatial resolution, though most of the scattering can be excluded by a suitable time-gating technique.
3. The hemoglobin oxygen-saturation of the cerebellum is higher than that of the cerebrum in untreated anesthetized rats.

Although further improvements in the technique should result in better resolution, we have demonstrated the feasibility of using near-infrared light to obtain in-vivo hemoglobin oxygen-saturation images.

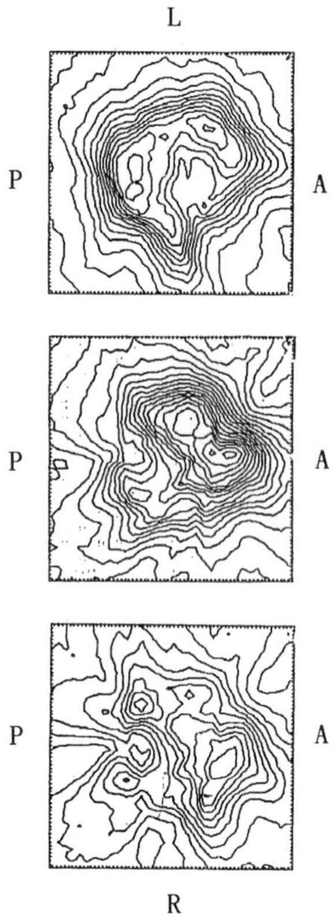

FIGURE 3. Hemoglobin oxygen-saturation images of the brain of a lightly anesthetized rat with right middle cerebral artery occlusion. Upper: time gate at 60 ps, middle: at 150 ps, and lower: at 300 ps. A: anterior, P: posterior, R: right, L: left

REFERENCES

Chance B, Leigh JS, Miyake H, Smith DS, Nioka S, Greenfeld R, Finander M, Kaufmann K, Levy W, Young M, Cohen P, Yoshioka H, and Boretsky R. Comparison of time-resolved and -unresolved measurements of deoxyhemoglobin in brain. Proc Natl Acad Sci USA 85:4971 (1988)

Delpy DT, Cope M, van der Zee P, Arridge S, Wray S, Wyatt J. Estimation of optical pathlength through tissue from direct time of flight measurement. Phys Med Biol33:1433 (1988)

Jobsis FF. Noninvasive, infrared monitoring of cerebral and myocardial oxygen sufficiency and circulatory parameters. Science 198:1264(1977)

Shinohara Y, Haida M, Kawaguchi F, Itoh Y, Yamashita Y, Takeuchi H. Hemoglobin oxygen-saturation image of rat brain using near infrared light. J Cereb Blood Flow Metab 11(Suppl 2):459(1991)

OPTICAL PROPERTIES OF NORMAL HUMAN INTRACRANIAL TISSUES IN THE SPECTRAL RANGE OF 400 TO 2500 NM

H.R. Eggert and V. Blazek*

Department of Neurosurgery, Kassel Community Hospital,
Kassel, Germany and
*Institute for High Frequency Technique, Aachen University of Technology,
Aachen, Germany

INTRODUCTION

Optical properties describe the physical interactions between electromagnetic radiation and matter dependent on the wavelength of the radiation. Incident power may be reflected, absorbed, scattered and/or transmitted. Reflection may consist of specular reflection and back scattering from inside the matter. The total amount of absorption and internal scattering may be expressed as extinction. According to the law of energy conservation, reflection, extinction and transmission add up to 1. For that reason, measuring optical properties of tissues must include at least two of these parameters or one of them must be excluded by the experimental design. Using a semi-infinite thickness of specimens, the influence of transmitted power may be neglected. According to the Kubelka-Munk-theory (Kubelka and Munk 1931, Kubelka 1948), the ratio of absorption and scattering can be calculated from reflection measurements assuming an ideally black background and a semi-infinite slice thickness (Blazek 1979). In reflection as well as in transmission measurements it has to be considered that the intensity of reflected and transmitted power depends on the solid angle of reflection or transmission (Hardy et al. 1956, Longini et al. 1968). This problem may be solved by using a two-beam spectral photometer with an integrating sphere as measuring instrument (Blazek 1975).

The knowledge of optical properties of normal intracranial tissues within a rather wide spectral range may influence the use of lasers as neurosurgical tools. It may be of some help for photodynamic therapy of brain tumors (Boggan et al. 1985, Cheng et al. 1986) as well as for in vivo measurements of cerebral blood flow and metabolism. The following contribution deals with the optical properties of normal brain tissue, meninges and arterial walls between 400 and 2500 nm measured with a two-beam spectral photometer using an integrating sphere as measuring instrument.

Optical Imaging of Brain Function and Metabolism, Edited by
U. Dirnagl *et al.*, Plenum Press, New York, 1993

MATERIALS AND METHODS

A two-beam spectral photometer with an integrating sphere of 60 mm in diameter (UV/VIS/NIR-Spectrometer, Lambda 9; Perkin & Elmer, Überlingen, Germany) was used in order to measure spectral reflection and transmission. The measuring range extended from 400 to 2500 nm. The instrument provides a wavelength accuracy of ± 0.2 nm (UV/VIS) respectively ± 0.8 nm (NIR), a wavelength reproducibility of ± 0.05 nm (UV/VIS) resp. ± 0.2 nm (NIR), a photometrical accuracy of ± 0.003 E at 1E, and a photometrical reproducibility of ± 0.001 E. The measuring and reference beam are calibrated by aiming the beams at reflection standards made of barium sulfate. Calibration and zero adjustment are carried out automatically. A spectral band width of 2 nm and a scan speed of 960 nm/min were choosen. Measured data were recorded on-line with a plotter and simultaneously saved at a floppy disc. By this means, computer assisted data processing became possible. For statistical calculations a spectral band width of 10 nm was used.

The spectral reflectance expressed as a percentage of incident power was measured in tissue samples of at least 5 mm in thickness. This thickness seemed to be sufficient to meet the Kubelka-Munk theory (Kubelka and Munk 1931, Kubelka 1948) and to approximate the intraoperative use of lasers, in which a semi-infinite slice thickness usually has to be assumed. The tissue samples were mounted at slides made of quartz glass (Suprasil^R). The ratio of spectral absorption and scattering was calculated according to the Kubelka-Munk theory as described earlier (Eggert and Blazek 1987).

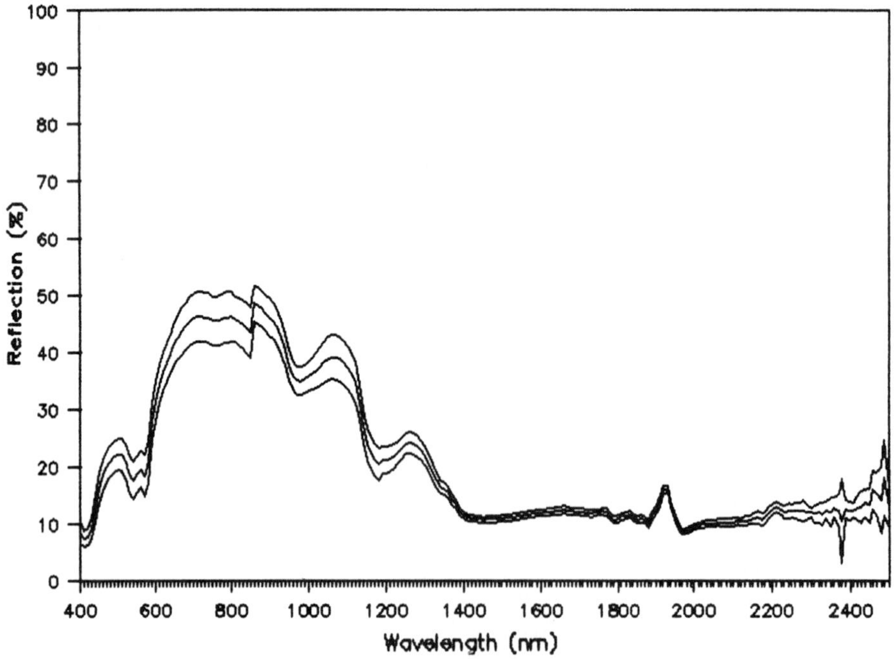

FIGURE 1. Frontal gray matter: relative levels of reflection plotted against wavelength, means ± SD.

In thin tissue slices, transmittance and reflectance were measured with the same device. The moist samples were put between two glasses providing a fixed distance of 0.8 mm between them. Care was taken to prevent the samples from drying out during the measuring procedure. From the measured transmittance and reflectance, the penetration depth at which the total optical power is reduced to 37 % was calculated according to Beers' law. Absorption coefficients were expressed as the reciprocal of penetration depth.

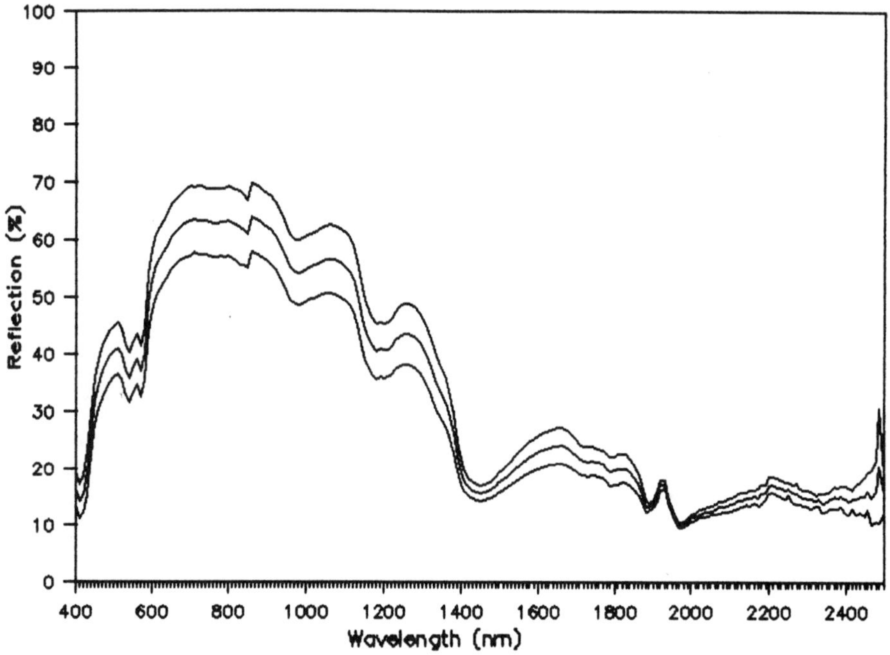

FIGURE 2. Frontal white matter: relative levels of reflection plotted against wavelength, means ± SD.

Normal gray and white matter, dura of the convexity, falx, tentorium, arterial walls of intra- and extracranial carotid artery and basilar artery were removed from 9 cadavers during autopsy. The mean age of the cadavers at autopsy was 70.7 years (range 54 - 85 years). Brains of patients with evidence of neurological disorders were excluded from the investigation. The average thickness of dura of the convexity was 0.43 mm (range 0.25 - 0.5 mm), of falx 0.43 mm (range 0.35 - 0.65 mm) and of tentorium 0.56 mm (range 0.3 - 1.0 mm). The walls of investigated arteries were between 0.05 and 1.4 mm in thickness, basilar artery 0.1 - 0.55 mm (mean 0.26 mm), intracranial internal carotid artery 0.05 - 0.4 mm (mean 0.26 mm) and extracranial internal carotid artery 0.5 - 1.4 mm (mean 0.80 mm).

RESULTS

Basically, in all investigated tissues the measured reflectance increased from 400 to 700 nm. From 800 to 1400 it decreased gradually (Fig. 1, 2). Correspondingly, absorption decreased from 400 to 700 nm and increased from 800 to 1400 nm. Because of an automatically performed change of the light source and the detector of the photometer a leap of up to 10 % reflection was recorded at 860 nm. Beyond 1440 nm, absorption decreased to a minimum at 1660 nm. The next maximum of absorption was found at 1970 nm. Maximum absorption was recorded at 410, 540, 580, 970, 1180, 1440 and 1970 nm. At wavelengths longer than 2170 nm the statistical dispersion of data increased significantly probably due to a rather poor signal to noise ratio (Fig. 3, 4)nI general, absorption was significantly higher in gray matter than in white matter. Maximum absorption was observed at 1970 nm in both, gray and white matter. In gray matter covered by intact arachnoid, absorption was up to 5 % higher than after removing the arachnoid between 400 and 1370 nm. At longer wavelengths, there were no major differences.

Scattering was usually higher in white matter than in cortical gray matter. In gray matter maximum scattering (41 %) was observed at 650 and 1060 nm (Fig. 5). In white matter, scattering reached the same level, but maxima were found at 580 and 1210 nm (Fig. 6).

Between 600 and 1130 nm white matter scattering seemed to be less than scattering of gray matter. However, because the scattering values are derived from reflection measurements which include the power scattered back out of the tissue these values might appear too low in tissues and at wavelengths with high internal scattering.

From measurements of spectral reflection and transmission in thin tissue slices penetration depth and absorption coefficients were obtained. As expected, the absorption coefficients of gray matter were definitely higher than those of white matter. The smallest values were found between 640 and 1080 nm, the highest at 1970 nm (Fig. 7). Maxima corresponded to the maxima revealed by reflection measurements of thick tissue specimens.

Arachnoid, convexity dura, tentorium and falx showed no major differences with respect to their absorption coefficients. Within the visible spectral range, the higher values were obtained in falx. This may be due to the higher vascularization of the falx (Fig. 8).

Similar observations were made in estimating the absorption coefficients of arterial walls. Only at 1440 nm the absorption coefficient of extracranial internal carotid artery was obviously higher than that of intracranial carotid artery and basilar artery (Fig. 9).

DISCUSSION

Due to the high chemical and structural complexity of biological tissues methods like photoacoustic spectrometry (De Tommasi et al. 1986) or reflection and transmission spectrometry (Blazek 1975, van Gemert et al. 1985) seem to provide more useful information about optical properties of these tissues than they can be obtained by measurements of molar extinction coefficients of single chromophores and water (Bayly et al. 1963, Parrish 1982). However, it has to be considered that these methods can only provide an approximation of optical tissue parameters since each method requires certain simplifying assumptions.

There are only few observations of optical properties of human brain tissue at few wavelengths in vivo and in vitro. The data obtained by our measurements are in rather good correspondence to values obtained with other different methods in vitro (Svaasand and Ellingsen 1983, Svaasand and Ellingsen 1985, De Tommasi 1986). However, three in vivo measurements performed in two patients at 630 nm (Wilson 1986) revealed penetration depths between 2.9 and 4.9 mm which are significantly higher than values calculated from our data (gray matter 2.2 ± 0.2 mm, white matter 2.4 ± 0.8 mm) as well as in other post mortem studies (Svaasand and Ellingsen 1983). It seems unlikely that this difference is caused by blood flow of living tissues because there is good evidence from theoretical considerations (Welch et al. 1980) as well as from experimentally obtained results (Doiron et al. 1982, Eichler et al. 1978) that optical properties are not influenced by blood flow to a degree that may significatly confound our measurements. On the other hand, there may be differences of blood and water content between specimens and living tissues. It became clearly evident by our measurements that within the visible spectral range optical properties of brain tissue are mainly influenced by hemoglobin. Between approximately 600 and 900 nm scattering due to myelin may play a major role especially in white matter (Svaasand and Ellingsen 1983). At wavelengths longer than 900 nm, absorption maxima of brain tissue coincide with the absorption maxima of water (Bayly 1963).

In conclusion, valuable information about optical properties of intracranial tissues may be obtained from reflection and transmission measurements using a two-beam spectral photometer with an integrating sphere as measuring instrument. There are rather high differences of optical properties between gray and white matter. Optical properties of brain tissue are obviously influenced by the hemoglobin and water content of the tissue.

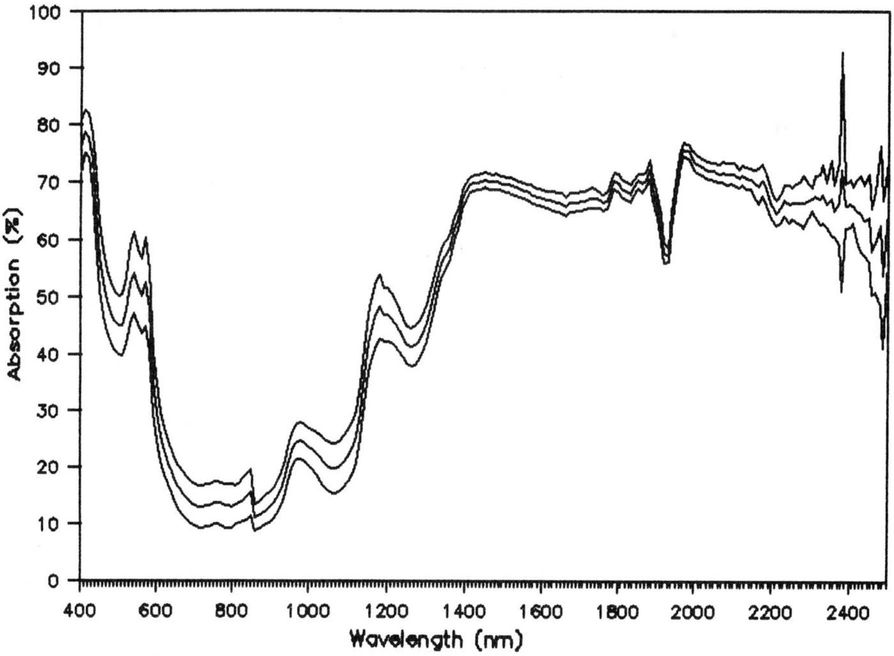

FIGURE 3. Frontal gray matter: relative levels of absorption plotted against wavelength, means ± SD.

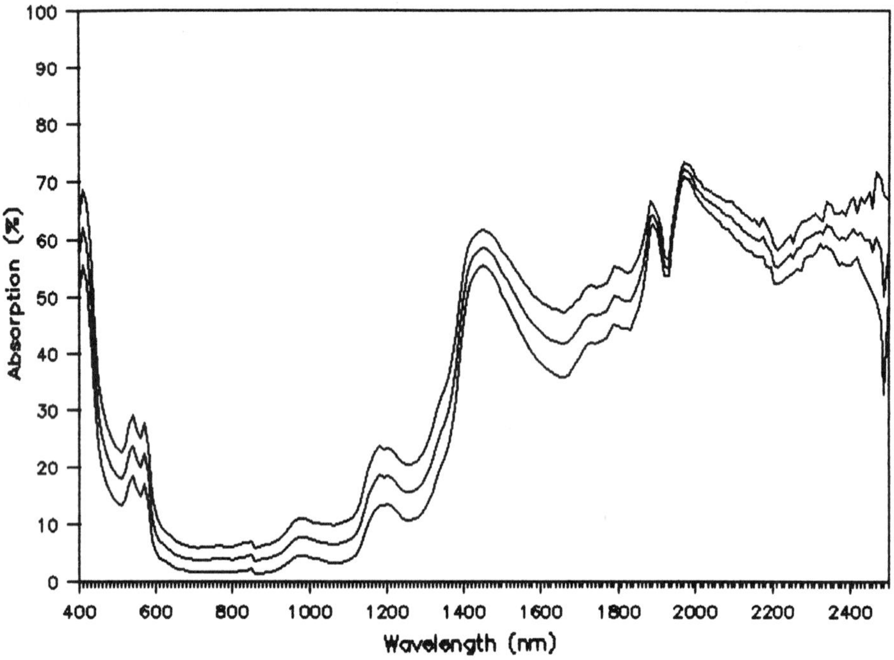

FIGURE 4. Frontal white matter: relative levels of absorption plotted against wavelength, means ± SD.

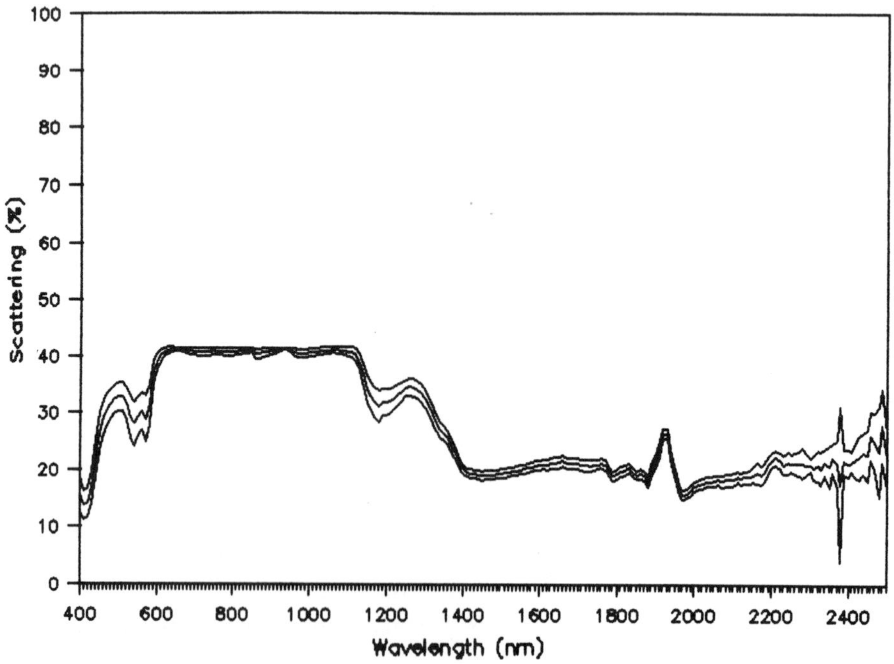

FIGURE 5. Frontal gray matter: relative levels of scattering plotted against wavelength, means ± SD.

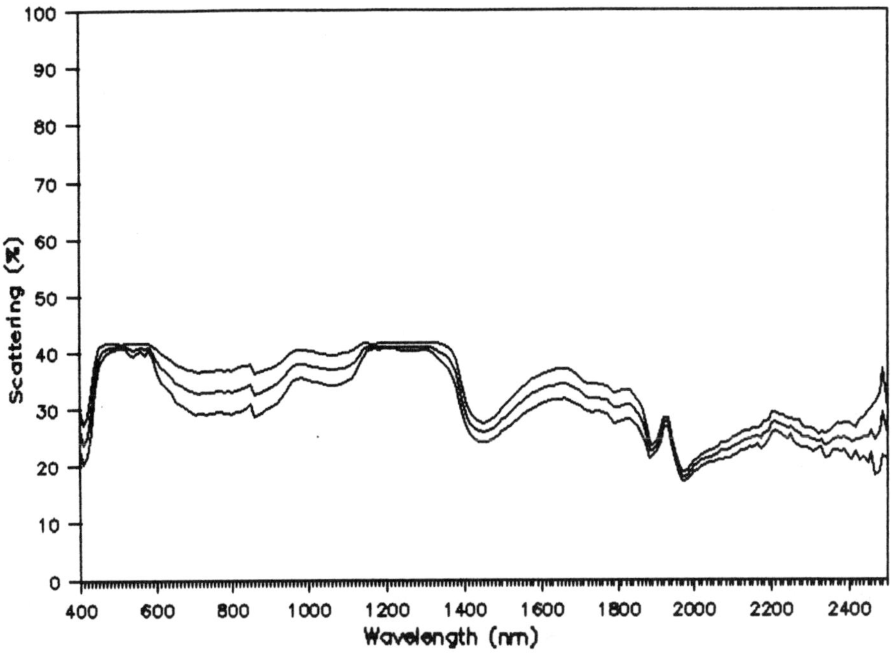

FIGURE 6. Frontal white matter: relative levels of scattering plotted against wavelength, means ± SD.

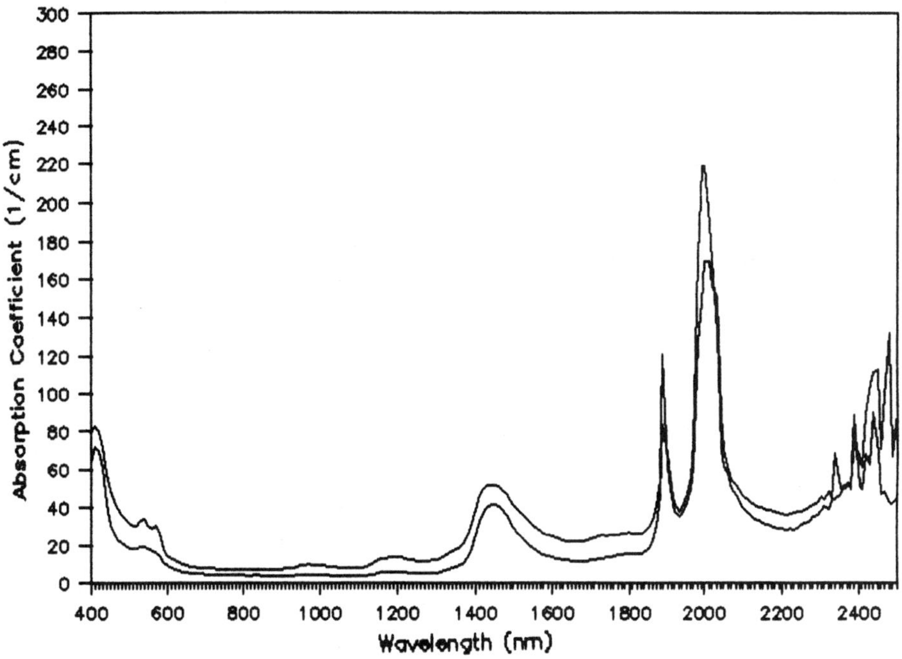

FIGURE 7. Frontal gray (upper line) and white (lower line) matter: absorption coefficients plotted against wavelength, means.

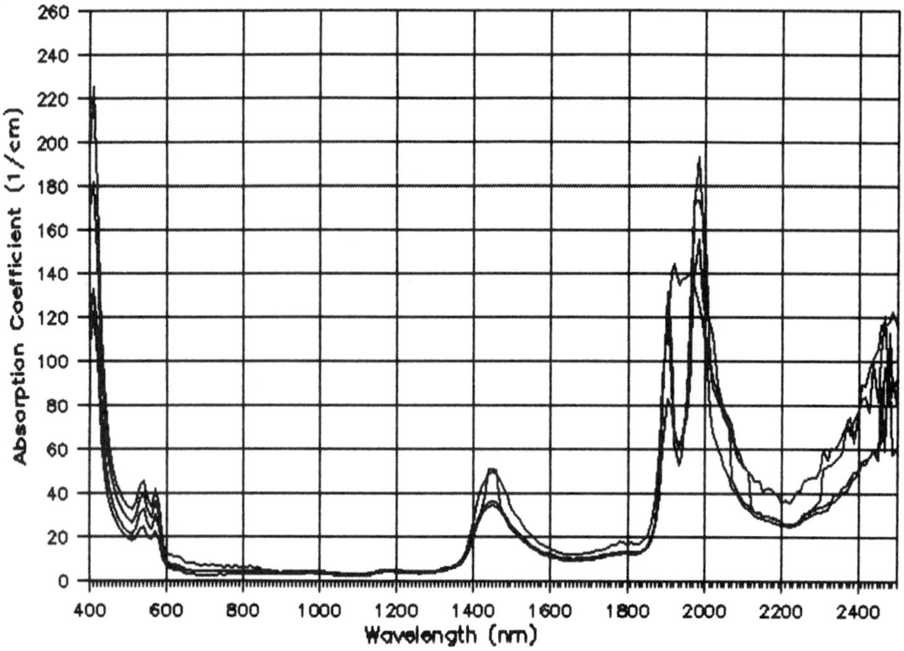

FIGURE 8. Arachnoid, convexity dura, tentorium, and falx: absorption coefficients plotted against wavelength, means.

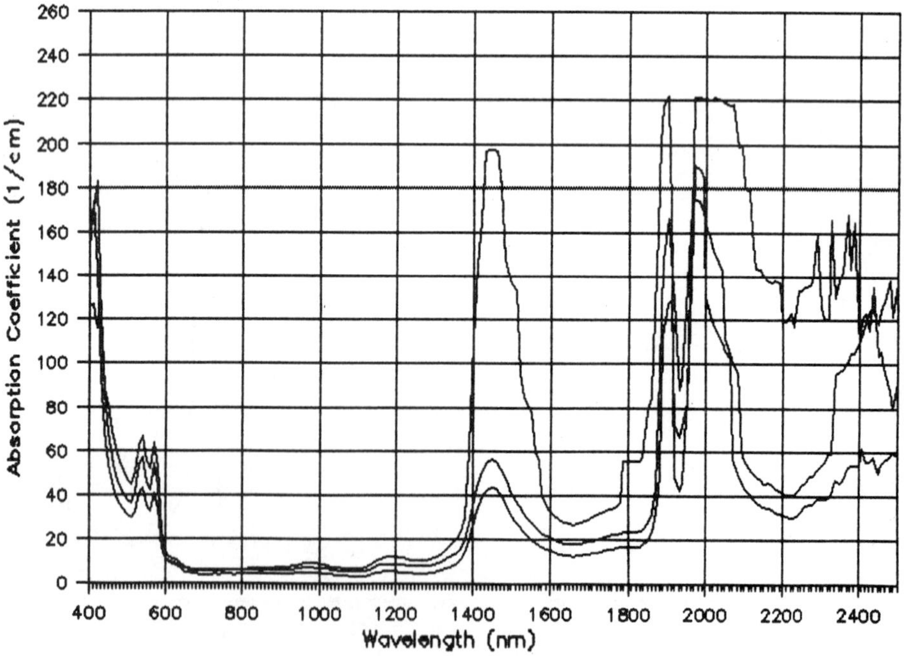

FIGURE 9. Arterial walls of basilar artery, extracranial and intracranial carotid artery: absorption coefficients plotted against wavelength, means.

REFERENCES

Bayly J.G., Kartha U.B., Stevens W.H. (1963) The absorption spectra of liquid phase H_2O, HDO and D_2O from 0.7 μm to 10 μm. Infrared Physics 3:211-223.

Blazek V. (1975) Reflexions-, Transmissions- und Absorptionsverhalten von biologischem Gewebe für elektromagnetische Strahlung im sichtbaren und nahen IR-Bereich. Biomed Technik 20:299-300.

Blazek V. (1979) Absorption and scattering properties of biological tissue in relation to monochromatic optical radiation. Biomed Technik 24:331-332.

Boggan J.E., Bolger C., Edwards M.S.B. (1985) Effect of hematoporphyrin derivative photoradiation therapy on survival in the rat 9L gliosarcoma brain-tumor model. J Neurosurg 63:917-921.

Cheng M.K., McKean J., Boisvert D., Tulip J. (1986) Photoradiation therapy: Current status and application in the treatment of brain tumors. Surg Neurol 25:423-435.

De Tommasi A., Occhiogrossi M., Vailati G., Baldarre L., Cingolani A. (1986) Evaluation of the Ar+ laser thermal effect in rabbit brain tissue by means of optical absorption coefficients: photoacoustic measurements. Acta Neurochir(Wien) 79:139-144.

Doiron D.R., Svaasand L.O., Profio A.E. (1982) Light dosimetry in tissue: Application to photoradiation therapy, in: "Porphyrin Photosensitization," D. Kessel, T.J. Dougherty, eds., Plenum Press, New York.

Eggert H.R., Blazek V. (1987) Optical properties of human brain tissue, meninges, and brain tumors in the spectral range of 200 to 900 nm. Neurosurgery 21:459-464.

Eichler J., Knof J., Lenz H., Salk J., Schäfer G. (1978) Temperature distribution in tissue during laser irradiation. Radiat Environ Biophys 15:277-287.

Hardy J.D., Hammel H.T. Murgatroyd D. (1956) Spectral transmittance and reflectance of excised human skin. J Appl Physiol 9:257-264.

Kubelka P., Munk F. (1931) Ein Beitrag zur Optik der Farbanstriche. Z Techn Phys 12:593-601.

Kubelka P. (1948) New contributions to the optics of intensely light scattering materials: Part I. J Opt Soc Am 38:448-457.

Longini R.L., Zdrojkowski R. (1968) A note on the theory of backscattering of light by living tissue. IEEE Trans Biomed Eng, BME-15:4-10.

Parrish J.A. (1982) Photobiologic considerations in photoradiation therapy, in: "Porphyrins and Photosensitization," D. Kessel, T.J. Dougherty, eds., Plenum Press, New York.

Svaasand L.O., Ellingsen R. (1983) Optical properties of human brain. Photochem Photobiol 38:293-299.

Svaasand L.O., Ellingsen R. (1985) Optical penetration in human intracranial tumors. Photochem Photobiol 41:73-76.

Welch A.J., Wissler E.H., Priebe L.A. (1980) Significance of blood flow in calculation of temperature in laser irradiated tissue. IEEE Trans Biomed Eng 27:164-166.

Wilson B.C., Muller P.J., Yanch J.C. (1986) Instrumentation and light dosimetry for intraoperative photodynamic therapy (PDT) of malignant brain tumours. Phys Med Biol 31:125-133.

van Gemert M., Verdaasdonk R., Strassen E.G., Schets G.A.C.M., Gijsbers G.H.M., Bonnier J.J. (1985) Optical properties of human blood vessel wall and plaque. Lasers Surg Med 5:235-237.

OPTICAL IMAGING OF THE FUNCTIONAL ARCHITECTURE IN CAT VISUAL CORTEX: THE LAYOUT OF DIRECTION AND ORIENTATION DOMAINS

T. Bonhoeffer[1] and A. Grinvald[2]

[1]Max-Planck Institut für Hirnforschung, Frankfurt, Germany
[2]The Weizmann Institute of Science, Rehovot, Israel

INTRODUCTION

Optical imaging of the functional architecture in cat visual cortex. The layout of direction and orientation domains.

It is a well established fact that the cerebral cortex and in particular the visual cortex in mammals is organized in a columnar fashion (Mountcastle, 1957; Hubel and Wiesel, 1965): neurons lying below each other from the pia to the white matter - despite some lamina-dependent differences - share important functional properties. A method which has proven to be very useful for investigating the functional architecture across a cortical area is the 2-Deoxyglucose (2-DG) technique (Sokoloff, 1977). It takes advantage of the fact that neuronal activity is concomitant with enhanced metabolic activity, which can be permanently marked with the radioactive glucose analog.

One important limitation of this powerful method is related to the fact that there are only two ways to mark 2-Deoxyglucose radioactively. This presents a severe experimental restriction since it limits the number of activity maps which can be recorded from a piece of cortical tissue. This hampers for instance clear answers to outstanding questions about the spatial relationships between different functional components in the cortex (for example the relationship between columns of different orientation preference or between orientation and ocular dominance columns in the visual cortex). Additionally, results can only be obtained *post mortem* and it is therefore impossible to use the acquired information to guide further *in vivo* experiments on the same animal.

Recently our lab has developed a new technique which allows to visualize metabolic activity *in vivo* (Grinvald et al. 1986; Ts'o et al. 1990) . This new technique, optical imaging of intrinsic signals, therefore has the advantage that one can obtain multiple activity maps from a single cortical area by presenting the animal with many different stimuli. From these data one can obtain a far more complete picture of the functional architecture of the cortex.

Optical Imaging of Brain Function and Metabolism, Edited by
U. Dirnagl *et al.*, Plenum Press, New York, 1993

Optical imaging of intrinsic signals makes use of the fact that the optic properties of cortical tissue change with its electrical activity. There are various sources for the optical signal which conveys information about cortical activity (Frostig et al. 1990). The major component of the intrinsic signal is caused by the fact that oxy-hemoglobin and hemoglobin have very different absorption spectra. This makes oxygen consumption by active tissue detectable optically. Beside the so called oximetry signal there are also other sources for intrinsic signals. They include activity dependent changes in blood volume, cytochrome-absorption and light scattering.

In our experiments we utilized these signals to obtain multiple maps from one cortical region to examine whether there are "directionality columns". Such columns have been reported in cat area 17 (Tolhurst et al. 1981; Payne et al. 1980) and recently also in area 18 (Swindale et al. 1987) but these results remained somewhat controversial. Our method allowed us to directly compare high-resolution activity maps obtained with a grating moving in one direction with that obtained with the same grating moving in the opposite direction. This comparison can resolve the above question of whether cells are sorted into columns according to preferred direction of motion or whether they are not.

We also wanted to clarify the detailed organization of orientation preference in cat visual cortex. There has been a great deal of debate about how orientation columns are spatially organized in cat visual cortex. The original proposition by Hubel and Wiesel was that they are "swirling stripes with many bifurcations and endings" (Hubel et al. 1977; Hubel et al. 1978). Later on von Seelen (1970) and Braitenberg and Braitenberg (1979) proposed that many of the data obtained by Hubel and Wiesel could also be explained by centers around which the different orientation domains are organized in a pinwheel-like fashion (each orientation appearing around the center twice). Löwel et al. (1987) challenged this interpretation by performing an extensive 2-DG study of area 17 and 18 in the cat from which they concluded that orientation is organized in stripes rather than in a pinwheel-like fashion. Subsequent data from monkey V1 (Blasdel and Salama, 1986) and area 18 of the cat (Swindale et al. 1987) again revived the idea of orientation being organized around centers. These data prompted papers by a number of theoreticians (Linsker, 1986; Götz, 1987; Durbin and Mitchison, 1990; Obermayer et al. 1990) who suggested that instead of orientation centers around which every orientation appears twice (as suggested by Braitenberg and Braitenberg) the actual organization might rather be one with every orientation appearing only once in the circle. These studies were based on work with monkeys, but also work on cat area 18 (Swindale et al. 1987) suggested that the organization there might be similar. The latter experiments had, however, the disadvantage that the maps obtained had a resolution of only 300-400 μm. Consequently the orientations between these data points could only interpolated and were thus burdened with some uncertainty.

Since all the previously performed experiments lacked either sufficient resolution (electrode recordings) or the benefit of having data from many different orientations (2-DG experiments) we wanted to use our new method to take a closer look at the organization of orientation preference in cat area 18. In particular we wanted to be able to distinguish between the three proposed models for the organization of orientation preference. (1) Parallel bands of orientation preference (Hubel et al. 1977; Hubel et al. 1978), (2) orientation centers with every orientation appearing twice around the center (Braitenberg and Braitenberg, 1979) or (3) orientation centers (appearing with two different handednesses, clockwise and counterclockwise) with every orientation only appearing around the center singularity once (Swindale, 1982; Linsker, 1986; Götz 1987).

MATERIAL AND METHODS

Animals

Cats were initially anesthetized with ketamine HCl (20mg/kg, IM) followed by sodium pentothal (20mg/kg, IV) and further injections as needed. A tracheotomy was performed and after paralysis (succinylcholine HCl, 20 mg/kg/h, IV) the animal was artificially respirated. EEG, ECG, expired CO_2 and rectal temperature were monitored throughout the experiment.

A retinoscope was used to determine the correct refractive power for the contact lenses which were put on the cats eyes.

Optical chamber

The skull of the cat was opened above area 18 by opening two half-circled holes whose "centers" lay at Horsley-Clarke A5 and 0.5 mm lateral of the mid-line. After a stainless steel chamber was cemented onto the skull, the dura was removed, the chamber filled with silicone oil and sealed with a round glass coverslip. This arrangement minimized the movement of the cortex due to heartbeat and respiration which otherwise severely hampered satisfactory optical imaging.

CCD camera and optics

Figure 1 shows the general setup used for our experiments. The surface of the cortex was illuminated by two adjustable light guides. These guides were attached to a Zeiss tungsten-halogen lamp. The light was passed through interference filters of different wavelengths. The filter most often used for visualizing the surface of the cortex and its blood vessels had a transmission maximum at 540 ± 15 nm whereas the filter for the optical imaging proper was a 605 ± 5 nm filter.

A slow scan CCD camera (Photometrics) recorded pictures from the exposed parts of the cortex. The pictures which in the current arrangement have a spatial resolution of 192 x 144 pixels were digitized and fed into DEC Microvax II computer.

We used two different kinds of lenses for our experiments. In some of the experiments standard 35 mm camera lenses were used; in most cases this was a 2.8/135 macro-lens (Pentax), in a few cases we additionally used a 2x converter in order to achieve a two-fold higher magnification. In this case the aperture of the lens was always set to 2.8 which provided us with a depth of field of several hundred microns.

In other experiments we used a "tandem-lens" arrangement which was developed in our laboratory (Ratzlaff and Grinvald, 1991). This device is in essence a microscope (with a rather low magnification) built out of two photographic lenses which together provide an unusually high numerical aperture. This system consequently has a very low depth of field (approximately 50 μm). It allowed us to eliminate virtually all artifacts caused by blood vessels on the surface of the cortex (Malonek et al. 1990).

Visual stimuli

Visual stimuli were displayed on a CRT screen (Mitsubishi) in 60 Hz non-interlaced mode at a distance of 25-45 cm from the animals eyes. The stimuli were generated by an IBM PC/AT with a Number Nine Graphics board. This system was programmed in a language, which allowed to specify the stimulus quality (e.g. bars, gratings, random dots) as well as parameters like color, spatial frequency, velocity etc.

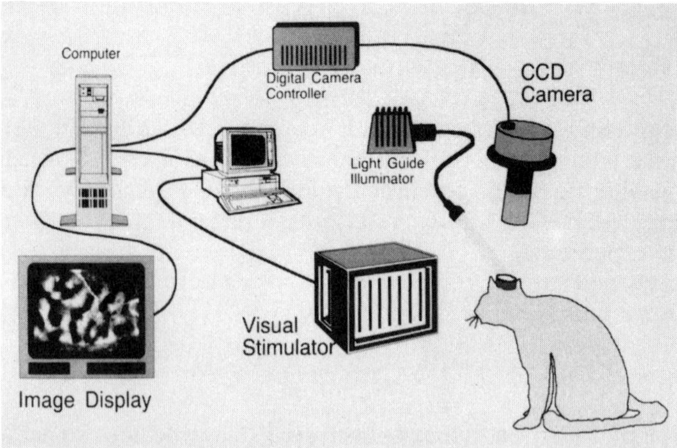

FIGURE 1. Setup of an optical recording experiment using a CCD camera. Pictures are taken from the cortex of the cat which is exposed in a sealed, oil-filled chamber. The cortex is illuminated with light of 630 nm wavelength. Images are acquired with the CCD camera while the cat is visually stimulated with moving gratings on a CRT-screen. These pictures are digitized in a camera controller which passes the data to the computer which controls the entire experiment. Data can then (in breaks of the experiment) be analyzed on this computer and shown on an image display or (during the experiment) they can be sent to a second computer (VAX 3100, not shown) for more detailed, quasi on-line analysis.

in very simple terms. The monitor was positioned so that it stimulated an angle of 40-60 degrees in the contralateral visual field. The stimuli consisted of gratings which had a spatial frequency of 0.15 cycles/deg and moved at a speed of 2.25 cycles/sec.

Data acquisition

For data acquisition we normally presented a particular stimulus to both eyes of the cat for 3 seconds and recorded 7 frames of 450 ms duration with the CCD camera. This period of data acquisition was followed by a 6 sec inter-stimulus interval after which a different stimulus was shown. We presented every stimulus 10 to 100 times randomizing the order of appearance in order to average out any stimulus dependent after-effects.

Data analysis

The first step for data analysis was to add the subsequent frames acquired for every visual stimulus. These pictures were then divided by a "blank"-trial in order to correct for uneven illumination. This blank trial was obtained by two different procedures. In some experiments the pictures for all different stimuli were summed together ("combination-blank"), in others we used the frames acquired during a trial in which the cat saw a blank screen. The resulting blank-adjusted pictures represented the activity map for the corresponding stimulus.

For complete analysis of the organization of orientation preference we calculated orientation maps in which the responses to all four (or eight) gratings were summed vectorially. The angle of the resulting vector was then displayed in a pseudo color code.

RESULTS

The first step in analysis of our data was to look at the activity maps that were produced by dividing the pictures acquired during one stimulus condition (e.g. a moving grating of oblique orientation) by the blank stimulus. These maps were calculated for all the stimulus conditions and served us as a first indication of the quality of the experiment. Figure 2 shows four activity maps that were recorded from the cortical area shown in Figure 4A. They were obtained with moving gratings of four different orientations. High activity leads to strong absorption of the red illuminating light; thus these regions reflect less light and appear darker in the picture. The strong intensity differences between the light and the dark areas in the Figure are deceptive. They are only obtained by appropriate scaling. The difference in reflected light intensity between the lighter and the darker areas in absolute numbers is minute; it is only in the order of 10^{-3}. By comparing Figure 2 A and B one can notice that the cortical regions activated by two orthogonal stimuli are interdigitating; regions strongly activated by one grating are weakly activated by the orthogonal one and vice versa.

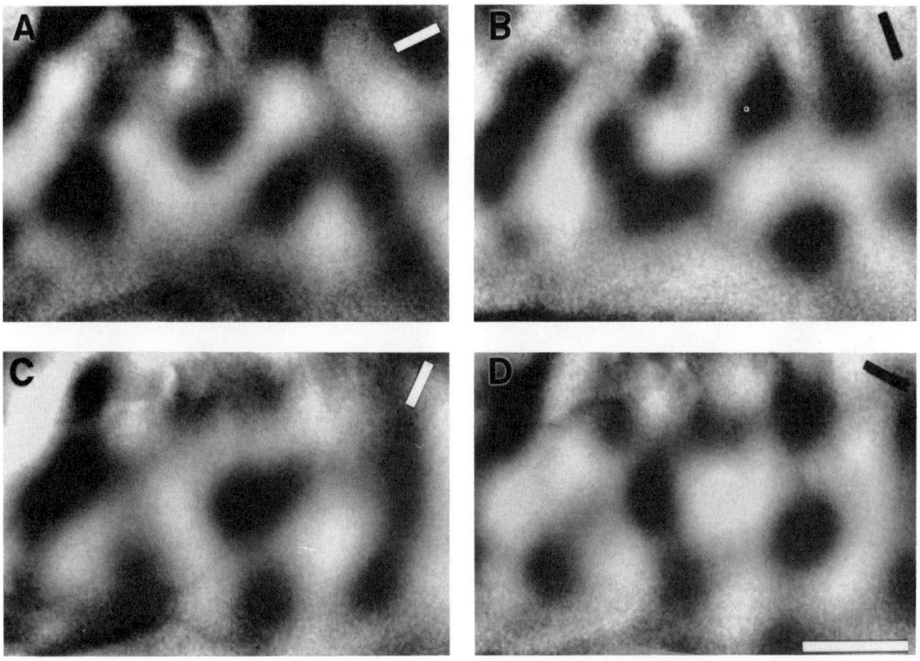

FIGURE 2. Optical imaging of iso-orientation columns in cat visual cortex. Panels A, B, C and D show four activity maps, each produced by a moving grating of a particular orientation (displayed in the upper right corner of each panel). Areas which appear dark in the picture are the regions in which the response to the respective stimulus is maximal. The characteristic patterns in all four maps is one of "beaded bands", dark regions of very strong activation (beads) connected by lighter "interbead-regions". The size of the mapping signal (i.e. the relative reflectance change DR/R in a black point minus the value of DR/R in a white point) in the four panels is 0.0015. Scale bar is 1 mm.

This confirms what is well-known from countless electrophysiological studies: cells in one location of the cortex respond very well to a moving grating of a particular orientation but hardly at all to a grating of the orthogonal orientation. This emphasizes once more the point that all the data gathered in optical experiments are in good agreement with conventional electrophysiological data. In some of our experiments we confirmed the optically obtained maps with conventional electrophysiological recordings. In every case we found an excellent agreement between the optical and the electrophysiological data.

The first question of the current study was to ask whether cells besides being clustered in orientation columns (cf. Figure 2) are also - as some investigators have proposed - clustered into direction columns (Payne et al. 1980; Tolhurst et al. 1981; Swindale et al. 1987). To this end we stimulated a cat with a grating moving in one direction and recorded the activity map from area 18. We then compared this map with the activity map obtained for a grating of the same orientation but the opposite direction of motion. This comparison told us whether the two gratings stimulate distinct cortical regions (implying that cells are grouped into separate directionality columns) or whether the same cortical area is activated by the two stimuli (implying that there are no directionality columns). An answer to this question can be found in Figure 3. The activity map for a vertical grating moving to the left is shown in Figure 3A, the map for the same grating moving to the right in Figure 3B. Scrutiny of these two figures already suggests that the two maps are very similar and, indeed, if one subtracts the two maps one is left with a patternless picture shown in Figure 3C. We performed analogous experiments in 16 hemispheres of 9 cats (with 4 or 8 orientations per experiment) and invariably found that maps obtained for a grating moving in one direction were indistinguishable from those obtained with a grating moving in the opposite direction. From these data we conclude that cells in area 18 are not grouped into directionality columns, but rather that cells preferring the same orientation but different directions of motion are intermingled.

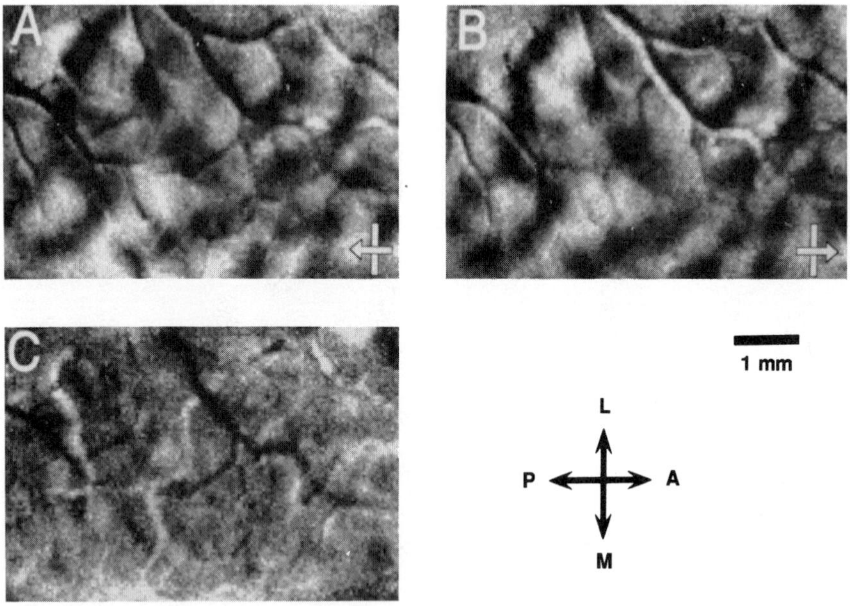

FIGURE 3. Maps obtained with two gratings having the same orientation but moving in different directions. A: activity map recorded with a vertical grating moving to the left. B: map produced by the same vertical grating but moving to the right. C: A patternless picture is produced by subtracting these two images from each other. This suggests that neurons in area 18 are not grouped in "directionality columns". Scale bar is 1 mm.

Data like the one presented in Figure 2 are equivalent to 2-DG pictures and only give a very incomplete picture of the organization of orientation in area 18. The unique advantage of optical imaging of intrinsic signals is that one can obtain many maps from one piece of tissue. We made use of this fact and combined the data from eight different orientations which then yielded a much more complete picture of the overall organization of orientation preference. To this end we took the eight different activity maps and (after multiplying the angles by two added them vectorially and coded the resulting angle (again divided by two) in color. Data of this kind are presented in Figure 4B which then gives a complete orientation map.

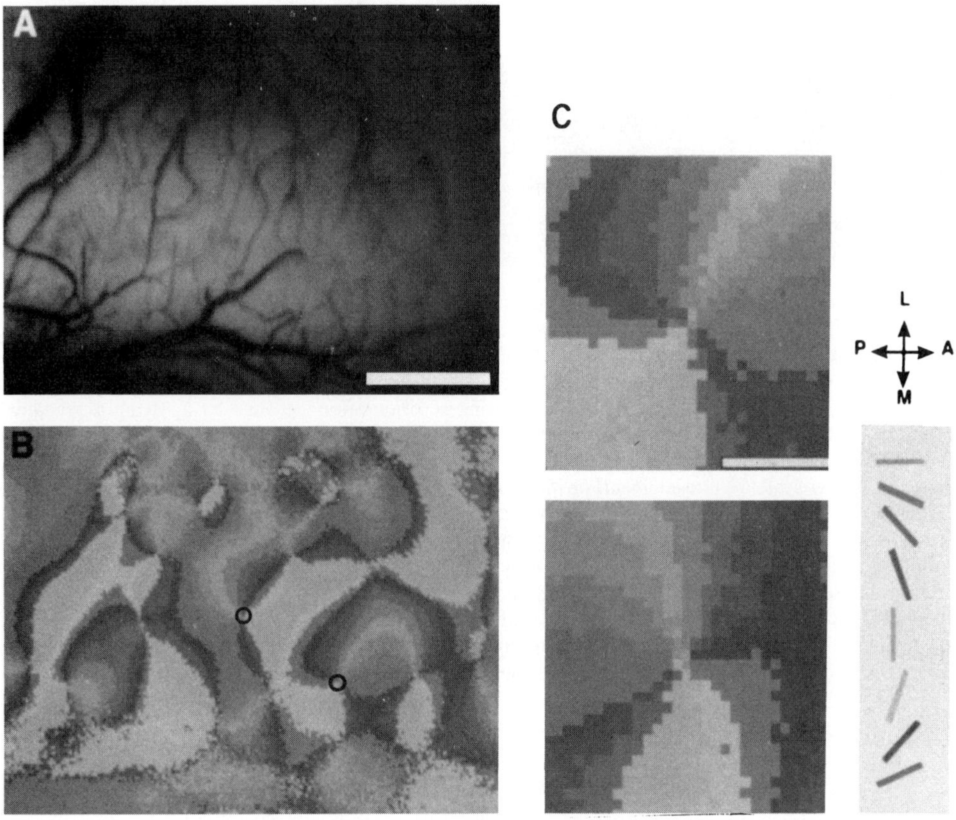

FIGURE 4. Pseudo-color-coded map of orientation preference. A: Cortical area from which this orientation map was imaged. Scale bar is 1 mm. B: Organization of orientation preference across this cortical area. Preferred orientations are color-coded according to the scheme to the right of the picture (regions preferring moving gratings of vertical orientation are coded in blue, those preferring horizontal gratings are coded in yellow, etc.). C: Close-up of two orientation centers. Two fundamental features of the orientation centers are illustrated in this Figure. (1) In all "pinwheels", each orientation appears once around the center. (2) The pinwheels exist in two forms: a clockwise form (upper part), in which the sequence of the colors yellow-green-blue-red is clockwise, and a counterclockwise form (lower part), in which the inverse is the case. Scale bar is 300 mm.
Note of the editors: Color laser prints of this figure can be obtained from the author upon request.

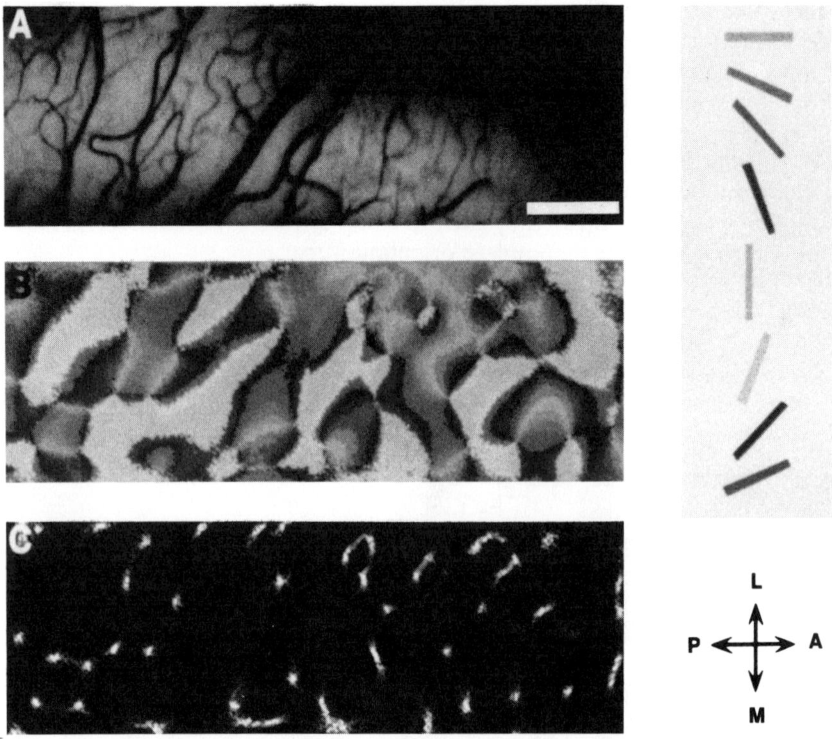

FIGURE 5. "Orientation centers" and the rate of orientation change across the cortical surface. A: An image of a long stretch of area 18 which was examined in this experiment. Scale bar is 1 mm. B: Orientation preference pattern of this piece of cortex. The color coding is again shown at the right. This map is derived from the responses to gratings of eight different orientations. C: "Fracture map": we screened our data for the locations with the strongest orientation change by applying a gradient operator to the orientation map. We calculated the two-dimensional derivative (taking the circular nature of the orientation data into account) for every point of the picture and displayed its absolute value on a gray scale. White elements in the maps show the regions of strong orientation changes. Evidently, the regions where orientation changes are strongest are the orientation centers manifesting themselves as small white patches. Note that there are no rapid changes along lines, except at the periphery of the map where the cortex is already folding down, and where the data are much noisier. The data presented in this figure were actually accumulated in two independent experiments performed approximately 4 hours apart on adjacent pieces of cortex. In panels A, B and C one can observe a slight break where the data of the two different experiments where pasted together.
Note of the editors: Color laser prints of this figure can be obtained from the author upon request.

A closer look at these orientation map reveals a number of noteworthy features in these data. The first one is that the overall appearance of this map is certainly patchy as opposed to banded as one might have expected from 2-DG studies and indeed also from Figure 2 of the present work. We found that in 19 out of the 21 hemispheres in which we performed this kind of analysis the appearance was clearly patchy. Only in two of the hemispheres we found the appearance of the map to be banded (both of those experiments were performed in a relatively posterior part of area 18 close to the vertical meridian).

A prominent feature of these maps is that orientation preferences are organized around points which we termed "orientation centers" (two of them are marked with black circles in Figure 4). Areas with different preferred orientation are ordered in a pinwheel-like fashion around the orientation centers. We found on the average 1.2 of these orientation centers per mm^2. Figure 3C shows two typical orientation centers at a higher magnification. Iso-orientation domains encircle the orientation centers. Every orientation appears once around an orientation center. In our experiments we found altogether 147 orientation centers and in all of them we found that every orientation appears around the center once. Figure 3C also shows that the "pinwheels" appear in two different forms. One in which the sequence of the colors (and thus orientations) blue, red, yellow and green is clockwise (Fig. C, upper part) and the other one having a counterclockwise sequence (Fig. 3C, lower part).

We also examined our orientation preference maps for locations were the rate of orientation change is very high. Such regions have been reported in monkey V1 (Blasdel and Salama, 1986) and have there been termed "fractures". Figure 5 shows a typical result as we obtained it in 21 orientation maps that we analyzed in this fashion. Figure 5A and B show the blood vessel pattern of the investigated region and the corresponding orientation preference map. Figure 5C then shows a picture in which we applied a two-dimensional gradient operator onto the map shown in Figure 5B. One can see many points in which the rate of orientation change is high. These points coincide with the orientation centers shown in Figure 5B (where orientation changes obviously are strong). Apart from point singularities we did not observe any other structures (e.g. lines) along which orientation changes very rapidly. This shows that in cat area 18 - in contrast to what has been reported for monkey V1 (Blasdel and Salama, 1986) - there are no fractures.

DISCUSSION

"Direction Maps"

Our experiments showed that the maps obtained with a grating moving in one direction are very similar to the ones obtained with the same grating moving in the opposite direction. In fact, when subtracting two images acquired with opposing directions a patternless picture resulted (Fig. 3). This suggests that cells in area 18 of the cat are not arranged in directionality columns. However, two alternative possibilities remain to be ruled out. (1) The directionality clusters could be so small that they lie below our technique's resolution of approximately 100 μm. (2) If there were clusters of direction preference which had opposite preferred direction of motion and they were to lie above each other in the cortex our method would not be able to tell the different clusters apart. We are, however, working on improvements of our apparatus which will allow three-dimensional imaging of the functional architecture of the cortex. We performed preliminary experiments (D. Malonek, T. Bonhoeffer, A. Grinvald, unpublished results) which seem to refute this possibility, at least for the upper 500 μm.

In area 17 Payne et al. (1980) and in area 18 Swindale et al. (1987) reported clusters of cells which all had similar direction preferences. The clusters described by both these groups had a size which was in the order of 500 μm and should thus be easily resolvable with intrinsic imaging. In none of our experiments were we able to find such areas in the cortex. The only remaining explanation for these contradicting results is the above mentioned possibility that there are clusters of cells having opposite direction preference lying above each other. However, Tolhurst et al. (1981) reported similar direction preference for cells across one whole cortical column reaching from the pia to the white matter. This sharply contrasts our findings. Altogether we find it unlikely that our results would be due to one of the two alternative explanations and we propose that there are indeed no direction columns in area 18 of cat visual cortex.

"Stripes" or "Patches" ?

In most of the experiments that we analyzed for the global organization of orientation we found that the distribution of iso-orientation domains was patchy rather than banded as one might have expected from 2-DG pictures. Only in two out of twenty-one hemispheres that we analyzed in such a fashion we observed a banded appearance. It should however be kept in mind that all the information we gather in our experiments is collected from a very restricted part of area 18 namely from a narrow strip of the lateral gyrus which is exposed to the surface. We can thus not rule out the possibility that in other regions and at other eccentricities the functional organization might be different.

2-Deoxyglucose pictures published in the literature seem to suggest that a region preferring the same orientation should be a band or slab rather than a patch (Hubel et al. 1977; Hubel et al. 1978; Löwel et al. 1987). And for example also Figure 3 of our data seems to forward such an interpretation. If one however looks at the organization of orientation preferences for all the orientations (which is not possible with the 2-DG technique) the picture is very different. The gray regions connecting the dark beads in the "band" for one particular orientation (cf. Figure 2) turn out to respond very vigorously to another orientation and should thus not be considered belonging to the same orientation "band" but to a patch of different orientation preference. We would therefore like to point out that the data that lead us to conclude that iso-orientation domains are patchy and not banded are not at variance with data from 2-DG studies. It is rather that richer data enable us to gain a more complete picture of the overall cortical organization.

"Pinwheels" around orientation centers

In our experiments we observed that the patches of orientation preference seem to be organized around points which we called "orientation centers" (Bonhoeffer and Grinvald, 1991). In these points regions of all orientation preferences border onto one another. The orientation preferences are distributed around the centers either clockwise or counterclockwise in a pinwheel-like fashion. Centers of this sort have also been observed by Swindale et al. (1987) using maps obtained from interpolation between multiple electrode recordings. Also the data of Blasdel and Salama obtained with voltage sensitive dyes in monkey visual cortex show pinwheels of the sort that we describe. However their angle maps were calculated from preprocessed data so that the interpretation of their maps is not straightforward. Moreover these authors suggested that not pinwheels but fractures are the salient features organizing the representation of orientation in monkey visual cortex.

It had been previously suggested from purely theoretical considerations (e.g. von Seelen 1970, Braitenberg and Braitenberg 1979) that the visual cortex could be organized around orientation centers instead of being a banded structure as originally proposed by Hubel and Wiesel. Braitenberg and Braitenberg (1979) showed that many of the data obtained by Hubel and Wiesel could very well be fit by a scheme of circular organization. Braitenberg and Braitenberg proposed in their work that these centers should be spaced at a distance of approximately 0.5 mm and that every orientation should appear twice around an orientation center. More recently different groups have - mostly on theoretical grounds - proposed an organization with orientation centers around which every orientation appears once. Götz (1987) based on data obtained by Blasdel and Salama (1986) in the monkey cortex proposed that V1 of the monkey cortex might be constructed in such a way. He and Swindale (1982) used the notion of handedness and the topological index to describe the different kinds of singularities. Singularities around which every orientation between 0 and 180 degrees appears twice (i.e. 360 degrees orientation change appear once in the 360 degrees around the singularity) have an index of 1 or -1 whereas orientation centers around which every orientation appears once consequently display an index of +1/2 or -1/2.

The topological index is the crucial distinction between the different models. We therefore inspected our data very closely with respect to the question whether most of our orientation centers have a topological index of +1/2 or -1/2, or whether it is most often +1 or -1. We found that virtually all of the singularities that we observed were of the type ±1/2. We conclude that our data obtained in area 18 of the cat favor theories which propose orientation centers with only one representation of all orientations around the center singularity.

Emergence of orientation centers

There are several models in the literature which predict the emergence of orientation centers for different reasons. Durbin and Mitchison (1990) and Obermayer et al. (1990) show that the "principle of continuous mapping" leads to orientation centers. Swindale (1982), Linsker (1986) and Götz (1988) use different models of self-organization to achieve a pattern of orientation preference which looks similar to the experimentally observed orientation maps. Götz's model is particularly interesting since it predicts (K.G. Götz, personal communication) that in early stages of the cats postnatal development there should be more centers with an index of 1 or -1 (which means that one should be able to find orientation centers around which every orientation appears twice) and only later on through the self-organizing process these centers each break up into centers with an index of +1/2 or -1/2. In order to experimentally test this theory it would thus be interesting to see whether in young kittens (or other mammals in which orientation tuning develops mostly postnatally) one indeed finds the centers around which every orientation appears twice which we rarely found in adult cats.

Fractures

Our data in cat areas 17 and 18 do not show the fractures that have been reported for monkey V1 (Blasdel and Salama, 1986). Since we did not do experiments in monkey visual cortex it has to remain unresolved what caused the difference between our and Blasdel and Salama's results. One possibility is the different algorithm the they used to compute the orientation preference maps. These authors computed the preferred angle from maps which themselves were already obtained by dividing one map by the orthogonal one. In this case, every single map already contains a mixture of information, and the resulting angle maps are difficult to interpret. Secondly, Blasdel and Salama did not use the macroscope with the extraordinarily low depth of field that we used. This macroscope minimizes blood vessel artifacts so that these could be considered another source of divergence between the two different results. And lastly, there is clearly the possibility that in cats the precise organization of the visual cortex is slightly different than in monkeys. But whatever the explanation for the different results, what can be concluded from our results is that fractures, i.e. lines of rapid orientation change are not present in the visual cortex of the cat. Thus, at least in the cat, fractures do not seem to play an important role in organizing the visual cortex.

OUTLOOK

We have described how one can make use of signals intrinsic to the brain to investigate maps of orientation preference in the visual cortex. Our technique is obviously not limited to the investigation of orientation maps nor is it limited to the visual cortex. Any region of the cortex which is lying on the surface so that it is accessible to optical methods can be examined with respect to parameters which are mapped in an orderly fashion onto the cortex. Maps in the auditory or the somatosensory cortex are very apt to be investigated with this technique. Furthermore imaging of intrinsic signals is a non-invasive method in the sense that the cortex itself is not touched during the experiments. This makes it particularly well suited for developmental experiments. More knowledge about the development of orientation preference maps and the emergence of pinwheels in young kittens will hopefully give some insight into the underlying, presumably self-organizing principles that govern the formation of maps the mammalian cerebral cortex. Yet further in the future one could imagine making use of intrinsic signals in the operating-theater during neurosurgical procedures to map an epileptic focus or to map important functional areas like for instance a speech area in order to spare it from an operation. However a lot of technical development will have to be done before this technique will be a standard tool in research and possibly in medicine.

ACKNOWLEDGMENT We thank Kaare Christian for developing the visual stimulator and Dae-Shik Kim for commenting on the manuscript.

REFERENCES

Blasdel, G.G. and G. Salama (1986) Voltage-sensitive dyes reveal a modular organization in monkey striate cortex. Nature 321: 579-585.

Bonhoeffer, T. and A. Grinvald (1991) Iso-orientation domains in cat visual cortex are arranged in pinwheel-like patterns. Nature 353: 429-431.

Braitenberg, V. and C. Braitenberg (1979) Geometry of orientation columns in the visual cortex. Biol. Cybern. 33: 179-186.

Durbin, R. and G. Mitchison (1990) A dimension reduction framework for understanding cortical maps. Nature 343: 644-647.

Frostig, R.D., E.E. Lieke, D.Y. Ts'o, and A. Grinvald (1990) Cortical functional architecture and local coupling between neuronal activity and the microcirculation revealed by in vivo high-resolution optical imaging of intrinsic signals. Proc. Natl. Acad. Sci. USA 87: 6082-6086.

Götz, K.G. (1987) Do "d-blob" and "l-blob" hypercolumns tessellate the monkey visual cortex. Biol. Cybern. 56: 107-109.

Götz, K.G. (1988) Cortical templates for the self-organization of orientation-specific d- and l-hypercolumns in monkeys and cats. Biol. Cybern. 58: 213-223.

Grinvald, A., E. Lieke, R.D. Frostig, C.D. Gilbert, and T.N. Wiesel (1986) Functional architecture of cortex revealed by optical imaging of intrinsic signals. Nature 324: 361-364.

Hubel, D.H., T.N. Wiesel, and M.P. Stryker (1977) Orientation columns in macaque monkey visual cortex demonstrated by the 2-deoxyglucose autoradiographic technique. Nature 269: 328-330.

Hubel, D.H., T.N. Wiesel, and M.P. Stryker (1978) Anatomical demonstration of orientation columns in macaque monkey. J. Comp. Neurol. 177: 361-380.

Hubel, D.H. and T.N. Wiesel (1965) Receptive fields and functional architecture in two non-striate visual areas (18 and 19) of the cat. J. Neurophysiol. 28: 229-289.

Linsker, R. (1986) From basic network principles to neural architecture: Emergence of orientation columns. Proc. Natl. Acad. Sci. USA 83: 8779-8783.

Löwel, S., B. Freeman, and W. Singer (1987) Topographic organization of the orientation column system in large flat-mounts of the cat visual cortex: a 2-deoxyglucose study. J. Comp. Neurol. 255: 401-415.

Malonek, D., D. Shoham, E. Ratzlaff, and A. Grinvald (1990) In vivo three dimensional optical imaging of functional architecture in primate visual cortex. Society for Neuroscience Abstracts 130.4.

Mountcastle, V.B. (1957) Modality and topographic properties of single neurons of cat's somatic sensory cortex. J. Neurophysiol. 20: 408-434.

Obermayer, K., H. Ritter, and K. Schulten (1990) A principle for the formation of the spatial structure of cortical feature maps. Proc. Natl. Acad. Sci. USA 90: 8245-8349.

Payne, B.R., N. Berman, and E.H. Murphy (1980) Organization of direction preferences in cat visual cortex. Brain Res. 211: 445-450.

Ratzlaff, E.H. and A. Grinvald (1991) A Tandem-lens epifluorescence macroscope: hundred-fold brightness advantage for wide field imaging. J. Neurosci. Methods 36: 127-137.

Seelen, W.v. (1970) Zur Informationsverarbeitung im visuellen System der Wirbeltiere. Kybernetik 7: 89-106.

Sokoloff, L. (1977) Relation between physiological function and energy metabolism in the central nervous system. J. Neurochem. 19: 13-26.

Swindale, N.V. (1982) A model for the formation of orientation columns. Proc. R. Soc. Lond. B 215: 211-230.

Swindale, N.V., J.A. Matsubara, and M.S. Cynader (1987) Surface organization of orientation and direction selectivity in cat area 18. J. Neurosci. 7(5): 1414-1427.

Tolhurst, D.J., A.F. Dean, and I.D. Thompson (1981) Preferred direction of movement as an element in the organization of cat visual cortex. Exp. Brain Res. 44: 340-342.

Ts'o, D.Y., R.D. Frostig, E.E. Lieke, and A. Grinvald (1990) Functional organization of primate visual cortex revealed by high resolution optical imaging. Science 249: 417-420.

MAPPING OF NEURAL ACTIVITY PATTERNS USING INTRINSIC OPTICAL SIGNALS: FROM ISOLATED BRAIN PREPARATIONS TO THE INTACT HUMAN BRAIN

B.A. MacVicar, T.W.J. Watson, F.E. LeBlanc, S.G. Borg, and P. Federico

Neuroscience Research Group
Faculty of Medicine, The University of Calgary
Calgary, Canada

INTRODUCTION

It has long been known, since the work of Hill and Keynes (1949), that the intrinsic optical properties of nerve tissue change with neuronal activity (reviewed in Cohen, 1973). There are several possible cellular mechanisms which could underlie activity-dependent changes in the optical properties of neuronal tissue. These include neuronal and glial swelling (MacVicar and Hochman, 1991), neurosecretion (Salzberg *et al.*, 1985) and the rapid reorientation of membrane dipoles induced by membrane depolarization (Stepnoski *et al.*, 1991). Neuronal activity can also alter the optical properties of tissue by inducing changes in the characteristics of intrinsic chromophores such as hemoglobin or cytochromes (Chance and Williams, 1955; La Manna *et al.*, 1987; Federico *et al.*, 1991; Frostig *et al.*, 1990; Grinvald *et al.*, 1986), or by changing local blood flow and tissue hemoglobin content (Frostig *et al.*, 1990).

Imaging of intrinsic optical signals has provided an excellent technique for mapping patterns of neuronal activity over large areas with high spatial resolution. This has provided an important tool for recent investigations into the functional architecture of the cortex. For example, several studies have shown that imaging of intrinsic signals can be used to study the functional architecture of cerebral cortex (Frostig *et al.*, 1990; Bonhoeffer and Grinvald, 1991; Grinvald *et al.*, 1986; Ts'o *et al.*, 1990). We have imaged intrinsic optical signals in three different preparations. This has allowed us to determine the different mechanisms by which these signals were generated and to examine the efficacy of this technique for mapping neural activity patterns on different scales. In one experimental series, we examined changes in light transmittance through hippocampal slices induced by electrically evoked synaptic activity. Synaptic activation has been shown previously to cause a decrease in the size of the extracellular space due to cellular swelling (Heinemann *et al.*, 1983; McBain *et al.*, 1990). Changes in light scattering are also associated with changes in cellular swelling (Andrew and MacVicar, 1992). We therefore determined whether optical techniques can be used to localize synaptically evoked cellular swelling in isolated slices.

Optical Imaging of Brain Function and Metabolism, Edited by
U. Dirnagl *et al.*, Plenum Press, New York, 1993

Another preparation that we have used is the isolated and perfused guinea-pig whole brain. This preparation, which was developed recently by Mühlethaler and Llinás (Mühlethaler and Serafin, 1990; de Curtis *et al.*, 1991), provides the mechanical stability of an *in vitro* slice preparation with a relatively intact synaptic circuitry. This preparation is also blood-free and therefore allows the examination of cellular mechanisms which are independent of blood components in the generation of intrinsic optical signals. We have examined whether it is possible to image patterns of synaptic activity in the isolated whole brain and whether the mechanisms by which the signal is generated are similar to those in the *in vitro* brain slice.

In the final series of experiments described in this paper, we used imaging techniques to map neural activity patterns in the human brain during neurosurgical procedures for the treatment of intractable epilepsy. Monitoring optical signals may provide a novel technique for delineating areas of increased activity in the human brain. During neurosurgery it is important to demarcate functional areas in the cortex and areas which generate seizure activity so that tissue retraction can be limited to epileptic tissue. We have therefore examined whether stimulus-evoked neuronal activity and seizure activity can be imaged in the human cortex.

METHODS

Tissue Preparation

The preparation of isolated brain slices and perfused guinea pig brains are described in detail elsewhere (de Curtis *et al.*, 1991; Federico *et al.*, 1992, MacVicar and Hochman, 1991). Hippocampal slices (500 μm) were prepared from 3-5 week old male Sprague-Dawley rats and were maintained using conventional methods. During experiments, slices were maintained at 21°C or 33°C in a chamber mounted on an inverted microscope. Brains were removed from 150-250 gm male Hartley guinea pigs under barbiturate anaesthesia (Somnotol) and were placed in a temperature regulated chamber. A vertebral artery was cannulated and arterial perfusion of dextran-containing artificial cerebrospinal fluid was commenced. Experiments were performed at 30°C.

Studies of exposed cortex of humans were performed on patients with intractable epilepsy during the course of intracranial surgery designed to identify and remove the seizure focus (MacVicar *et al.*, 1990). Ethical approval and informed consent was obtained prior to the operations in all cases. Electrocorticography, combined with recordings using a temporal lobe depth electrode, was routinely employed as part of the intraoperative exploration to delineate the area from which the seizure activity originated.

Imaging Techniques

The procedures and system that we employed in these experiments is described in detail in MacVicar and Hochman (1991). In all experiments a series of digitized images (512 x 480 pixel array) was acquired by averaging frames from a charge-coupled device camera (CCD; Dage-MTI or Cohu) at video rates (30 hz) using an Imaging Technology 151 image processor controlled by a 486 computer. Changes between control images and subsequent images were analyzed by pixel-by-pixel subtraction and displaying the difference images. Tissue was illuminated using a voltage-regulated tungsten lamp attached to an inverted microscope (Zeiss IM-35) for slice experiments, or to an operating microscope (Zeiss OPMI-1) for experiments on human cortex or isolated guinea pig brains. Gain and black level of the camera were set at high levels so that small gradations in image intensity would span the full dynamic range of the input range of the A/D boards of the image processor.

RESULTS AND DISCUSSION

Hippocampal Slices

Synaptic activation of the Schaffer collateral pathway to the CA1 region of the hippocampus consistently induced an increase in light transmittance in the synaptic region, the stratum radiatum (Figure 1). The increase gradually developed over 3 to 5 sec and took approximately 20 sec to recover. The magnitude of the signal was dependent on the stimulation rate. An optical signal was detected at stimulation rates of 2 Hz and the magnitude of the signal reached a plateau at 15 Hz (MacVicar and Hochman, 1991).

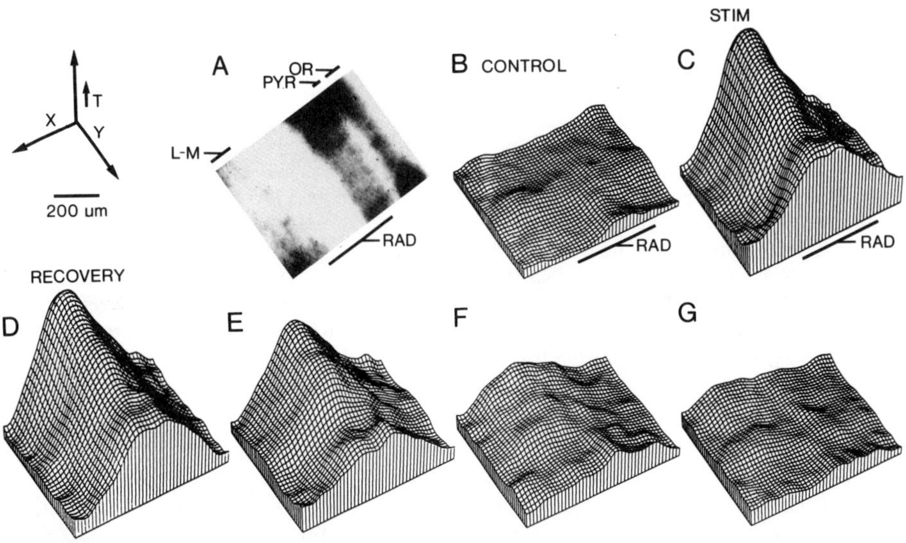

FIGURE 1. Schaffer collateral stimulation in a hippocampal slice evoked an increase in light transmission through the stratum radiatum. Averages of 128 video frames were obtained during 10-Hz synaptic activation. Each average image took 4 sec to acquire and there was a 1 sec delay between the acquisition of averages. The control average image is shown in A. The layers within CA1 in the control image are labelled as Or (oriens), Pyr (pyramidal), Rad (radiatum), and L-M (lacunosum-moleculare).B-G represent the subtraction of image 1 from images 2-7 respectively. The magnitude of the differences, representing increased transmittance, are illustrated as 3-D surface plots where the z-axis (labelled T) represents increased transmittance. The length of the z-axis bar represents $\Delta T/T$ of 1.2%. B) Stimulation of the Schaffer collaterals caused an increase in light transmittance in the radiatum. After cessation of the stimulation, light transmittance recovered to control levels (D-G) within 20 sec. Figure reproduced with permission from MacVicar and Hochman, 1991.

The wavelength dependency of the signal was determined by imaging the slices at different wavelengths using interference filters (20 nm bandpass). The largest increases in light transmission during 10 Hz stimulation were observed at wavelengths greater than 650 nm (Figure 2). There was no significant peak at these longer wavelengths. The smaller magnitude of the signals observed below 600 nm was most likely due to the poor transmittance of light through tissue at shorter wavelengths. This pattern of wavelength dependency is consistent with light scattering changes that could be caused by cellular swelling. Increased swelling is believed to decrease light scattering by cells, which would lead to increased light transmittance (Cohen, 1973).

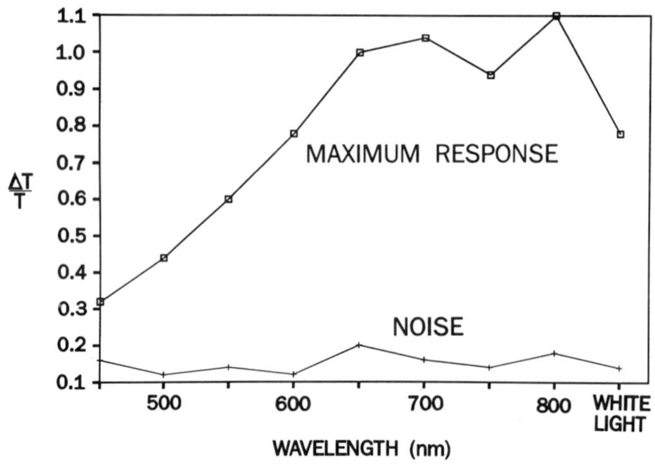

FIGURE 2. Wavelength dependency of the intrinsic optical signals in hippocampal slices. An experimental protocol was followed similar to that described in Figure 1. The changes in light transmittance were observed at different wavelengths (every 50 nm from 450 to 800 nm) from 10 Hz Schaffer collateral stimulation. The response observed in white light is also plotted. Light illumination was changed to compensate for the increased transmittance at longer wavelengths. After each stimulation trial, the optical signal was observed to recover to control levels. The noise level (measured in unstimulated controls) was fairly constant but ΔT increased and stayed constant at wavelengths >600 nm. Figure reproduced with permission from MacVicar and Hochman, 1991.

There are other observations concerning intrinsic optical signals which also indicate that the signals are most likely generated by cellular swelling. Increased light transmittance was observed in slices when hypo-osmotic solutions were superfused and decreased transmittance occurred in hyper-osmotic solution (Andrew and MacVicar, 1992). The increase in light transmittance in hypo-osmotic solution was correlated with the increase in field potential amplitude that occurred when the extracellular space was decreased during cell swelling. How can neuronal activity induce swelling? Synaptic transmission was critical because intrinsic optical signals were blocked by blocking synaptic transmission with either a 0 Ca^{++}-EGTA solution or an excitatory amino acid antagonist, kynurenic acid (MacVicar and Hochman, 1991). Synaptic activity most likely induces the signal by increasing extracellular K^+ since high K^+ (20 mM) caused an increase in light transmission through slices even after synaptic transmission was blocked in 0 Ca^{++}-EGTA solution (Andrew and MacVicar, 1992).

How might high extracellular K^+ cause the cellular swelling that is detected as increased light transmission through the slice? High K^+ has been shown to induce swelling of astrocytes due to KCl uptake and water movement (Walz, 1987). A Na-K-2Cl cotransporter is thought to underlie these ion movements. Activity-dependent shrinkage of the extracellular space in the optic nerve is also believed to result from Cl^- dependent swelling of astrocytes (Ransom *et al.*, 1985). We found that the optical signals were sensitive to Cl^- transport inhibition with furosemide or reduced extracellular Cl^- (MacVicar and Hochman, 1991). It is possible, therefore, that intrinsic optical signals in isolated slices result from changes in light scattering that are secondary to K^+-evoked cellular swelling (potentially of astrocytes).

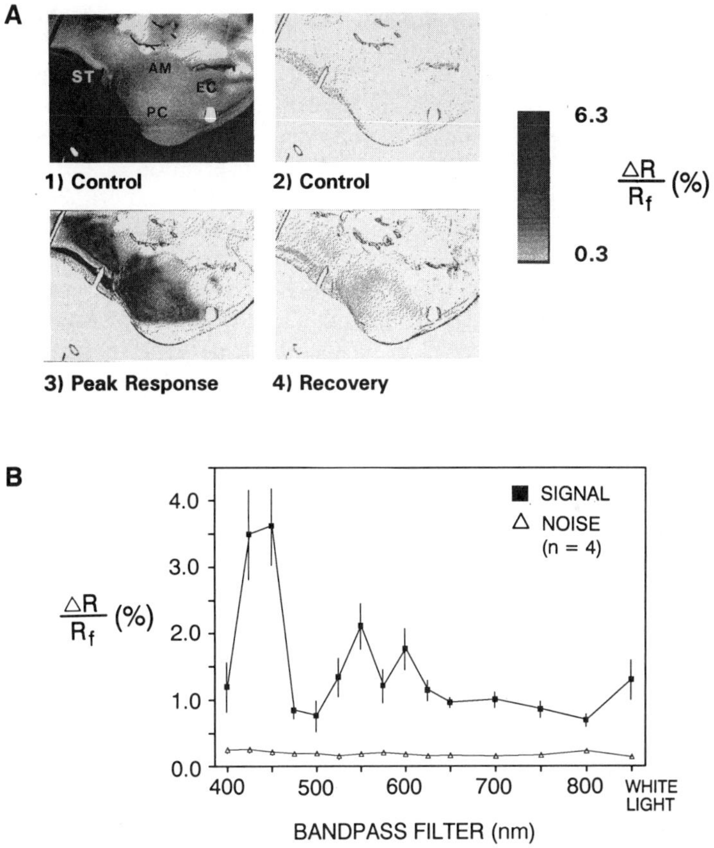

FIGURE 3. A. Stimulation of the lateral olfactory tract decreased light reflectance in perfused guinea-pig brain. Panel A1 shows a control unsubtracted image acquired at the start of the experimental series. A control subtracted image acquired immediately prior to stimulation is shown in A2. Stimulation of the lateral olfactory tract (5 Hz, 15V, 0.1 msec for 18 sec) decreased tissue reflectance in the olfactory cortex. Panel A3 shows an image acquired 9 secs after stimulation was stopped. Following this peak, the change in tissue reflectance recovered over the next 81 sec (Panel A4). Abbreviations: amygdalar cortex (AM), entorhinal cortex (EC), piriform cortex (PC), and stimulating electrode (ST). B. Peak intrinsic signal magnitude (mean ± S.E.) at different wavelengths following stimulation of the lateral olfactory tract in 4 different brains. In order to insure that the baseline reflectance was equal at all wavelengths, the intensity of light illumination was adjusted to compensate for increased reflectance at longer wavelengths. Since R_f was kept constant at all wavelengths, it was possible to compare ΔR at different wavelengths. At 400 nm, the F stop had to be lowered from 4.0 to 2.8 in order to allow enough light to enter the CCD camera to saturate the image. Thus, the peak response at 400 nm was compared to a 450 nm control trial carried out with an F stop value of 2.8. The noise level in unstimulated controls was fairly constant in all cases. Note the peaks at 425-450, 550 and 600 nm.

Isolated Perfused Guinea Pig Brain

Synaptic activation in the isolated whole brain always elicited a decrease in tissue reflectance in areas in which field potentials could be recorded (Federico *et al.*, 1992). Repetitive electrical stimulation (1 to 20 Hz) of the lateral olfactory tract evoked a progressive decrease in tissue reflectance in the main projection areas of this pathway, the pyriform cortex, olfactory tubercle, amygdala and entorhinal cortex (Figure 3A). The decrease in tissue reflectance took approximately 3 sec to become discernible above the noise and took over 40 sec to recover following the end of stimulation. We compared the map of the evoked intrinsic optical signals to the distribution of the evoked field potential amplitudes in order to ascertain the reliability of optical maps for predicting neuronal activity. In each experiment the relationship was linear (Federico *et al.*, 1992).

The isolated whole brain is blood-free so that it is obvious that intrinsic optical signals detected using this preparation are generated independently of blood-related components (ie. increased hemoglobin content or decreased oxygen binding to hemoglobin) (T'so *et al.*, 1990). We repeated the observation of intrinsic signals at different wavelengths using interference filters, as in the slice experiments, in order to examine the wavelength dependency of these signals. Above 650 nm, the graph was again flat (Figure 3B). However, there were significant peaks at 450, 550 and 600 nm. These peaks correspond to the absorption maxima of tissue cytochromes which exhibit increased light absorption in the reduced form (Chance and Williams, 1955).

The wavelength dependency of the intrinsic optical signals in the perfused whole brain indicates that there may be another factor contributing to the generation of optical signals besides changes in light scattering. It is conceivable that the increased neuronal activity caused by afferent stimulation increased metabolic demand in the active tissue. This, in turn, would cause a net reduction of tissue cytochromes if oxygen delivery stayed constant. Vascular reactivity is probably compromised in the perfused brain preparation so that the normal increases in blood flow which occurs during activity most likely would not be observed.

In conclusion, imaging of intrinsic optical signals provides a reliable map of evoked neuronal activity in the isolated guinea pig brain. The mechanisms underlying the generation of these signals may be multifactorial with reduction of tissue cytochromes and changes in light scattering both contributing.

Human Cortex

We wished to determine whether it was possible to image activity-dependent intrinsic optical signals in the human cortex. These studies were performed on patients with intractable epilepsy during the course of intracranial surgery designed to identify and remove the seizure focus (MacVicar *et al.*, 1990). During the standard intracranial operative procedure for the treatment of these patients, the exposed cortex of awake patients was stimulated to localize cortical areas committed to sensory, motor and speech functions. In all patients, direct stimulation of the brain induced a decrease in the reflectance of the cortical surface (Figure 4A). The decrease in reflectance progressively increased in magnitude during the stimulation period (15 sec) and gradually recovered to control levels within 40 sec following cessation of stimulation. In all patients, these changes were repeatable in subsequent trials.

In some patients, cortical stimulation evoked localized afterdischarges of seizure activity that were measured with EEG electrodes. Afterdischarges outlasted the stimulus train for up to 30 sec. Associated with these afterdischarges was a large decrease in light reflectance in the same cortical locus where the afterdischarges occurred (Figure 4B). When seizure afterdischarges occurred after the stimulation was stopped, the magnitude of the decrease in tissue reflectance continued to increase in the seizure focus (Figure 4B). Optical signals were also detected in adjacent cortical sites as the seizure activity spread. This is in contrast to control situations when afterdischarges were not elicited by cortical stimulation and the change in tissue reflectance recovered to control levels directly after the cessation of stimulation.

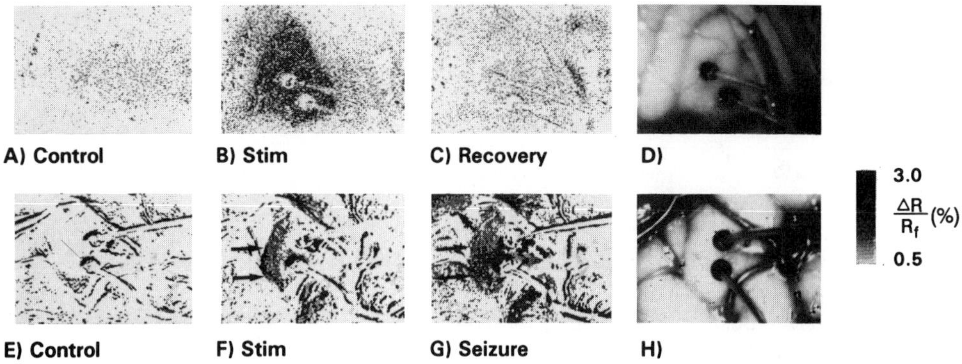

Figure 4. Direct stimulation of the human cortex evoked a decrease in the reflectance of light from the surface of the brain. Images were obtained by averaging and the subtractions were accomplished as described in the previous figure legends. A) Subtraction of a control image from an unstimulated frame showing the noise levels. B) Direct stimulation by bipolar electrodes (100 Hz, 10 V, 0.1 msec) induced a gradual decrease in the intensity of the reflected light from the surface of the cortex. C) Following the cessation of stimulation reflectance returned to control levels within 80 sec. D) The control image of the brain and the stimulating electrodes are shown. E) In another patient direct stimulation of the cortex induced a seizure afterdischarge which was recorded using depth electrodes. The control subtraction is shown illustrating the movement noise in this experiment that can be seen as outlines of the electrodes and blood vessels. F) Stimulation induced a decreased reflectance in the activated gyrus (illustrated by arrows). G) When a seizure afterdischarge followed the stimulation period, the reflectance signal did not recover to control levels as it did in seizure free recordings (C). Instead it increased in intensity in the stimulated gyrus (arrows) and spread to adjacent areas (double arrow, upper left corner). H) The control image showing the electrode placement on the cortex.

These results indicate that increased neuronal activity in the awake human cortex is associated with decreases in light reflectance and, with appropriate computer analysis, can be used to map areas of stimulus-evoked activity and seizure discharge.

CONCLUSIONS

The experiments described in this chapter support the utility of imaging techniques that detect intrinsic optical signals for mapping neuronal activity patterns. They also provide some clues regarding mechanisms underlying the generation of these signals. Reduction of tissue cytochromes appears to contribute to increased light absorption and changes in cellular swelling may contribute to decreased light scattering. In the slice, stimulus-evoked changes in light transmission are confined to the regions of synaptic activation. The spatial resolution of the imaging system used in our laboratory appears to be in the range of 50 μm. Gradients in the intensity of the signal in the radiatum are discernible over this scale (Figure 1). In the perfused brain it is possible to map neuronal activity patterns over a much larger scale of several mm to cms. Resolution of activity on this scale is impossible to obtain using conventional microelectrode techniques. EEG recordings can be used for mapping neuronal activity over larger areas, however, fine resolution is lost. Imaging of intrinsic signals from the human cortex indicates that this type of technology may be useful for delineating functional areas and patterns of abnormal activity.

ACKNOWLEDGEMENTS: Supported by the Medical Research Council of Canada (MRC) and the Foothills Hospital Research and Development Fund. P.F. is a recipient of studentships from the MRC and Alberta Heritage Foundation for Medical Research (AHFMR). T.W.J.W. is an AHFMR Clinical Fellow. B.A.M. is an MRC Scientist and an AHFMR Scholar.

REFERENCES

Andrew, R.D. and MacVicar, B.A. (1992). Imaging cell volume changes in rat hippocampal slices. Soc. Neurosci. Abstr. 18 (in press).

Bonhoeffer, T. and Grinvald, A. (1991). Iso-orientation domains in cat visual cortex are arranged in pinwheel-like patterns. Nature 353: 429-431.

Chance, B. and Williams, G.R. (1955). A method for the localization of sites form oxidative phosphorylation. Nature 176: 250-254.

Cohen, L.B. (1973). Changes in neuron structure during action potential propagation and synaptic transmission. Physiol. Rev. 53: 373-418.

de Curtis, M., Paré, D. and Llinás, R.R. (1991). The electrophysiology of the olfactory-hippocampal circuit in the isolated and perfused adult mammalian brain *in vitro*. Hippocampus 1: 341-354.

Federico, P., Borg, S.G. and MacVicar, B.A. (1991). Stimulation-evoked changes in intrinsic optical signals in the guinea pig isolated whole brain. Soc. Neurosci. Abstr. 17: 443.13.

Federico, P., Borg, S.G., Šalkauskus, A.G. and MacVicar, B.A. (1992). Mapping neuronal activity patterns by imaging intrinsic optical signals in the guinea-pig isolated whole brain (submitted).

Frostig, R.D., Lieke, E.E., Ts'o, D.Y. and Grinvald, A. (1990) Cortical functional architecture and local coupling between neuronal activity and the microcirculation revealed by *in vivo* high-resolution optical imaging of intrinsic signals. Proc. Nat. Acad. Sci. 87: 6082-6086.

Grinvald, A., Lieke, E.E., Frostig, R.D., Gilbert, C. D. and Wiesel, T.N. (1986). Functional architecture of cortex revealed by optical imaging of intrinsic signals. Nature 324: 361-364.

Heinemann, U., Neuhaus, S. and Dietzel, I. (1983). Aspects of potassium regulation in normal and gliotic brain tissue. IN: Current problems in epilepsy. I. Cerebral blood flow, metabolism and epilepsy (Baldy-Moulinier, M., Ingvar, D.H., Meldrum, B.S. eds), pp 271-277. London.

Hill, D.K. and Keynes, R.D. (1949). Opacity changes in stimulated nerve. J. Physiol. 108: 278-281.

La Manna, J.C., Sick, T.J., Pikarsky, S.M. and Rosenthal, M. (1987). Detection of an oxidizable fraction of sytochrome oxidase in intact rat brain. Am. J. Physiol. 253: C477-C483.

McBain, C.J., Traynelis, S.F. and Dingledine, R. (1990). Regional variations of extracellular space in the hippocampus. Science 249: 674-677.

MacVicar, B.A. and Hochman, D. (1991). Imaging of synaptically evoked intrinsic optical signals in hippocampal slices. J. Neurosci. 11: 1458-1469.

MacVicar, B.A., Hochman, D., LeBlanc, F. and Watson, T. (1990). Stimulation evoked changes in intrinsic optical signals in the human brain. Soc. Neurosci. Abstr. 16: 309.

Mühlethaler, M. and Serafin, M. (1990). Thalamic spindles in an isolated and perfused preparation *in vitro*. Brain Res. 524: 17-21.

Ransom, B.R., Yamate, C.L. and Connors, B.W. (1985). Activity-dependent shrinkage of extracellular space in rat optic nerve: a developmental study. J. Neurosci. 5: 532-535.

Salzberg, B.M., Obaid, A.L. and Gainer, H. (1985). Large and rapid changes in light scattering accompany secretion by nerve terminals in the mammalian neurohypophysis. J. Gen. Physiol. 86: 395-411.

Stepnoski, R.A., LaPorta, A., Raccuia-Behling, F., Blonder, G.E., Slusher, R.E. and Kleinfeld, D. (1991). Noninvasive detection of changes in membrane potential in cultured neurons via light scattering. Proc. Natl. Acad. Sci. 88: 9382-9386.

Ts'o, D.Y., Frostig, R.D., Lieke, E.E. and Grinvald, A. (1990) Functional organization of primate visual cortex revealed by high resolution optical imaging. Science 249: 417-420.

Walz, W. (1987). Swelling and potassium uptake in cultures astrocytes. Can. J. Physiol. Pharmacol. 65: 1051-1057.

FIBER OPTIC IMAGING OF SUBCORTICAL NEURAL TISSUE IN FREELY BEHAVING ANIMALS

D.M. Rector, G.R. Poe, and R.M. Harper

Department of Anatomy and Cell Biology and
The Brain Research Institute,
University of California at Los Angeles, Los Angeles, CA, USA

INTRODUCTION

Changes in voltage potentials have been used as representative measures of activity from single neurons, as well as from large populations of neurons. However, electrophysiological techniques are limited in the number of individual cells that can be recorded simultaneously. Correlates of neural activity are also measured in two or three dimensional space with procedures such as Positron Emission Tomography or Magnetic Resonance Imaging. However, the spatial resolution of these procedures is insufficient to observe single neurons or small groups of cells, and the temporal resolution is inadequate to examine many functional aspects.

Intracellular calcium and voltage-sensitive dyes enable the use of light and video technology to record the activity of cell populations in vitro with great precision. This procedure can also examine exposed brain surface regions. However, the use of dyes requires damaging laser light and substantial equipment, and is potentially toxic to the tissue.

Imaging of neural tissue has taken a significant step forward through the application of high sensitivity CCD technology to record minute changes in reflected light from cells at video frame rates. Thus, correlates of neural activity can be measured via changes in reflected light without using dyes (Cohen 1971, Grinvald et al. 1977) and functional maps of the cortex can be made using this technique (Grinvald et al. 1986, Blasdel 1989, Ts'o et al. 1990).

Our goal was to measure changes in reflective properties of subcortical neural tissue in freely behaving animals. To accomplish this task, we devised a fiber optic probe attached to a miniaturized CCD camera. During one day of surgery, the probe tip can be place in any desired location in the brain. One week after recovery, video images can be collected over a period of months with the connection of a cable from the camera to the digitization hardware. The camera presented here incorporates several improvements to a device which was previously described (Rector and Harper, 1991).

Optical Imaging of Brain Function and Metabolism, Edited by
U. Dirnagl *et al.*, Plenum Press, New York, 1993

MATERIALS AND METHODS

A schematic diagram of the probe and camera is illustrated in Figure 1. A rigid coherent fiber optic bundle (Fig. 1a) (Schott Fiber Optics; Southbridge, MA) is attached to a 128 x 128 pixel CCD array (b) (Sanyo Corp, Compton, CA; part LC9943). The lower panel of Figure 1 illustrates a view from the peripheral end of the probe. The optic bundle is composed of 6 to 25 μm fibers, which conduct light from the brain structure to the array. Glass cladding surrounds each fiber to prevent light from transferring between fibers. A cooling chamber (c) with coolant ports (d) is affixed to the CCD chip with silicon heat conducting paste to transfer the heat of the CCD chip to the cooling chamber. A light emitting diode (LED) attached at point (f) illuminates the target through flexible optic fibers (g) attached around the perimeter of the image probe. The LED emits a narrow-bandwidth light of operator-selectable frequencies. Timing and amplification circuitry (h) are attached to the side of the CCD chip. A miniature gold BNC connector (e) carries the video output. Electrodes (m) attached at the peripheral end of the probe acquire electrophysiological signals. Note the arrangement of flexible illuminating fibers and electrodes around the bundle perimeter in the lower panel of Figure 1.

One objective was to examine brain structures deeper than the cortex. The probe can be cut to any length and polished with diamond lapping techniques, allowing examination of structures at any depth. To minimize light loss, the fiber bundle was cemented to the CCD array, eliminating the need for a light focusing lens. A rigid probe also provides a stable mount for illuminating fibers and recording electrodes.

Cats were prepared under general anesthesia (25 mg/kg intravenous sodium pentobarbital) for surgical placement of electrodes. Typically, electrodes to record electrocardiographic (ECG), diaphragm electromyographic (EMG), electroencephalographic (EEG), and eye movement (EOG) activity were inserted. The fiber bundle was lowered stereotaxically to the site of interest. The imaging device was then attached to the skull with dental cement.

A schematic diagram of the system with associated peripherals is shown in Figure 2. Light from the LED was transmitted down the flexible fibers, reflected by the target tissue, and transmitted through the bundle to the CCD array. The video signal from the camera was amplified, recorded on video tape, and digitized by a frame grabber (Progressive Peripherals, Denver, Colorado). The resulting images were downloaded to a video computer (Commodore Business Machines, West Chester, PA) and subjected to digital averaging, subtraction, and pseudocoloring. Sequences of images were also replayed by the computer in a time lapse video fashion.

Although the camera can output frames every 16 ms, the transfer rate of the framegrabber was limited by the parallel port of the computer. This limitation restricted transfer of images to one frame every 3 seconds. We triggered the framegrabber to digitize the video signal to particular electrophysiological events, such as the rise of the cardiac R wave and respiratory onset. Triggering on these events minimized the potential for vascular or movement artifact. In addition, the frames could be digitized in synchrony with a stimulus pulse for correlation of images with evoked potentials.

We placed the imaging device over the dorsal hippocampus of an adult cat. Electrophysiological activity (including electrical activity from recording electrodes which were attached to the fiber bundle) was simultaneously gathered, amplified on a polygraph, and stored onto digital media on a second acquisition computer (A/D computer, Fig. 2). The video recorder stored synchronizing signals from the analog computer, ECG, and the video image. Bipolar electrodes were also placed in the contralateral hippocampus to evoke the crossed Schaeffer collateral CA1 response.

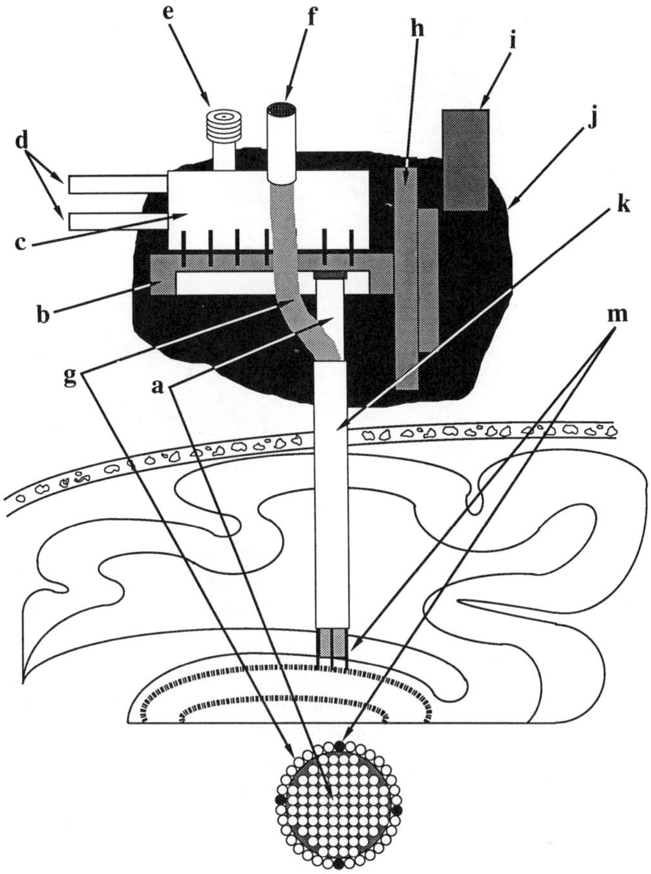

FIGURE 1. Outline drawing of imaging device. A straight coherent fiber bundle (a), reaching the brain target, is attached to the CCD array (b). A stainless steel chamber (c), through which a coolant enters and exits through 14 gauge tubing (d), is attached to the CCD chip. The video signal exits the device through a miniature gold BNC connector (e). An LED attached at point (f) illuminates the target through the flexible fiber optics (g) which are attached around the perimeter of the probe tip. Emitted light is reflected by the tissue and then transmitted to the CCD sensor. Video timing circuitry and amplification is achieved by components attached to the side of the CCD chip (h). Control voltages and synchronizing pulses are conducted through a 9-pin connector (i). The entire device is epoxy dipped (j) to protect the components. A thin wall stainless steel tube (k) surrounds the image probe and flexible fiber optics to provide stability and to prevent breakage. Recording electrodes (m) are also attached around the circumference of the fiber bundle. The bundle enters the skull through an opening which is sealed by bone wax. Internal wiring connections are not shown to simplify the diagram. The drawing is not to scale; the imaging device is a 2 cm cube.

A week following surgical recovery, an LED was selected for emission at 700 nm. Thirty images were gathered from the dorsal hippocampus over a 100 s period when the cat was alert and awake. Thirty additional images were synchronized to each pulse of a 0.5 ms, 20- to 50- μV electrical stimulation train. Images for each condition were averaged, and the differences in each pair calculated by subtraction. For each pixel of the resulting difference image, a multiplication factor was applied to amplify the changes.

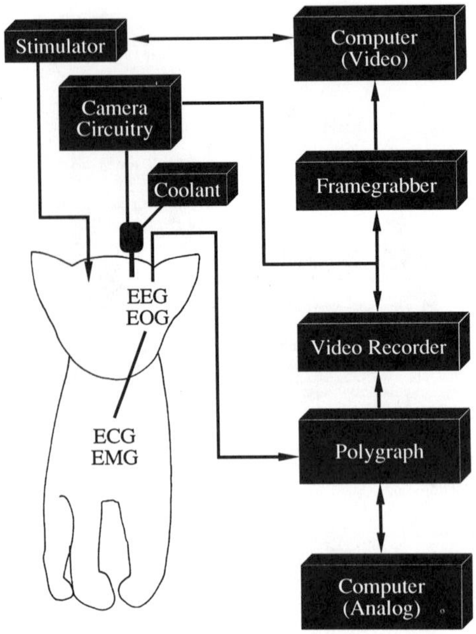

FIGURE 2. Schematic outline of components used in optical and electrophysiological acquisition. The electrophysiological signals (ECG, diaphragmatic EMG, EEG, EOG) are passed to polygraphic amplifiers via a shielded cable and to an acquisition (A/D) computer and audio channels of the video recorder. Reflected light is transferred to the imaging device via the fiber bundle. Video signals from the CCD are passed through a cable to the camera circuitry, and then to a frame grabber. The video signal is stored on a video recorder, together with synchronizing signals from the Schmitt trigger electronics and a time code from the A/D computer. The video computer acquires images through the frame grabber synchronized to respiratory, cardiac or stimulation events. A liquid nitrogen canister pumps coolant through the cooling chamber of the camera.

RESULTS

Figure 3 is an illustration of a typical evoked response elicited by stimulating the contralateral Schaeffer collaterals. A small but detectable difference occurs between the pre-stimulation (Fig. 3a) and during-stimulation (b) averages displayed as absolute reflectance levels. The difference image (c) was multiplied by a factor of 10 to amplify the changes. Ninety-five percent of the pixels were significantly different using ANOVA procedures. The grey scale (d) shows the percent change for the difference image. The difference image revealed a one percent overall decrease in reflectance during stimulation; however, isolated groups of pixels changed as much as ten percent in either direction creating the pattern observed in (c).

DISCUSSION

Modifications to the previously described camera reduced the device size and improved the image quality. A considerable size reduction was achieved by replacing a Plexiglas housing with an epoxy coating. Attachment of the video timing and amplification circuitry reduced noise introduced by transmission of the timing signals through long cables. A loop in the optic bundle was formerly required to minimize detection of stray light through the glass cladding which was used to illuminate the tissue. We now illuminate the tissue more evenly and efficiently through flexible fibers placed around the probe tip. Since the cladding is no longer used to illuminate the tissue, the loop was eliminated, thus enhancing contrast and reducing potential distortion of the image.

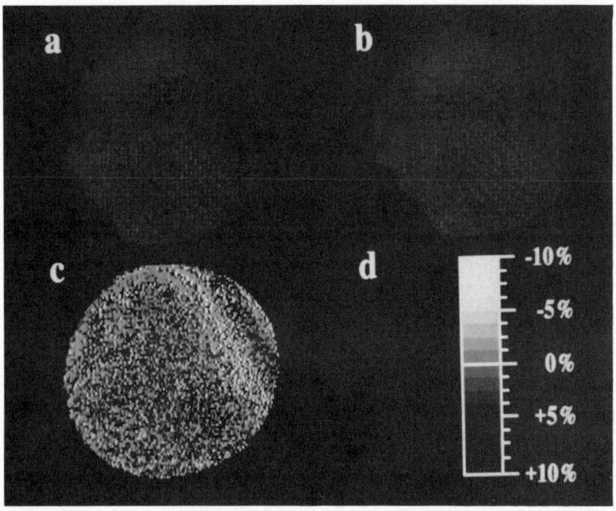

FIGURE 3. Averages of 30 dorsal hippocampal images synchronized to the cardiac R wave and inspiratory offset prior to (a), and during (b) Schaeffer collateral stimulation on the contralateral side of the hippocampal CA1 region. The averaged images (a) and (b) are displayed in absolute reflectance levels. The differences between the images are shown in (c). We observed an average change of one percent with isolated changes as great as ten percent in both directions. The grey scale in (d) illustrates the percent change in the difference image (c).

Many of the limitations of the current system result from trivial hardware issues. For example, the limitation imposed by the transfer rate through the parallel port of the computer could be easily overcome by more sophisticated frame grabbers, which store images in real time directly to memory and then to a mass storage device. Smaller CCD chips which include timing and amplification circuitry on the wafer with the CCD element will reduce the size of the device to allow implantation in smaller animals; such chips are just now becoming available.

Many cellular processes may contribute to the change in reflectance measured by the camera. For example, reflectance changes may be a function of neuronal cell body or glial swelling which accompanies neural activity and will alter the reflective properties of the tissue slice (MacVicar and Hochman 1991). Increases in neural activity may also alter local blood perfusion, creating localized changes *in vivo* for functional mapping of activity (Raichle et al. 1976, Baron et al. 1982, Frostig et al. 1990, Belliveau et al. 1991). In the guinea pig isolated whole brain preparation, changes in reflectance accompanies neural activity in the short wavelength region (blue end of the spectrum) which may be associated with changes in cytochrome oxidase activity (Federico et al. 1991). Measurements at these wavelengths, however, require specialized optics and CCD sensors.

The incorporation of a fiber optic probe with a miniaturized CCD camera provides many advantages. The technique allows recording from structures anywhere in the brain while the animal is freely behaving and drug free. The mounting technique allows recording for extended periods of time, even for periods of several months. Thus, the procedure is of great value for developmental and behavioral studies.

ACKNOWLEDGEMENTS: This research was supported by HL 22418. D.R. is supported by a predoctoral fellowship NIDR DE 07212. G.P. is supported by a Howard Hughes Medical Institute Predoctoral Fellowship. We thank Schott Fiber Optics for providing high resolution image conduit.

REFERENCES

Belliveau, J.W., Kennedy, D.N., McKinstry, R.C., Buchbinder, B.R., Weisskoff, R.M., Cohen, M.S., Vevea, T.J., Brady, J.M., and Rosen, B.R. (1991), Functional mapping of the human visual cortex by magnetic resonance imaging, Science, 254:716-719.

Blasdel, G.G. (1989), Visualization of neuronal activity in monkey striate cortex, Annu. Rev. Physiol., 51:561-581.

Baron, J.C., Lebrun-Grandie, Ph., Collard, Ph., Crouzel, C., Mestelan, G. and Bousser, M.G. (1982), Noninvasive measurement of blood flow, oxygen consumption, and glucose utilization in the same brain regions in man by positron emission tomography: concise communication, J. Nucl. Med., 23(5):391-399.

Cohen, L.B. and Keynes, R.D.(1971), Changes in light-scattering associated with the action potential in crab nerve, J. Physiol., 212:259-275.

Federico, P., Borg, S.G. and MacVicar, B.A. (1991), Stimulation-evoked changes in intrinsic optical signals in the guinea-pig isolated whole brain, Neurosci. Abstr., 17(2):1114.

Frostig, R.D., Lieke, E.E., Ts'o, D.Y. and Grinvald, A. (1990), Cortical functional architecture and local coupling between neuronal activity and the microcirculation revealed by in vivo high-resolution optical imaging of intrinsic signals, Proc. Natl. Acad. Sci. U.S.A., 87:6082-6086.

Grinvald, A., Salzberg, B.M. and Cohen, L.B. (1977), Simultaneous recording from several neurons in an invertebrate central nervous system, Nature, 268:140-142.

Grinvald, A., Lieke, E., Frostig, R.D., Gilbert, C.D. and Wiesel, T.N. (1986), Functional architecture of cortex revealed by optical imaging of intrinsic signals, Nature, 324:361-364.

MacVicar, B.A. and Hochman, D. (1991), Imaging of synaptically evoked intrinsic optical signals in hippocampal slices, J. Neurosci., 11(5):1458-69.

Raichle, M.E., Grubb, R.L., Gado, M.H., Eichling, J.O. and Ter-Pogossian, M.M. (1976), Correlation between regional cerebral blood flow and oxidative metabolism, Arch. Neurol., 33:523-526.

Rector, D.M. and Harper, R.M. (1991), Imaging of hippocampal neural activity in freely behaving animals, Behav. Brain Res., 42:143-149.

Ts'o, D.Y., Frostig, R.D., Lieke, E.E., and Grinvald, A. (1990), Functional organization of primate visual cortex revealed by high resolution optical imaging, Science, 249:417-420.

OLFACTORY INFORMATION PROCESSING IN INSECTS REVEALED BY REAL-TIME OPTICAL IMAGING OF INTRINSIC SIGNALS

E.E. Lieke

Institut für Tierphysiologie
Abt. Neurobiologie
Berlin, Germany

INTRODUCTION

When honeybees search for nectar in the field they use visual cues like the color and shape of flowers to approach their target. The moment they are close to their target they have to find answers to several questions: (1) Is this really the right flower species? (sometimes several flowers have a very similar color and shape), and (2) Is this object already producing nectar, or is it still at a premature stage? These questions can be answered by probing the scent the flower produces.

Bees can learn both visual (von Frisch, 1914) and olfactory (von Frisch, 1967) cues in order to differentiate flowers. Under restricted laboratory conditions bees do not learn visual cues. They do, however, learn odors very fast and precisely (Menzel, 1983).

Nevertheless, odors are not unique as food markers. Several kinds of social behaviors are affected by different pheromones which are sensed via the same sensory pathways. The mandibular pheromone or queen substance guides the drones for mating and suppresses maturation of the worker bees. Guard bees at the entrance of the hive produce an alarm pheromone when the beehive is attacked by an enemy.

Therefore, odors can have very different meanings and elicit very different types of behavior. As we are interested in the physiological modifications caused by learning, we would like to know how the bee brain processes odor information.

The antennal lobes (AL) are the primary olfactory neuropiles of insects (see figure 1, top left). They are organized into about 170 glomerular substructures (Pareto, 1972; Christensen and Hildebrand, 1988) containing sensory input neurons, projection or output neurons and local interneurons. The complex firing pattern of insect AL output neurons is thought to result from the stimulus-specific temporal-spatial distribution of excitation in this neuropil. In this respect the primary afferent processing is reminiscent of the processing found in the mammalian olfactory system (see e.g. Boeckh et al., 1990). Both neuropiles show a mass convergence of many input neurons onto relatively few output neurons (in the case of the honeybee 30,000 to about 200). They also process olfactory information modulated in time and categorize olfactory stimuli.

Optical Imaging of Brain Function and Metabolism, Edited by
U. Dirnagl *et al.*, Plenum Press, New York, 1993

The vertebrate olfactory bulb has been modeled extensively on how the network performs these difficult tasks (see Kauer, 1991 for a review) based on intracellular and electro-olfactogram recordings as well as optical recordings in salamanders (Orbach and Cohen, 1983; Kauer, 1988) and rats (Catarelli and Cohen, 1989). For the insect AL only data from single-cell recordings have been reported until now (see e.g. Christensen and Hildebrand, 1987a; Waldrop et al., 1987; Boeckh et al., 1983, 1984, 1987). Therefore, network models for the insect AL are rare and quite immature, although there is a belief that parallels to the olfactory bulb of vertebrates can also be found at this level. Techniques that allow the electrophysiological investigation of the insect AL in space and time might help to bridge this gap.

Optical recording of neural activity offers the ability to study a parallel working neural system by means of parallel multisite recording with high time and spatial resolution. The optical recording technique was successfully applied to honeybees in *in vivo* preparations as well as in brain slices. This article shows that intrinsic optical changes which can be used to monitor electrical activity indirectly occur in this tissue.

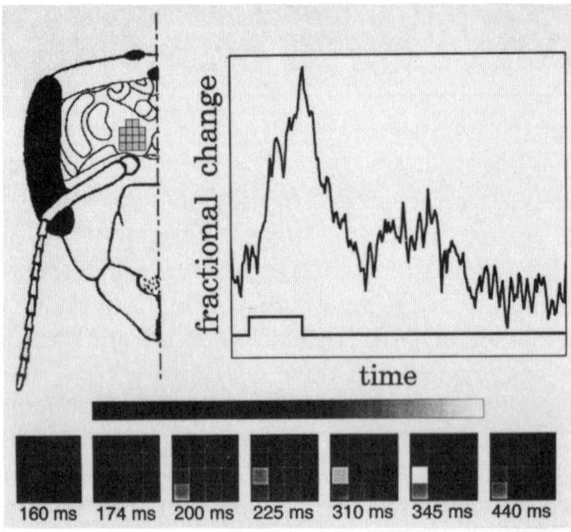

FIGURE 1. Recording site, response to an olfactory stimulus. The top left panel depicts the bee head and some parts of the brain. The position of the antennal lobe is indicated by a grid representing the field of view of the photodiodes. The top right panel shows the time course of the intrinsic optical signal elicited by a 200 msec carnation stimulus (average of 30 trials) at the site of maximal response. The bottom panel shows the spatial distribution of the odor response at different times. Frame 6 displays most clearly the site of maximal response.

OPTICAL SIGNALS ELICITED BY AN ODOR

The sensory cells sitting on the antennae have broad response profiles (Vareschi, 1971; Akers and Getz, 1991). Therefore, natural odors should excite many sensory cells and the stimulus-specific excitation pattern should be encoded primarily in an across-fiber pattern (Boeckh et al., 1984). It is also known that honeybees distinguish odors very well with large overlapping component profiles (Vareschi, 1971; Akers and

Getz, 1991). Such a tuning should be modeled within the early stages of olfactory information processing.

Figure 1 (top right) shows the activity elicited by stimulation with carnation in the AL of a worker bee. The activity starts with some delay, increases during the stimulus and returns to its baseline after cessation of the stimulus. The time course of the response is complex - at least two components produce the rising phase.

An interesting question about olfactory information processing is whether odor "images" are encoded as across-fiber patterns in a widely distributed network or in spatially neighboring cells forming well defined patterns of hot spots in the glomerular organization of the AL. The spatial distribution of the response (figure 1, bottom) shows that the second assumption is more likely. However, the size of the activated area is time-dependent - starting from a small initial patch the activity spreads somewhat laterally. After cessation of the stimulus the activity contracts back into the initial patch and finally vanishes.

CONFIRMATION OF THE BIOLOGICAL ORIGIN OF THE INTRINSIC SIGNAL

In the course of using these intrinsic signals in further studies on the temporal-spatial distribution of excitation in the AL, two features have to be shown: (i) evidence must be obtained that they are not stimulus artifacts, and (ii) their origin should be investigated. Both prerequisites can be acquired easiest in brain slices because of the improved manipulative access to the cell environment.

One possible artifact might be that the optical signals arise from the production of gas bubbles caused by the electrical stimulation. The data shown in figure 2 test this option. The threshold for eliciting an optical response can be well defined and is very narrow (between 60 and 70 μAmpere in the preparation shown). When increasing the stimulus strength only the response amplitude changes. A stimulus current about 50% above the threshold already resulted in a saturation of the response amplitude. Therefore, the slope of the response amplitude vs. stimulus intensity function (figure 2B) is very deep, while that of the gas bubble production should be much more shallow.

In the same preparation the wavelength dependence of the response amplitude was also tested. Figure 2C shows the response amplitude of the intrinsic signals elicited by saturating stimulus currents at several wavelengths. All intensity changes are normalized to the background intensity at the same wavelength. The result indicates that the intrinsic signal has at least two components: one is mostly wavelength-independent, and one strongly wavelength-dependent with a maximum at about 500 nm. When assuming that the magnitude of the wavelength-independent signal at 500 nm is the same as for the other wavelengths, the additional signal is about 2.5 times larger than the wavelength-independent signal. The wavelength-independent signal probably arises out of some physical changes of the tissue like change of scattering properties. Stepnoski et al. (1991) have shown that such a light scattering signal is linearly proportional to the change of membrane potential. The wavelength-dependent signal is probably an absorption change of some activity-dependent protein. Candidates for such an enzyme are cytochromes and/or flavoproteins. The possible sources of the intrinsic signals have not yet been investigated in detail, however, fake sources like movement of the preparation or electrolysis can be excluded.

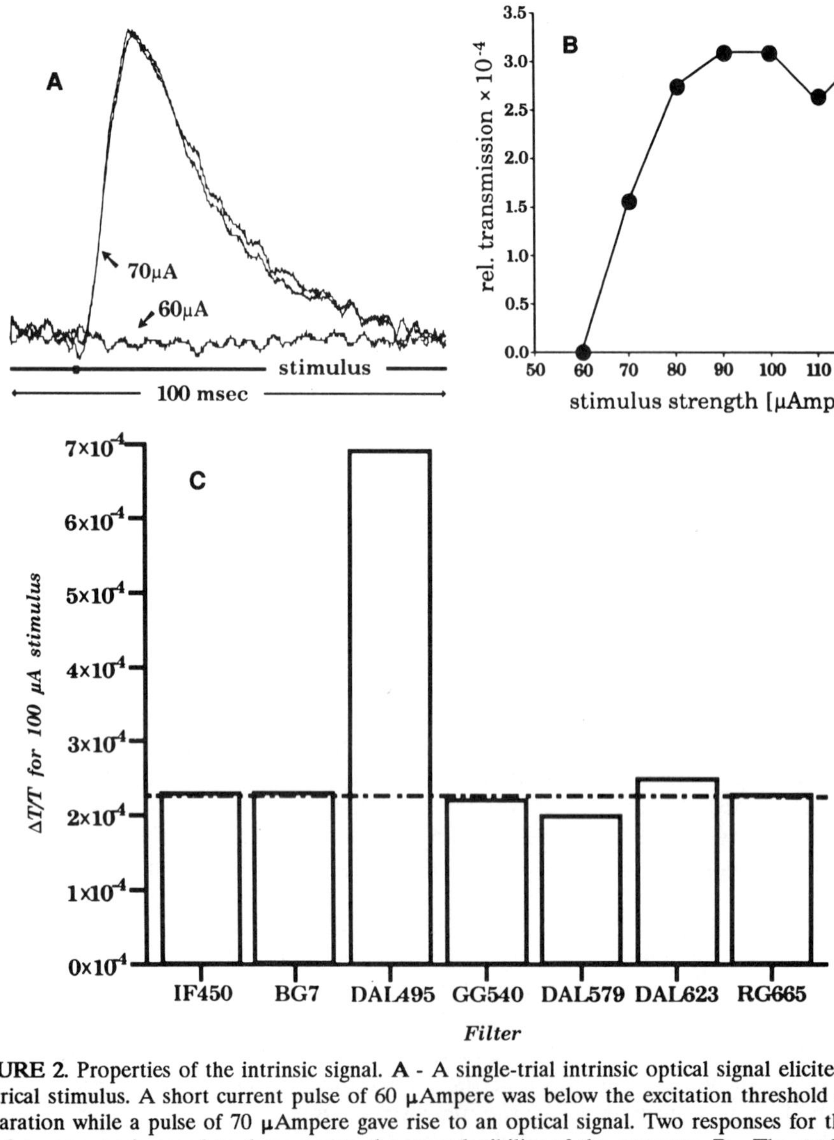

FIGURE 2. Properties of the intrinsic signal. **A** - A single-trial intrinsic optical signal elicited by an electrical stimulus. A short current pulse of 60 μAmpere was below the excitation threshold for this preparation while a pulse of 70 μAmpere gave rise to an optical signal. Two responses for the later stimulus are superimposed to demonstrate the reproducibility of the response. **B** - The peak signal amplitude is plotted as a function of the stimulus strength. A stimulus only 50% above the threshold already resulted in a response saturation. This result clearly excludes that the optical signal could be caused by gas bubbles. **C** - The peak signal size is shown for several wavelengths. The intrinsic signal has at least two components - one is mainly wavelength-independent, and one wavelength-dependent with a maximum at about 500 nm. The dash-dotted line indicates the mean value of the wavelength-independent signal. This result also excludes possible movement artifacts producing the optical signal.

GENERAL NETWORK PROPERTIES OF THE INSECT ANTENNAL LOBE

Many of the AL local interneurons have been found to be GABA immunoreactive and thus appear to sculpt olfactory processing by inhibitory connections (Schäfer and Bicker, 1986; Christensen and Hildebrand, 1988; Itagaki and Hildebrand, 1990). By blocking GABAergic chloride channels with picrotoxin, inhibitory pathways in this neuropil should be unmasked.

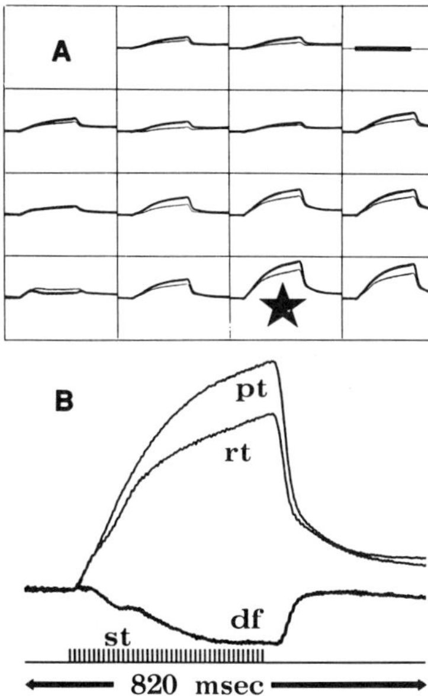

FIGURE 3. Inhibition in the antennal lobe. The figure shows the average optical responses (30 trials) elicited by 400 msec electrical bursts. **A** - responses before (thin traces) and after (bold traces) picrotoxin treatment (10^{-5} molar, 5 min). The signal amplitudes were normalized to the background intensities at each position. The response amplitude in most places increased after picrotoxin treatment. **B** - magnified traces from panel A (see asterisk in A for position) and the calculated difference. The regular trace (*rt*) starts with a fast component followed by a slower one. After picrotoxin treatment (*pt*) the time course of the signal becomes more continuous. The difference signal (*df*) shows a delay of about 10 msec and two following phases which correspond to the fast and slow component of the regular trace. This difference can be interpreted as the time course of the inhibition unmasked by blockage of GABA-sensitive chloride channels. *st* - stimulus.

Bursts of electric stimuli applied to the antennal nerve before and after picrotoxin treatment (10^{-5} molar, 5 min) elicited different optical signals (see figure 3A). As expected, the signal amplitudes increased in most places after the picrotoxin treatment. However, the signal modifications are much more complex as can be seen better in the magnified traces in figure 3B. While the response from the untreated slice (*rt*) starts with a steep phase followed by a shallow one, this transition becomes more continuous after the picrotoxin treatment (*pt*). This change in the time course of the signals is also visible in the calculated differences (*df*) of the raw data. The difference signal starts with a flat phase of about 10 msec duration indicating no modification of the time course during this period. This lag is followed by a fast and a slow alteration. The time course of the signals before the treatment looks much like that elicited by an odor (see figure 1). Thus the effects seen by stimulation with electrical currents are likely to represent common features within this neuropil.

The raw data (figure 3A) should be suitable for an extraction of spatial information about coarse neuropilar organization in the AL. Figure 4 displays the spatial distribution of the picrotoxin effect at several different times. The time of the first panel corresponds with the transition from the steep to the shallow slope of the signal, i.e., the transition from the fast to the slow inhibition (see figure 3B).

The central region of the AL with a predominance of axons over neuropil in the optical image of the whole AL, displays a larger effect than peripheral glomerular regions. In addition, the overall distribution of the inhibition appears inhomogeneous in the glomerular neuropil with steep profiles of higher or lower inhibition. Ratio images calculated for later times yield less inhomogeneities.

As shown in the magnified traces (figure 3B) the overall inhibition comprises at least two components. These may be generated by different neuronal mechanisms. However, one of these might not result from local inhibitory circuits but rather from inhibitory collaterals of the output neurons looping back on its own input or on the input of a neighbor. Evaluation of the spatial distribution of the inhibition (figure 4) suggests that some part of this interaction is running orthogonal to the main stream of the information pathway. The known connectivities in the AL of insects are in accordance with such an interpretation (Boeckh et al., 1983, 1984; Christensen and Hildebrand, 1987b; Homberg et al., 1989). Local patches of information processing units can thus be produced as in the olfactory bulb of mammals.

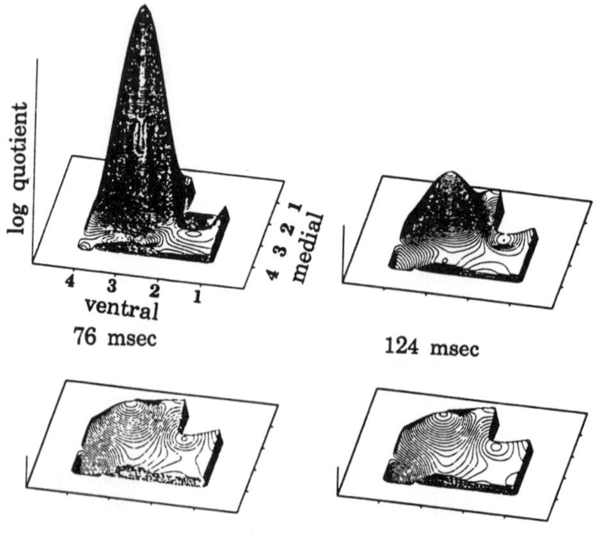

FIGURE 4. Spatial distribution of the picrotoxin effect. The logarithm of the quotient *'response after picrotoxin/regular response'* is displayed at its spatial position. The spatial distribution 76 msec after response onset (first panel) is extremely inhomogenous. This time corresponds to the transition from the fast to the slow component of the regular response (see figure 3B). Ratio images calculated for later times yield less inhomogeneities.

REFERENCES

Akers, R. P. and Getz, W. M. (1991) A test of identified response classes among olfactory receptor neurons in the honeybee worker. Chem. Senses (in press).

Boeckh, J., Ernst, K. D. and Polz, H. (1983) Convergence of inputs from different sense organs upon neurones of the central antennal pathway in insects. In: Horn, E. (ed) Multimodal convergences in sensory systems. G. Fischer, Stuttgart, pp. 143 - 148.

Boeckh, J., Ernst, K. D., Sass, H. and Waldow, U. (1984) Anatomical and physiological characteristics of individual neurones in the central antennal pathway of insects. J. Insect Physiol. 30, 15 - 26.

Boeckh, J., Ernst, K. D. and Selsam, P. (1987) Neurophysiology and neuroanatomy of the olfactory pathway in the cockroach. Ann. NY Acad. Sci. 510, 39 - 43.

Boeckh, J., Distler, P., Ernst, K. D., Hösl, M. and Malun, D. (1990) Olfactory bulb and antennal lobe. In: Schild D. (ed) Chemosensory information processing. Springer, Berlin, pp. 201 - 227.

Catarelli, M. and Cohen, L. B. (1989) Optical recording of the in vivo piriform cortex responses to electrical stimulation of the lateral olfactory tract in the rat. Chemical Senses 14, 577 - 586.

Christensen, T. A. and Hildebrand, J. G. (1987a) Functions, organization, and physiology of the olfactory pathways in the lepidopteran brain. In: Gupta, A. P. (ed) Arthropod brain: its evolution, development, structure, and functions. J. Wiley and Sons, New York, pp. 457 - 483.

Christensen, T. A. and Hildebrand, J. G. (1987b) Male-specific, sex pheromone-selective projection neurons in the antennal lobes of the moth Manduca sexta. J. Comp. Physiol. 160, 553 - 569.

Christensen, T. A. and Hildebrand, J. G. (1988) Frequency coding by central olfactory neurons in the sphinx moth Manduca sexta. Chem. Senses 13, 123 - 130.

Frisch, K. von (1914) Der Farbensinn und Formensinn der Biene. Zool. J. Physiol. 37, 1 - 238.

Frisch, K. von (1967) The dance language and orientation of bees. Cambridge University Press.

Itagaki, H. and Hildebrand, J. G. (1990) Olfactory interneurons in the brain of the larval sphinx moth Manduca sexta. J. Comp. Physiol. [A] 167, 309 - 320.

Kauer, J. S. (1988) Real-time imaging of evoked activity in local circuits of the salamander olfactory bulb. Nature 331, 166 - 168.

Kauer, J. S. (1991) Contributions of topography and parallel processing to odor coding in the vertebrate olfactory pathway. TINS 14, 79 - 85.

Menzel, R. (1983) Neurobiology of learning and memory: the honeybee as a model system. Naturwissensch. 70, 504 - 511.

Orbach, S. H. and Cohen, L. B. (1983) Optical monitoring of activity from many areas of the in vitro and in vivo salamander olfactory bulb. A new method for studying functional organization in the vertebrate central nervous system. J. Neurosci. 3, 2251 - 2262.

Pareto, A. (1972) Die zentrale Verteilung der Fühlerafferenzen bei Arbeiterinnen der Honigbiene, Apis melifera L. Z. Zellforsch. 131, 109 - 110.

Schäfer, S. and Bicker, G. (1986) Distribution of GABA-like immunoreactivity in the brain of the honeybee. J. Comp. Neurol. 246, 287 - 300.

Stepnoski, R. A., LaPorta, A., Raccuia-Behling, F., Blonder, G. E., Slusher, R. E. and Kleinfeld, D. (1991) Noninvasive detection of changes in membrane potential in cultured neurons by light scattering. Proc. Natl. Acad. Sci. USA 88, 9382 - 9386.

Vareschi, E. (1971) Duftunterscheidung bei der Honigbiene-Einzelzell-Ableitungen und Verhaltensreaktionen. Z. vergl. Physiol. 75, 143 - 173.
Waldrop, B., Christensen, T. A. and Hildebrand, J. G. (1987) GABA-mediated synaptic inhibition of projection neurons in the antennal lobes of the sphinx moth, Manduca sexta. J. Comp. Physiol. 161, 23 - 32.

INFRARED IMAGING OF BRAIN FUNCTION

A.M. Gorbach

Institute of Higher Nervous Activity and Neurophysiology
Russia Academy of Sciences
Moscow, Russia, and
National Institutes of Health
Bethesda, MD, USA

INTRODUCTION

The aim of this article is to describe in detail a technique to record the thermal response of the animal brain cortex using an infrared (IR) optical system (Budko et al., 1985). We compare this method, which we call thermoencephaloscopy (TES), with other techniques for studying the cortex. Results obtained using TES are analyzed in relation to results from other neuromapping methods. Previous reports of our methodology were published primarily in Russian or in journals with limited circulation.

Recently a new concept of brain organization invoking multiple representation and various kinds of network organization has gained great popularity in neurophysiology. The previously dominant methodological concept of investigating individual neurons has been replaced by investigation of the complex behavior of a microcircuit containing many neurons. Optical methods of investigating brain functions can be useful for the study of such microcircuits because they allow combining the study of the function of individual cells and the cooperative behavior of variously sized clusters of higher interconnected neurons.

We developed TES because other methods for studying brain function are either invasive, limit movement of the animal, or do not have sufficient spatial and temporal resolution. Other optical techniques share some of the advantages of the IR method, but they require illumination of the brain surface.

BIOPHYSICS OF HEAT EMISSION OF THE BRAIN

All bodies, the human body included, continuously emit IR radiation, which travels at the speed of light in the form of photons, or quanta of energy.

Thermography, or more correctly, radiometry, is defined as the technique for measuring radiant energy traveling in the form of electromagnetic waves (Glossary, 1987). Thermography, which is used in medical practice, is based on recording of the intrinsic IR radiation of man and is the measurement of natural emission from the human body.

Optical Imaging of Brain Function and Metabolism, Edited by
U. Dirnagl *et al.*, Plenum Press, New York, 1993

Any object above absolute zero Temperature will radiate electromagnetic energy to an extent governed by its radiant emission. A body upon which electromagnetic radiation falls may transmit, reflect, or absorb all of the incident radiation or energy. If it absorbs all the incident radiation, it is known as a "black body" or full radiator.

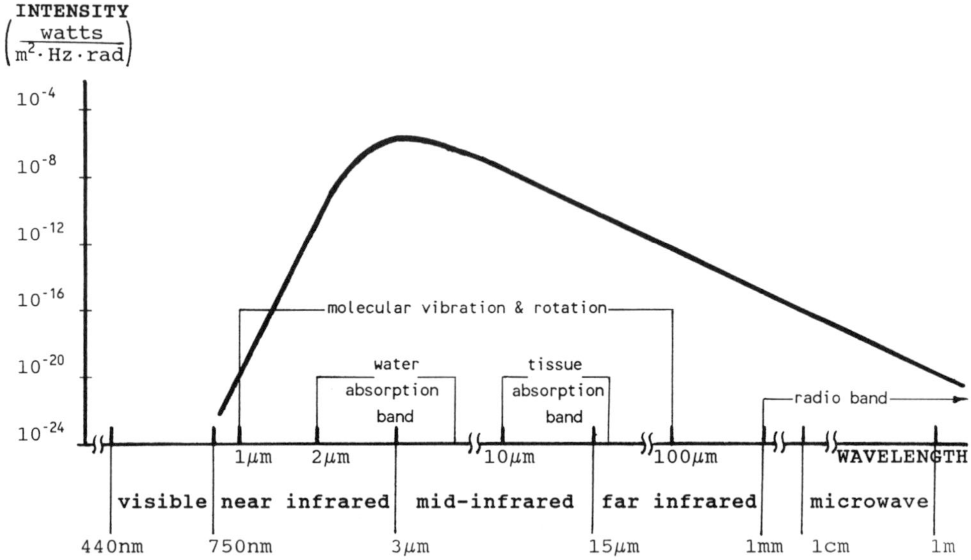

FIGURE 1. A qualitative illustration of the intensity of the black body radiation (1000 K) as a function of wavelength.

The distribution of the radiation is a function of both the wavelength and the temperature. Figure 1 illustrates the relative position of the electromagnetic waves of the spectrum of interest and of black body radiation.

Recording radiation at considerable distances from the biological object is possible because the radiation maximum at a temperature of 300 K corresponds to a window of atmospheric transparency in an average IR wavelength range of 8-15 μm. The atmosphere is also relatively transparent for passage of IR radiation in a range of 3-5 μm, where the maximal sensitivity to temperature contrast is achieved with less radiation intensity in comparison to the first window. However, the band at 3-5 μm has a significant absorption region at 4.2 microns. When imaging outdoors on a humid day, one sees significant "fog" due to this absorption band while using a 3-5 μm system. The 8-15 μm region is relatively insensitive to humidity.

The IR radiation can be detected using a device called a thermoviser. The radiation sensed by the thermoviser includes thermal energy emitted from all surfaces in the field of view and the reflection of the IR radiation from surrounding surfaces. The effect of reflection in the process of measurement is usually disregarded or reduced by a number of methodological means.

The depth of the heat sources which radiate in these wavelength ranges may reach 100 μm (Gulyaev et al., 1985). Using the distribution of temperature fields on the surface of the head or skull, the processes which are linked to the activity of the deepest heat sources, for example, in the brain cortex can be assessed. However, in this context the inverse heat conduction problem must be considered, since the usual or "direct" problem in heat conduction is to determine the temperature using the known distributions of thermophysical properties. For this purpose we will briefly discuss the main fundamentals of heat exchange of the head of man or animals with the ambient environment as well as internal heat exchange.

Since thermal currents are caused by temperature gradients between the ambient environment and the living organisms, the temperature on the surface of the latter is determined by the equilibrium between heat production and its loss. It is possible to determine four important processes which occur in the coverings of bioobjects of interest to us: heat radiation, heat conduction (linked to the direct transmission of kinetic energy of molecules from the warm surface of the body to the cool air), heat convection (heat transport by moving air), and evaporation. By means of the first three pathways the body can both disseminate heat into and receive heat from the environment. It is known that in man under agreeable temperature, 3% of the heat is carried away from the body surface by conduction, 12% of the heat by convection, 25% by evaporation, and 60% by radiation (Ivanov, 1990).

For neurophysiological tasks the temperature dynamics of the coverings of the head or skull is the source of noise in relation to generated responses of brain origin. Thus, it is necessary to achieve an equilibrium with the environment before thermal imaging is carried out. Reduction of the influence of air convection and evaporation on the process of measuring thermal radiation is achieved by a number of methodological approaches described below.

We will examine the main properties of internal heat exchange in the cranial bone and brain of organisms. In addition to the processes of heat generation in the brain, which we will discuss later, heat exchange here is characterized by heat conduction (of tissues, fluids, and bones) and convection of heat by the blood.

The flow of heat by conductive means is a function of the physical properties of the tissue, including the tissue density, the thermal constant, and the specific heat (or heat capacity). The relation between total protein and total water in a tissue is a major factor in determining its specific heat. The value of specific heat is slightly lower in white matter (0.86 cal/g °C) than in gray matter (0.88 cal/g °C) (Cooper and Trezek, 1972).

The heat capacity of blood, 0.93 cal/g °C, is close to that of water. The thermal constant of soft tissues is practically equal to the thermal constant of water. This suggests that physiological changes in the water content in tissues will not be substantially reflected in their heat conduction (Ivanov, 1990). Such a hypothesis is justifiably applicable to the brain cortex if we consider that the functional activity of the latter is linked first of all to a change in the velocity of local brain blood flow and, to a lesser degree, to a change of its momentary volume.

The process of heat convection by blood depends not only on the caliber and type of the blood-carrying vessels where heat exchange occurs, but also on the temperature gradients between the blood and tissues, and the direction, volume and velocity of the blood flow in the vessels (Lemons et al., 1987). This presupposes the possibility of functional activation of the brain cortex of mechanisms which control the convection process, in addition to mechanisms of physiological regulation of the activity of heat sources. The effectiveness of the process of heat convection depends on the heat conduction between generators of heat and the walls of the blood vessels as well as between these vessels and the heat carrier (blood).

Heat diffuses intramurally through blood slowly, compared with the rate at which the fluid carrying heat moves through the cardiovascular system. A number of data indicate that an increase in the velocity of the blood flow can lead to a twofold increase in heat transmission by living tissue (Grayson, 1952). On the other hand, blood has a rather high thermal capacity, which suggests that the fluid has a good ability to absorb large quantities of heat for subsequent transport.

The appearance of temperature gradients on the surface of the skull as a result of heat emission by the brain cortex is facilitated by the low heat capacity of bone. Flat bones have the lowest heat capacity (0.3 cal/g °C) of any tissue (Ivanov, 1990). We recall that when supplied with the same quantity of heat, a body with a low heat capacity increases its temperature considerably more than a body with a large heat capacity. Heat exchange inside the body between organs, tissues, and fluids has by far not been exhaustively studied, which does not allow assessment of a number of its important characteristics, including time characteristics.

It is convenient to categorize heat currents in the brain as global and local. By global we mean a heat current which is directed, as in all heated physical bodies, from the center to the periphery in all directions. Local currents can be formed with the appearance of local temperature gradients caused by local heat production of the brain. On the grounds of theoretical and experimental studies, three basic processes of local heat production of the brain can be considered:

1. Heat emission in membranes of neurons linked to synaptic transmission (Kusano and Tasaki, 1990) and generation of action potentials (Ritchie and Keynes, 1985; Tasaki et al., 1989).
2. Heat emission linked to metabolic heat production of neurons and glia (Ivanov, 1990).
3. Heat emission linked to changes of general blood flow, temperature changes in the arterial blood flowing to the brain, and changes in local blood flow (Baker et al., 1973; Baker, 1982; Falk, 1990; Ivanov, 1990).

Heat currents originating from these sources modulate the global heat current which, running through brain membranes, is projected toward the external coverings of the head of the bioobject and, becoming transformed into electromagnetic waves, may be measured by the thermovisor. Thanks to the relative stability and inertia of the global heat current, its contribution may be subtracted from the total signal as a constant component, which enables assessment of local heat currents.

Therefore, the main purpose of our study was to obtain maps of temperature changes which reflect the spatial and temporal distribution of local sources in the brain cortex arising from sensory and motor stimulation, as well as in different states of the animal and human subject.

METHOD

Part of the energy radiated from the bone or scalp surface can be recorded at a considerable distance from the head using a special optical scanning system and an IR detector. We used the standard AGA-780 IR imager (Agema, Sweden), with the lens placed at a distance of 10 to 75 cm from the head. The objective was focused on the scalped dorsal surface of the animal's skull (Fig. 2).

For precise focusing, pieces of aluminum foil were temporarily fixed on the dorsal surface of the animal's skull or on the skin of a human subject's head. Depending on the position of the animal's head, limitations of the camera inclination (because of the inoperability of this model at angles greater than 45 °), and the best surveying site for the head surface of interest, its image completely or partially fell into the lens through an IR mirror located at an angle of 45°, or without it.

An opticomechanical system of rotating mirrors in the IR imager carried out x-y scanning of the heat image and transmitted its elements successively to the IR detector. The imager used for this study had a 40 msec scanning time (25 scan/sec) and a differential thermal sensitivity of 0.2 °C, which is not sensitive enough for analysis of the heat generated by the cerebral cortex. To increase the signal-to-noise ratio, several procedures were adopted: synchronized summation of a series of images during 4 to 9 repetitions of the stimulus, in some cases 4 successive IR scans were averaged over a period of 160 msec, and the heat images were filtered by digital high-frequency filters.

To detect weak thermoreactions against the background of considerable initial temperature nonhomogeneities of the brain, one of the background scans (most frequently the first), taken as the reference, was subtracted from each scan. By the methods indicated above the differential temperature sensitivity of the method was increased, during evaluation of spatial nonhomogeneities of the IR scans, to 0.01-0.015 °C and, for more extensive details of the temperature relief, even lower.

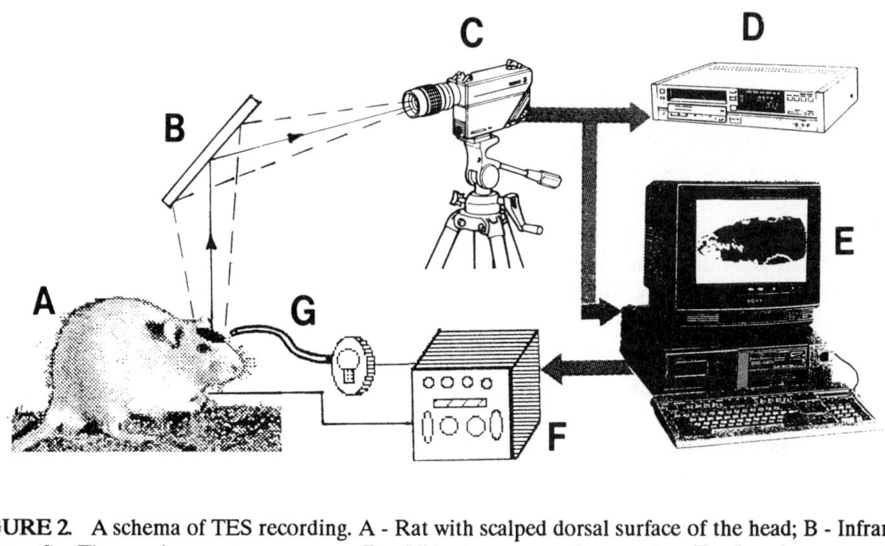

FIGURE 2. A schema of TES recording. A - Rat with scalped dorsal surface of the head; B - Infrared mirror; C - Thermovisor scanner camera; D - Videotape magnetic recorder; E - Imaging processing computer; F - Computer controlled stimulator; G - Fiber optic

The thermoviser is focused on the skull of the animal through an IR mirror. The type of lens and the distance between the camera and cortex determine the spatial resolution. The camera output is digitized with a frame grabber, filtered and processed in the computer and displayed on the monitor. The computer also provides control over the stimulator and the timing of the experiment. The visual stimulus was delivered to the animal's eye by a fiber optical cable.

Besides IR scans, we also investigated thermograms: dependence of temperature at a given point or region on time. In the latter case, temperatures at 25 or 49 points were averaged, resulting in an even greater increase of sensitivity, up to the limit determined by the specialized computer (Numelec, France) and the best working range (2 °C) of the IR scan: 0.002 °C. Instrumental spatial resolution was 70-100 μm/pixel in experiments on animals and in human investigations 0.6-1.3 mm/pixel.

With a combination of hard- and software it was possible to record a series of 72-216 IR scans in the computer and transfer it to a magnetic tape recorder with a frequency of 1.3 to 25 frames/sec depending on the aims of the experiment. Most frequently, recordings of 4 IR scans in 160 msec were followed immediately by a pause of 600 msec, after which the next 4 scans were recorded. Each IR scan consisted of 10,880 pixels (128 x 85 elements), and of this total number, from 4,000 to 9,000 pixels were related directly to the skull. The time taken to obtain one scan was 40 msec, and measurement of the temperature at each individual point of the IR image took about 2.4 μsec. The temperature at each point was coded in the computer by a discrete number and it could be shown on the display or as a hard copy in the form of gradations of brightness or color.

To analyze the temperature dynamics, animation programs were developed, making it possible to examine a process at an accelerated rate and in real time. We also used differentiation of the image with respect to time. This permits visualization of the propagating fronts of temperature waves. As a rule, before presentation of the stimulus a series of 24 background IR scans was recorded, after which the computer activated the stimulator, and a further 48-192 scans were recorded at the same rate, some during and some after the action of the stimulus.

The experiments were conducted in a lightproof and soundproof chamber with a monitoring window in acute experiments on mature albino rats and rabbits, anesthetized with pentobarbital (40 mg/kg body weight), but more frequently, immobilized with D-tubocurarine (1.5 mg/kg, artificial ventilation). In the latter case, possible air currents from the artificial ventilation tube were directed away from the dorsal surface of the brain with the aid of simple paper screens. As test objects we used white rats, rabbits, cats, monkeys, and humans. During animal experiments, the body temperature was kept constant, the scalp was removed beforehand, and the head fixed in a halter. A number of experiments on animals were carried out under conditions of partial mechanical immobilization and also with free movement during sleep.

In the process of the experiments air currents and radiation from the instruments inside the room were minimized. Temperature calibration of the entire measurement system was usually carried out before and after the experiments with the aid of a standard calibrator for the IR camera, within the working sensitivity range of the instrument. Here by virtue of the methods described above for processing of the images, main attention was not given to measurement of the absolute temperature values, but rather to the sensitivity of the system to temperature changes on the scalp.

Initial data processing consisted of examination of the data of the entire experiment, selection of images without artifacts and determination of the most interesting areas of temperature activity from a physiological standpoint. Secondary data processing consisted of the use of different methods for analyzing the images of interest and creation of thermograms for the "zones of interest" and for the control zones. The parameters of these thermograms, as well as the characteristics of the thermal foci on the images (localization, number, and size of the foci, distinctness of their boundaries, behavior in time), were the subject of physiologicalal analysis and statistical processing.

RESULTS

1. Thermomapping of the Sensory Cortex

Altogether 38 acute experiments were conducted on mature Albino rats immobilized with D-tubocurarine and artificially ventilated. With each type of stimulation, 4-9 responses were recorded.

Photic Monocular Stimulation. A light guide, the far end of which was illuminated through a slide by a noiseless modulator tube, was brought up to the eye at a distance of about 30-50 mm. Geometric shapes (circles, triangles) flashing for 5-60 sec, creating an intensity of illumination of 15-20 lx at pupil level, were used. The experiments were carried out under conditions of dark or low scotopic adaptation. Because of the considerable hypermetropia of the albino rat eye, the image was focused on the retina by means of +35D lenses, placed in front of the eye. Refraction was tested by projection ophthalmoscopy. The stimuli subtended an angle of between 6 and 26°.

In response to the action of local illuminated shapes on the eye, circumscribed foci of heating developed in the primary visual cortex, and their location depended on the projection of the stimulus on the retina (Shevelev et al., 1986g). Foci of heating in the visual cortex were distinctly localized, and were fully formed 3-10 sec after the beginning of stimulation, when they measured up to 2-4 mm in the anteroposterior direction and 3 mm in the transverse direction (Fig. 3, frame D). Temperature contrast of the responses in the center of the zone usually varied from 0.03 to 0.1 °C, although it could reach 0.14 °C (Shevelev et al., 1986c, 1989a).

The early and fairly local response in field 17 of both hemispheres had a precise location and clear-cut boundaries, which usually persisted for over 50 sec until the recording had finished and never lasted less than 10-20 sec (Budko et al., 1985). Typical fluctuations of the initial latent period were from under 160 msec to 3 sec, whereas the peak latent period under these conditions could vary from 2-8 to 18-35 sec. The last two values, moreover, do not relate to the first peak latency, but to the time taken by the response to reach its absolute maximum, which was between 4.5 and 35 sec. After maximal heating had been reached, a slow return to the original temperature began, and the process was complete after 1-4 min, depending on the strength of the temperature response.

Besides local heating in the primary projection areas, during photic stimulation a local change of temperature also took place in the frontal cortex: in area 10, and also in areas 2 and 3.

The temporal parameters of the temperature responses in the primary projection area depended on parameters of the stimulus such as its brightness, size, duration of action, focusing on the retina and the number of repetitions of flashes in a regular series (Shevelev et al., 1986d; 1987b; 1989a; Volovik et al., 1990).

Acoustic Stimulation. For acoustic stimulation, clicks (0.2 msec, intensity 30-40 dB above the level of human audibility) were applied binaurally (a loud speaker in the midline of the head). Since the auditory cortex in rats lies on the lateral surface of the brain, we were unable to record its direct temperature response to clicks in the present investigation. However, in other parts of the cortex, on its dorsal surface, distinct temperature responses to sound could be seen (Shevelev et al., 1986g). For instance, local heating by 0.01-0.15 °C took place in the right motor cortex, after a short initial and peak latent periods (2.7 sec). It is interesting to note that cooling developed at this time in the left visual projection area followed by prolonged heating by 0.035 °C.

Cutaneous Stimulation. For cutaneous stimulation both natural stimuli (bending a group of vibrissae) and electrodermal stimulation were used. A single electric pulse had a duration of 0.3 msec, with an intensity of 2-2.5 times threshold for twitching the limb of the unimmobilized animal, which is 2-3 times below the pain threshold.

During weak electrical stimulation of the skin in the region of the vibrissae a local temperature reaction in the form of heating (Fig. 3, frame A) developed in the contralateral somatosensory and motor cortex. Heating began in these regions after a very short latent period (under 1 sec) and reached a maximum of +0.07 and +0.04 °C, after which the temperature returned fairly rapidly (by the 20th-25th sec) to the background level. Later, after a latent period of 3-4 sec and peak latency of about 20 sec local heating of the ipsilateral motor cortex and of the somatosensory area developed and reached 0.05 °C.

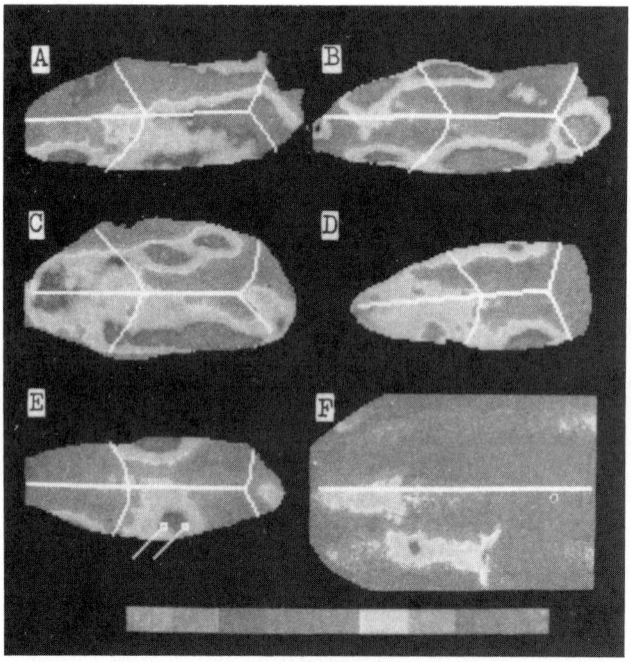

FIGURE 3. Thermomaps of the rat (A-D) and cat (F) dorsal surface of the cortex during reaction to different kinds of sensory stimulation. In each map, the frontal part is at left and the right hemisphere is at the top. The temperature difference between dark and light is 0.1 °C. The heated zones are shown as dark patches. A: Weak electroskin stimulation in vibrissae region from right side. B: The same stimulation as in case A plus a 2 sec weak light flash in left eye (conditioning). C: Right hind leg electrically stimulated at pre-pain level. D: Adequate 2 sec light flash stimulation of right eye. E: Direct electrocortical stimulation in occipital region of left hemisphere under 60 mg/kg nembutal. F: Moving of the right foreleg due to touching it with a stick. Note of the editors: Color laser prints of this figure can be obtained from the author upon request.

Electrodermal stimulation in the region of the thigh led to rapid heating (at the rate of +0.023 deg/sec), increasing the temperature to 0.088 °C in the contralateral somatosensory area (Fig. 3, frame C). After rapid initial heating the temperature of this area continued to rise slowly, to reach +0.13 °C after 25 sec. Virtually simultaneously with this response local heating of the motor cortex contralateral to the side of stimulation took place, although much more slowly and also in the ipsilateral motor cortex.

The main result of this investigation was the discovery of multiple local temperature patterns in the cerebral cortex during sensory stimulation, in the form of foci of heating and/or cooling (Shevelev et al., 1986g; 1988; 1991c; Gorbach et al., 1989a; Gorbach and Tsicalov, 1990a). Some of them clearly coincided with the corresponding cortical projection areas (primary and secondary), but some appeared in distant areas of the cortex. The order of development, its duration, the intensity and sign of temperature contrast, and also the geometric characteristics of the thermal effects depended on the modality and parameters of sensory stimulation.

We note that studies of the temperature reaction of the brain of animals to various sensory stimuli are known (Serota and Gerard, 1938; Melzack and Casey, 1967; Baker et al., 1973). However, using point-type sensors only with limited leads, it is impossible to assess whether cortical responses to stimulation are local.

This new approach to the study of cortical activity has enabled preliminary estimates to be made of the size and, in some cases, the shape of the temperature patterns, and to follow their formation and interaction in different zones of the cortex (Shevelev et al., 1986e; 1992a; 1991d; Gorbach and Tsicalov, 1990b; George et al., 1991).

2. Thermomapping of the Motor Cortex

The study was performed on non-anesthetized cats without myorelaxation. In the preliminary experiment the animal was scalped and under local anesthesia three screws were fixed to the skull with plastic, and with these screws the skull was immobilized in a halter. The body and hind paws of the cat lay in a hammock, the forepaws hung freely down not touching the mount.

Touching the surface of the right-forepaw with a long wooden rod, which the cat did not see, induced a reaction of bracing the paw on the mount. The movement lasted approximately 1 sec (according to the mechanogram), but full relaxation of the extensors occurred at the end of the 2nd sec (according to the electromyogram). For recording of weak temperature effects the stimulation was repeated 5 times at intervals of 4 min. Through a mirror mounted above the skull at an angle of 45°, the lens of a thermovisor was directed toward and focused on the anterior third of the dorsal surface of the cat's skull.

The main result of the experiments was the detection of local temperature changes, linked to the performance of a natural reflex movement of the paw, in the area of the anterior sigmoid gyrus (motor cortex) of the cat (Shevelev et al., 1986b). As can be seen from Fig. 3, frame F the spatial distribution of the warm and cool foci proved to be rather complex. The predominant effect during and after performance of the movement (lasting up to 2 sec) was local cooling with a peak of up to -0.03 °C in the 6th sec in a small part of the motor cortex, corresponding to the motor projection of the forepaw. After cooling, warming reached a peak of +0.04 °C above the mean background level or 0.07 °C from peak to peak by the 21st sec. Simultaneously, in a small zone of the ipsilateral hemisphere symmetrical to the area described above, antiphase change of temperature developed.

The effects associated with movement proved to be local: at a distance of 3-5 mm to the side from their focus the amplitude of the thermoreaction declined by more than fourfold i.e., to the level of the background temperature fluctuations. In the control zone located between two main foci of the reaction, no significant differences were detected. By the 11th-12th sec after the beginning of the movement a sharp local cooling (by -0.11 °C) was observed in the frontal lobe of the cortex along the midline.

The fact that a considerable number of temperature "events" occur after the movement is finished indicates the inertia of the heat-producing mechanism, possibly because of the speed of metabolic and hemodynamic processes (Shevelev et al., 1989a).

3. Thermomapping During Direct Cortical Stimulation

Trials were conducted on four white rats immobilized using 1 mg/kg D-tubocurarine during artificial pulmonary ventilation. A dose of 60-160 mg/kg Nembutal i.p. was administered to the animal at the close of the experiment. Bipolar spherical silver stimulating electrodes were implanted on the cortical surface in the parietal region through two trepanned openings measuring 0.3 mm in diameter and placed 2-2.5 mm apart, aligned in an anterior posterior direction. Both single square-wave pulses of current with an amplitude of 1-50 V and duration of 300 μsec, and regular trains of 24 such pulses at a rate of 1 Hz, were applied.

A local thermal response, with heating next to the stimulating electrodes and at the corresponding site on the opposite hemisphere (Fig. 3, frame E) appeared within the first 160 msec when a single 10 V, 300 μsec electrical stimulus was applied to the cortex of the animal immobilized with myorelaxant (Tsicalov et al., 1987). The magnitude of the heated site did not exceed 1-2 mm initially (Fig. 3, frame E). The heated sites extended to 3-4 mm gradually, with intensified response, and became oval in shape. Several local zones of thermal response with a similar oval shape and dimensions (3-4 mm) had already formed by the 22nd sec in the anterior cortical regions and the parietal cortex.

The heating process took 2-5 sec to peak; cortical temperature rose fastest in the stimulation zone and the symmetrically opposite zone of the contralateral cortex. After reaching its peak, the temperature and configuration of zones of response remained virtually constant for 10-20 sec; temperature then began to decrease, reaching the original level is 2-3 min. The shift in temperature, averaged over 49 elements in the response zone, did not exceed +0.15 °C.

The data we obtained were consistent with the results of a rat brain temperature study obtained with the aid of point temperature sensors during the process of direct stimulation of the cortex (LaManna et al., 1989).

Cerebral thermal responses produced by direct electrical stimulation have the distinguishing feature of developing slowly both in time and space, encompassing many parts of the cortex and persisting for long periods. The fact that the thermal responses with the shortest latency and very low temperature thresholds are located in the vicinity of the stimulating electrodes and subsequently in the diametrically corresponding point in the other hemisphere would reinforce our deduction on the fundamentally neuronal origin of the responses recorded, potentiated by direct cortical stimulation and transmission of excitement to the opposite hemisphere by connections travelling through the corpus callosum. The dependence of these responses on anesthesia would support this view. Hence, the reduction in magnitude and number of heat build-up zones correlated with the initial Nembutal-induced suppression of polysynaptic interneuronal connections, restricting irradiation of excitation from the stimulation zone. Deeper anesthesia leads to localization of the thermal response in the stimulation zone only and a substantial reduction in level of response (Tsicalov et al., 1987).

On the strength of the correspondence of the main results described above with widely known facts in neurophysiology on the reactions of the brain cortex of animals during the process of sensory, motor, and direct electrical stimulation, the data presented may serve as a physiological control of the TES method according to the results described below.

4. Thermomapping During Conditioning (Associative Learning)

In this study, two different types of stimulation were given: initially, electrical skin stimulation (unconditional stimulation below the threshold for pain at the right side of the vibrissae) was given alone, then later it was combined with photic stimulation (conditional stimulation was given to the central part of the retina of the left eye). The combined stimulations were given simultaneously 9 times each. After conditioning with the combined stimulation, the animal received photic stimulation alone.

The thermal images obtained after photic stimulation were identical to the image previously obtained with electrical skin stimulation (Fig.3, frame B): that is, activation of somatosensory and motor cortex and cerebellum (Shevelev et al., 1986f; Gorbach and Tsicalov, 1991).

5. Dynamic Thermomapping of the Cortical Spreading Phenomena

For detailed visualization and study of wave processes in space and time (spreading process) it is essential to obtain information at many points in the medium along which the wave spreads. The main limitations of spatial and/or temporal resolution and relative invasiveness of the modern neuroimaging techniques often seem not to apply to TES study of the brain cortex.

We visualized spreading phenomena which encompass a large area of the cortex in two different neurophysiological models: i) spreading depression (SD) in rats and ii) evoked activity waves after regular visual stimulation in rats.

Spreading Depression

Temperature reactions of the brain at the onset of an SD wave in the cortex of the greater hemispheres were studied in 70 rats in acute experiments (under nembutal anesthesia or with the use of curare and artificial respiration), as well as in 8 experiments on alert, mobile rats. In the acute experiments the animals were scalped under local anesthesia, and their heads were fixed in a stereotactic instrument. For introduction of a microinjector needle, a trepanation opening 0.2 mm in diameter was made. The animals were scalped one week before the experiment. The opening for the injector was only indicated, and during the experiment the needle was introduced into the cortex through the thinned bone.

For stimulation of an SD wave a microinjection of 1 μL 2% KCl solution was applied, which induced one SD wave in the stimulated hemisphere. Recording of the temperature maps of the brain was carried out after 5 or 10 sec, taking into account the slow character of the process under study. A mode utilizing construction of derivatives and temperature graphs and profiles along different cross–sections was used. For analysis of the wave reactions, programs of amplitude field construction were also used.

Against a background of a general shift of temperature in the stimulated hemisphere, a temperature reaction (Fig. 4) in the form of thermowaves slowly spreading along the cortex from the injection site was observed (Tsicalov et al., 1985; Shevelev et al., 1986c; Gorbach et al., 1989b). The main parameters of SD thermal wave were:

-latent period: a KCl microinjection caused heating of the cortex near the site of action after 20–40 sec;
-rate of distribution: experiments with simultaneous recording of constant potential and temperature have shown that the rates of distribution of separate SD waves and potential waves are identical;
-amplitude: the mean amplitude of the wave was 0.49 ± 0.07 °C;
-duration of temperature waves: in the first instance the wave duration at 1/2 of its amplitude comprised 132 ± 2 sec, in the second 121 ± 3 sec;
-refractoriness: with reduction of the time interval between injections to 2 min the next temperature wave most often did not develop, or its formation was delayed for 1.5–2 min.

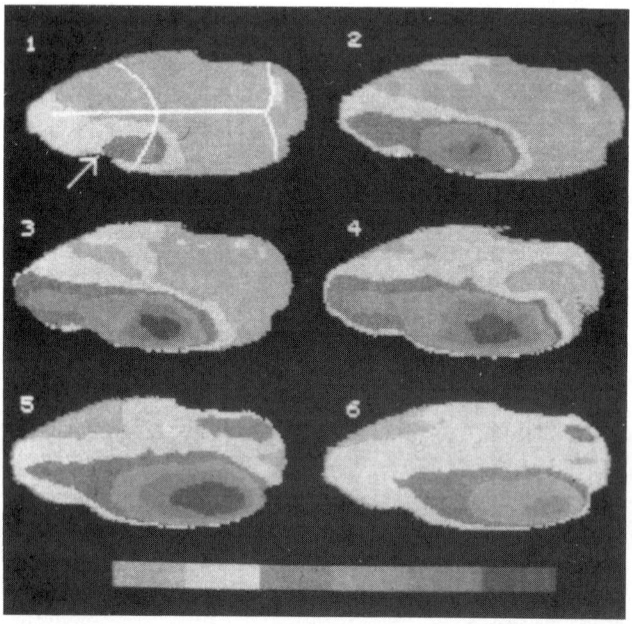

FIGURE 4. Thermomaps of the rat dorsal surface during moving of the Leao's Spreading depression wave over the cortex. Frontal part is on the left side, occipital on the right. White lines on first map are skull sutures. White arrow marks place of KCl micro injection-induced Spreading depression. Gray scale between dark and light is 1.0 °C between dark and light. The heated zones are shown as dark patches. Time interval between each map is 30 sec. Note of the editors: Color laser prints of this figure can be obtained from the author upon request.

The SD thermal wave completely corresponds in many parameters (velocity of spread, latent period, refractoriness) to analogous parameters of other known manifestations of SD.

The strong reaction found in all alert animals in the form of a decrease of temperature in the stimulated hemisphere may be explained by a reduction of the cerebral blood circulation (Duckrow, 1991). In anesthetized animals the thermal manifestation of SD may be explained by an increase of cerebral circulation and metabolism, which is in accord with the commonly accepted view.

In confirmation of the results of the described experiments a considerable heat production in the brain cortex of the rat (LaManna et al., 1989) and on isolated amphibian retinas (Tasaki and Byrne, 1991) connected to the SD phenomenon was found by other groups. There is an interesting correlation between the main temperature characteristics of SD which we recorded with the TES method and by the above-mentioned authors with the aid of point sensors.

Evoked Activity Waves after Regular Visual Stimulation

Investigations were made during nine acute experiments on curarized white rats. The animals were kept at a steady temperature and artificially respirated with the head partially fixed in a frame.

The spreading wave activity was found to exist in the cortex both in spontaneous and evoked thermoactivity. During the spontaneous activity, thermowaves are aperiodic and can spread with linear, circular or arc-shaped trajectories mono- and transhemispherically (Shevelev et al., 1987a). During evoked activity, waves were induced by single visual stimulus or trains of visual stimuli (at intervals of 1.5-3 min) or by initiating one of the flashes, and spread through the cortex along trajectories of some basic types: global rotation, local rotation, transhemispheric and linear spreading (Shevelev et al., 1989b; 1991b). The characteristics of thermowaves were: amplitude 0.005-0.1 °C; length 10-15 mm; duration 1.2-11.4 sec; velocity 1-33 mm/sec; extent of pathway 2-56 mm.

Evoked activity waves appeared with a high degree of probability (0.92) during 26 sec after the flash. The number of moving waves rose after the flash, to peak by 7-8 sec after the stimulus. Waves arose in 50% of cases in the contralateral visual cortex (areas 17 and 18a), spreading to the midline and crossing to the ipsilateral hemisphere (areas 17, 18a, and 7). Local waves moving along a circular trajectory were also discovered in the contralateral visual cortex. Several types of waves of differing trajectory also arose in the ipsilateral visual cortex. With the same means of processing and visual monitoring, such waves were not observed on a temperature calibration instrument and on a dead animal.

If we assume that the thermowaves reflect some informational processes in the brain, then apparently it is specifically the local changes of their characteristics which may be indicators of the interaction of waves with nervous tissue and reflect and reveal their significant, dynamically changing functional heterogeneity.

It might be postulated that we are seeing the projection of a wave process on the cortical surface occurring in one of the subcortical thalamic or hypothalamic nuclei (if we take this nucleus as an uninterrupted environment). Callosal connections could also be playing a part in transhemispheric passage if we accept that the shorter among these are also the most effective (Shevelev et al., 1989a).

The question of whether these waves are self-operating fluctuations can apparently only be answered after clarification of their internal mechanisms and their inclination towards autowave variables. Spreading depression is an autowave process, i.e., it arises in an active medium with restoration, and is refractory. With sufficient homogeneity of the functional state of the cortex, SD spreads through it in the form of a circular wave from the site of its origin. When obstacles are created along its path, the wave goes around them and can begin to reverberate around a damaged section of considerable dimensions. On the basis of a single spreading wave a spiral wave of SD may form. It was shown that local areas of warming (by 0.2-0.3 °C) symmetrical to the site of stimulation were found in the opposite hemisphere (Rusinov et al., 1990). Latent excitation foci during local chemical action were found on the dynamic thermal maps in regions remote from the stimulation site in the same or the contralateral hemisphere. On the basis of encountering two temperature waves which ran from two foci of chemical effect, the authors showed that in the cerebral cortex a stable auto-oscillatory process may form. Long-lasting temperature oscillations with a period of 1-3 min occurred in the new stable temperature focus against the background of a general warming by several tenths of a degree. In the opinion of the authors, such an auto-oscillatory process demonstrates the possibility of forming a stable excitation focus outside of the areas of the primary perturbation of the brain.

As we can see from the results, temperature waves of evoked visual stimulation have a number of properties which differentiate them from temperature SD wave processes. These are the variability of all main wave characteristics (amplitudes, velocities, forms) along its line of travel. Often rapid thermowaves develop against a background of higher–amplitude local and diffuse temperature shifts.

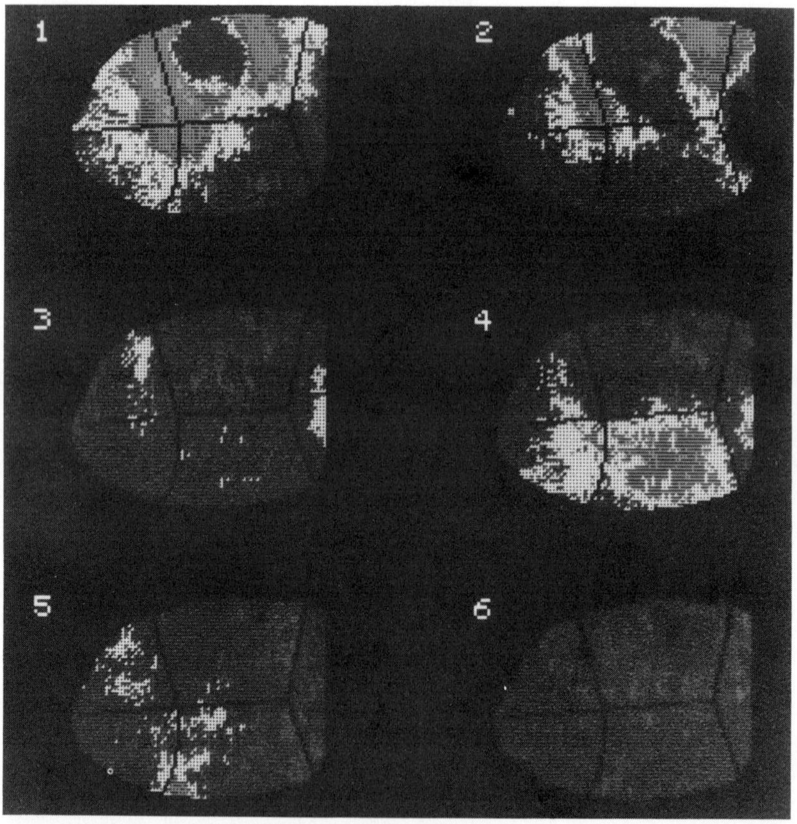

FIGURE 5. Thermomaps of the rat dorsal surface produced by local epidural application of penicillin crystals. In each map, the frontal part is at the left and the right hemisphere is at the top. Black lines on the maps mark the skull sutures. The temperature difference is 0.3 °C. 1 - 3 : ictal episodes; 4 - 5 : interictal episodes; 6 : start of the ictal episodes. Time interval between maps: 1-2: 8 sec; 2-3: 28 sec; 3-4: 40 sec; 4-5: 50 sec; 5-6: 70 sec. Note of the editors: Color laser prints of this figure can be obtained from the author upon request.

Common to all waves presented is the fact that, in addition to significant differences in the main characteristics of the spreading processes, one can isolate the general traits of the spreading brain waves:

- the presence of anterior and posterior fronts
- the movement of two fronts in a certain direction
- the monotonous shift of phases at all intermediate points on the trajectory of the front movement
- reproducibility of phenomena.

A wave can be defined as a propagated disturbance of activity which moves as a pulse wherein the shape of its temporal change is characteristic and similar for each point in its path. The presence of a monotonous shift of phases at all intermediate trajectory points serves as a criterion for regarding this process as waves.

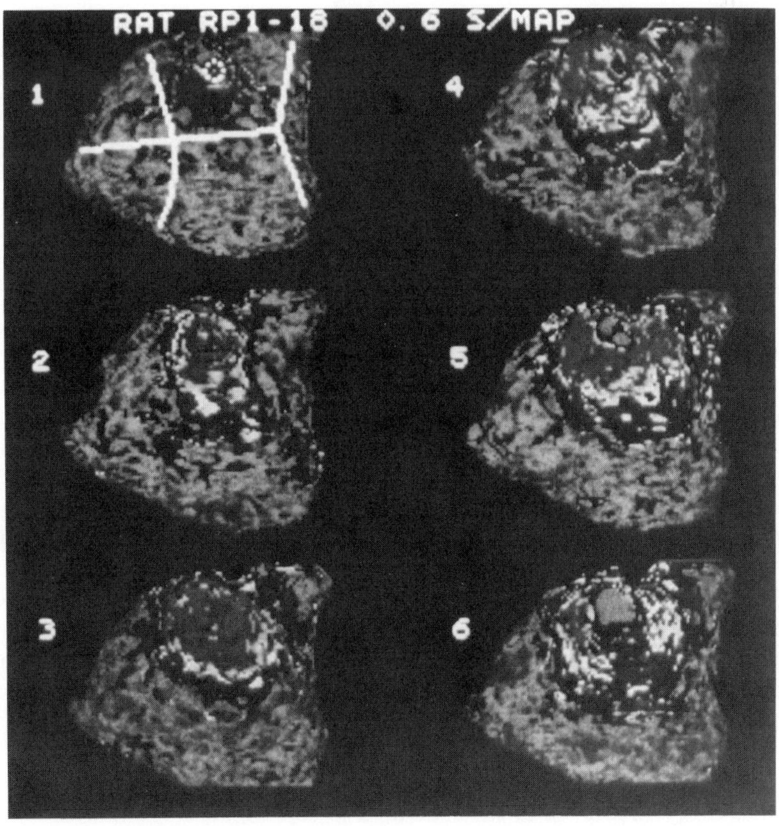

FIGURE 6. Local concentric thermowave around penicillin focus on the rat's cortex. In each map, the frontal part is at the left and the right hemisphere is at the top. White lines on the first map mark the skull sutures. The temperature difference between dark and light is 0.1 °C. The heated zones are shown as dark patches. Time interval between maps - 0.6 sec. Note of the editors: Color laser prints of this figure can be obtained from the author upon request.

6. Thermomapping of the Cortex with Experimental Epilepsy

The spatiotemporal dynamics of the cortical thermoactivity produced by local epidural application of penicillin crystals 0.5mm diam. was studied through unopened scalped skull bones of four rats. Temporal TES resolution in this work was 2 maps/sec (128x85 pixels), instrumental spatial resolution was 150 μm/pixel and temperature differential sensitivity was 0.008 °C.

After appearance of interictal spikes, monitored by steady potential and electroencephalogram measurements, the primary and secondary foci of thermoactivity were revealed in both hemispheres (Shevelev et al., 1993). The primary focus was more local and had dipolar temperature distributions (Fig. 5, frame 1,2,3): a cool region surrounded a relatively hot region and vice versa, or local cool and hot regions were found near the focus. The secondary foci were more diffused and uniform in temperature (Fig. 5, frame 4,5,6). The temperature increase during spikes was about 0.05 °C, and during ictal episodes was up to 0.5 °C.

Concentric thermowaves spreading from the primary focus with a velocity of 2-4 mm/sec and amplitude about 0.01 - 0.05 °C were found associated with spikes (Fig. 6). It was found that the foci were stationary in the first 6-8 sec after the start of the ictal episodes; then the foci disappeared, reappearing at the same place after 70 sec.

An increase of temperature in the rat brain cortex of 0.1 °C in response to individual seizure bursts in experimental epilepsy induced by intravenous pentylenetetrazol administration in small doses was found with the aid of point temperature detectors (LaManna et al., 1989).

We have shown that high spatial resolution optical thermoimaging can be used to visualize dynamic spatial patterns of the epileptic foci in real time.

7. Thermomapping During Sleep

After the animals were scalped and bipolar ball–type silver electrodes were implanted in the occipital region of the cortex of both hemispheres, the animals were placed in a confining box (30 x 20 cm), in which they were kept for 2 days before the experiment. The experiments were carried out in a semi–dark chamber with visual monitoring of the animals' behavior. We recorded the electrocorticogram (ECoG) over the course of the entire experiment.

The lens of the thermoviewer was focused on the dorsal surface of the skull by means of a slanted mirror at a distance of 30 cm. When the animal moved, the thermal imaging camera was moved, following relocations of the head, and the lens was refocused. For analysis of the thermal fields in the computer, portions of the recordings were introduced which corresponded to different phases of the wakefulness cycle, and these were assessed by correlation with the constantly running recording of the ECoG. In these experiments rather slow temperature dynamics are usually studied; therefore, thermomaps of the brain were recorded every 1–5 sec.

In the transition from the quiet wakefulness state, characterized by cortical asynchrony (Fig. 7, frame 1), to slow-wave sleep (SWS1 phase), characterized by 500 μV ECoG sleep spindles, interspersed with brief periods of cortical desynchronization, a cooling of both hemispheres by 0.3–0.5 °C occurred (Fig. 7, frame 2). In the SWS2 phase of slow-wave sleep, characterized by sleep spindles and near 50% slow wave activity on ECoG (Fig. 7, frame 3-4), the left hemisphere became colder, while the right warmed by 0.1–0.2 °C (Fig. 7, frames 3–4). During the REM sleep phase, characterized by cortical asynchrony, the thermal asymmetry begun in phase SWS2 intensified and became 0.3–0.4 °C warmer in the right hemisphere (Fig. 7, frame 5). After excitation

(Fig. 7, frame 6), a rapid warming was observed in which the left hemisphere warmed more quickly and the initial temperature distribution was restored.

In the REM sleep phase we also noted periodic general temperature fluctuations with an amplitude of approximately 0.1 °C and lasting 10–15 sec. A stronger asymmetry was observed in the parietal and occipital regions than in the frontal region.

It has been shown that brain temperature is extraordinarily sensitive to changes in the level of wakefulness (McGinty and Szymusiak, 1990). Reduction of brain temperature during the SWS1 and SWS2 phases correlates well with the data on the reduction of regional metabolic activity and regional cerebral flow both in animals and in man (Ramm and Frost, 1983; Madsen and Vorstrup, 1991). The interhemispheric asymmetry of brain activity in the SWS phase has only been noted thus far in dolphins and man. In dolphins, which exhibit unihemispheric sleep, cortex temperature drops selectively in the sleeping hemisphere (Koval'zon and Mukhametov, 1983). Studies of the regional cerebral hemodynamics of man have revealed that the right hemisphere becomes involved in the sleep cycle before the left (Meyer et al., 1981).

Interhemispheric differences in the rapid-sleep period were also found in man (Meyer et al., 1987; Maquet et al., 1990). It was shown that in REM sleep and dreaming cerebral blood flow values greatly increased, particularly in right temporoparietal regions in subjects experiencing both visual and auditory dreaming.

The fact of elevated activity of the right brain hemisphere in rats during the process of REM sleep, which we discovered, supports the hypothesis of interhemispheric specialization, which proposes that REM sleep seems to be a right hemisphere function (Sawaya and Ingvar, 1989). The results of this series of experiments show that the TES method can detect precisely changes in the functional states of the brain, although the results are preliminary and the experiments should be repeated using a larger number of animals.

In addition to known regularities, new results were obtained with the TES method which characterize the dynamics of interhemispherical asymmetry in catalepsy (Petrova and Markov, 1989), in acute stress (Kuznetsova et al., 1989a), in the state of animal hypnosis in the rabbit (Shevelev et al., 1989a), as well as in other experimental states in animals (Volovik et al., 1988; Kuznetsova et al., 1989b).

8. Evoked Thermofields of the Human Brain

The greater part of the surface of the human scalp, unlike that of animals, is relatively thin and poorly vascularized because of its lack of muscles. Motivated by this fact, we decided to undertake a noninvasive investigation of the dynamic temperature fields of the human brain during response to photic stimulation of the retina.

Four healthy adult men with normal vision were investigated. The subjects sat in an armchair with their chins supported in a lightproof chamber. To minimize involuntary movements of the head, it was gently fixed in place by two restraints which rested lightly against the left and right margins of the frontal bone. The objective of the receiving camera of the thermovisor was aimed from a distance of 35-75 cm toward the occipito-parietal region of the head (P3-C3 according to the 10-20 system of EEG derivations), which had been shaved 2-3 days before the experiments, or had a naturally bald area. Parameters of photic stimulation were: circular flash (mat screen, visible monocularly with the right eye, subtending an angle of 6°20', at a distance of 77 cm), duration 5 sec, intensity of illumination at the pupil level 18 lx, with general intensity of illumination of 1.5 lx against a white background (K = 0.7) or 0.33 cd/m^2. The observer fixed the center of the circle with his gaze. A weak warning sound was given

5 sec before the beginning of photic stimulation to prevent involuntary movements in response to the switching on of the light.

The interval between recording of two successive maps was 760 msec, and spatial resolution was 0.6-1.3 mm/pixel. Synchronous summation of effects during 2 and 4 repetitions of the stimuli was carried out. The measurements began with recording of a series of 24 background temperature maps (18.2 sec), after which the computer switched on the photic stimulation and recorded 7 temperature maps during stimulation (5 sec) and 41 maps (31.2 sec) in the after-period. The whole series of measurements lasted 54.7 sec. From each temperature map the computer subtracted one of the background maps (differential method). The interval between successive stimulations was 4 min. Allowing for repeated accumulations the series lasted from 8 to 15.6 min.

The investigation revealed, through the intact scalp and skull, a local change of temperature (Fig. 8) in the human brain in response to the action of light on the retina, and enabled us to determine the amplitude, spatiotemporal characteristics, and changes of temperature in the control zone as well as to statistically assess this response (Shevelev et al., 1986a; 1986d).

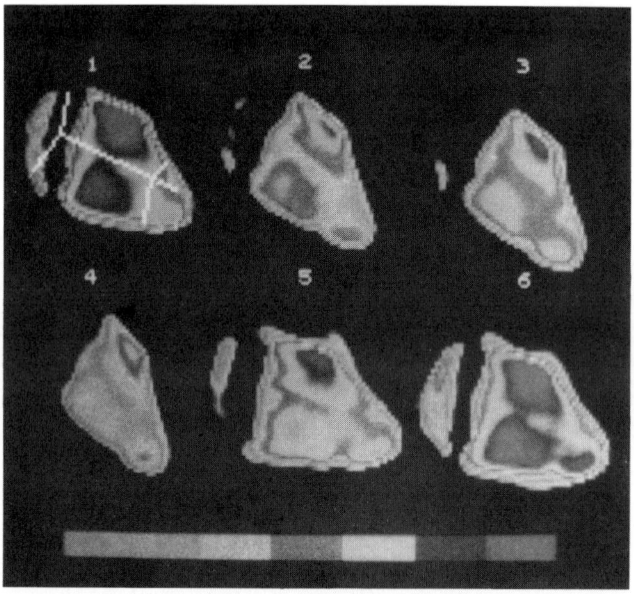

FIGURE 7. Temperature dynamics and right side asymmetry on the dorsal surface of the brain during sleep cycle of the free moving rat. White lines on first map mark the skull sutures. Frontal part is on the right side of each map, right hemisphere is up (mirror images). The heated zones of the cortex are shown as dark patches. Relative grey scale is about 1.0 °C. 1 : awake rat; 2 - 5 : are II, III, IV and REM stages of sleeping, monitored by EEG recording; 6 : after awaking. Note of the editors: Color laser prints of this figure can be obtained from the author upon request.

Analysis of the thermograms showed that the amplitude of the temperature response to light rose between +0.08 and +0.14 °C above the peak of initial cooling, or between +0.05 and +0.1 °C above the mean background temperature level in that zone.

A second zone of heating could be seen in the right hemisphere. It was located close to the outline of the head. Inasmuch as we were limited to the use of only one thermovision camera, assessment of the form and sizes of the foci in the other hemisphere was impossible. The time course of the response in this zone was close to that of the temperature response in the left hemisphere. Zones of temperature response on both the right and left sides did not coincide with the temperature distribution on the scalp before the stimulus. The temperature response to light, according to different estimates of its amplitude, was 3-5.5 times above the maximal fluctuations of temperature in the background and in the control zone. It is important to note that the fluctuations of temperature did not coincide in phase in the response focus and in the control zone. Analysis of the distribution of temperatures at individual points of the temperature map lying in the focus of response to light showed that they rose from the mean background level, taken as 0 °C, up to +0.112 °C.

These results indicate for the first time that temperature activity of the human brain evoked by adequate sensory stimulation can be investigated quantitatively and noninvasively. Although the bone and soft tissues of the scalp act as a heat filter, greatly weakening and obscuring the temperature contours of the brain both in time and in space, the results of the present investigation indicate that this filtration is passive in character. All that can be asserted unequivocally is that temperature changes are initiated by processes taking place in cortical neuronal structures. In fact, there are no other information systems in the brain capable of responding selectively and locally to sensory stimulation.

Encouraged by these results, we performed a series of experiments on a healthy subject to investigate the thermofoci of the brain during the process of mental arithmetic and a series of mathematical operations.
Preliminary analysis of the complex temperature mosaic that arose on the skin surface of the head revealed no focuses of activity which were repeated from experiment to experiment. Attempts to localize the brain hemisphere drawn into interictal epileptic activity in two patients (children aged 11 and 14 years with epileptic syndrome) also ended in failure. We think the reasons for this failure were inadequacies of the head fixation system and the impossibility under our laboratory conditions to immobilize the patients. Despite these setbacks, it seems that the effect discovered in this investigation justifies an optimistic assessment of the possibility that the method may be used to study human brain activity under normal and pathological conditions.

Nerve cells in the human body are distinguished by a high metabolism and thermoproduction per unit mass. Calculation of such energy expenditures revealed that neuronal mass consumes approximately 0.12 W/g; here 1 kg of neuron mass uses energy which would be sufficient to power a 120 W incandescent lamp (Ivanov, 1990). The author also showed that heat from a neuron can be carried away by a large volume of capillary circulation (minute blood volume exceeds the neuronal volume by approximately 45 times) with minimal drops of temperature between blood flowing in and out through the capillaries (0.03 °C).

Biophysical assessment of thermoproduction linked to blood flow and heat conduction showed that the duration of a thermal response to sensory stimulation in a human head should be greater than the duration in experiments in animals with a thin skull (Shevelev et al., 1989a). However, in our data the time characteristics are similar. Perhaps it is possible to explain this result by the mechanism of direct venous convection, which is characteristic for the human brain (Falk, 1990).

In addition to the mass transfer of heat into the vessels of the trunk and extremities by the blood, heat is also conducted from the brain through the tissues of the skull by means of conduction and direct venous convection with dispersal of heat. If one takes into account that the human brain releases approximately 12.4 kcal per hour (Stolwijk, 1970), then 3.9 kcal per hour (32% of all heat production) passes into the layer of skull tissues by means of conduction.

The amount of heat emitted by the brain by means of direct venous convection to the scalp is unknown. The external venous trunks of the skull are linked to the internal venous basin by means of diploetic and emissary veins. Traversing through the orifices in the skull, the emissary veins link up the veins of the external head coverings with the veins of the skull cavity, enabling an outflow of blood volumes from the internal venous basins of the skull to the cutaneous superficial veins. The possibility of a reverse blood flow from the outside to the inside of the cranium has also been shown. Not only the large veins are drawn into the functioning of this largely theoretical mechanism, but also a rather impressive number of small and microscopic veins over

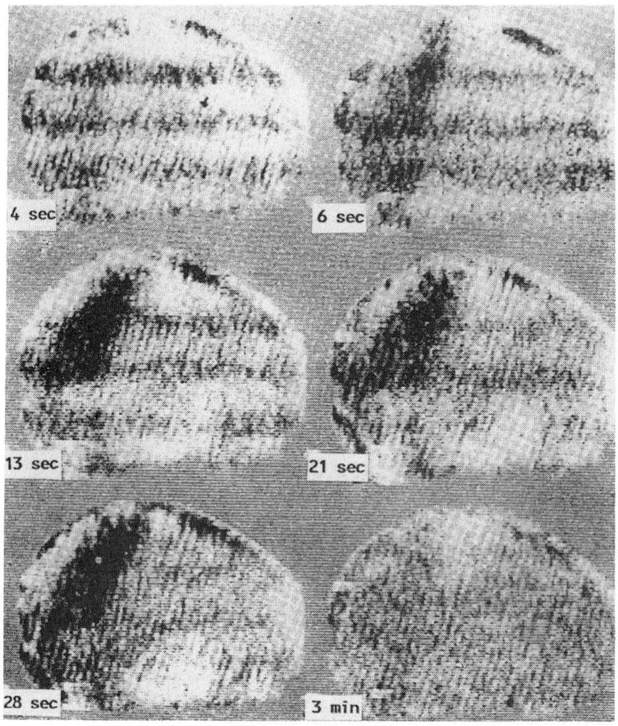

FIGURE 8. Series of temperature maps of occipitoparietal region of human head (viewed from behind and a little to the left) during (4 sec) and after (6 sec-3 min) action of photic stimulus on right eye (differential method). Shaded areas indicate regions of raised temperature. Time of recording each map after beginning of photic stimulation shown in seconds and minutes.

the entire surface of the scalp (Cabanac and Brinnel, 1985). Therefore, it is possible that the TES response of the brain to sensory stimulation is linked to the heat outflow given off by the cortex through the blood, even if partially by means of the primary venous convection mechanism. The amplitude variability of TES responses, form, and localization (within the limits of adequate stimulation of the hemisphere) may be explained by the anatomic variability of the position of the areas of emissary vein egress from the skull. Further comparison of these zones with TES foci might shed light on the participation of selective brain cooling in sensory and motor activity in man as well as on the "radiator" theory of brain evolution (Falk, 1990).

COMPARATIVE ANALYSIS OF TES WITH THE NEWEST METHODS OF NEUROVISUALIZATION

A comparison of different functional imaging and mapping techniques illustrates the relative advantages and disadvantages of each approach. Unit electrical recording, for example, provides excellent temporal resolution, but the mapping potential is limited due to the number of electrodes one can place within a grid for simultaneous recording. This problem can be eliminated with the use of the multichannel optical recording method utilizing a voltage-sensitive dye (Grinvald et al., 1988; Nakashima et al., 1992). However, the effects of photodynamic damage and dye-bleaching during optical experiments limit increasing the signal-to-noise ratio of this method and does not permit its use over a long period of time. Intrinsic signal optical mapping often requires the averaging of 100 images to adequately resolve the signal from the noise (Grinvald et al., 1988). In addition, measurements are difficult to carry out without a cranial window and they cannot be calibrated against a physical source.

The TES method is less noisy then the intrinsic optical method and can generate useful imagery with as few as 6-10 averaged scans (Shevelev et al., 1989a). It is relatively non-invasive, since no contact and no light sources are necessary to record intrinsic signals of the brain. Moreover, measurements are recorded at a distance from the object through the intact skull or through a cranial window, and may be calibrated relative to known temperature standards.

All the above-listed optical methods provide information on the processes occurring in the brain cortex at an average depth of 100-400 microns. However, thanks to the thermal conductivity properties of brain tissues, TES enables assessment of the activity of deep structures as well. Therefore, in principle, on the basis of analysis of latent periods and the spatial pattern of thermal responses as well as the inverse tasks in thermal conductivity, in the future TES may also be used for three-dimensional visualization.

The TES method may be more spatially precise than EEG, MEG, isotope, and PET, which have poorer than 50 μm resolution. It is more dynamic than MRI, PET, isotope clearance, and visible and near-infrared spectroscopy (due to averaging requirements). Table 1 provides a brief (non-statistical) comparison between several imaging techniques for cortical mapping. This table is intended to illustrate the relative advantages of using TES to gather data on rapid cortical responses with high resolution and comprehensive spatial mapping.

At present, the use of thermal mapping provides an answer to the questions "where" and "when" changes in cortical activity took place, but it cannot separate neuronal processes mediated by blood flow and metabolism. This is the essential limitation of TES in comparison to other methods of neuromapping. Assessment of possible heat sources in the cortex and their modeling revealed that on average for different physiological models with some activation of the elementary volume of the cortex approximately 90% of the amplitude of the TES signal was created by change in local blood flow in the brain; the contribution of metabolic thermal production and thermal production linked to electrogenesis comprises 9% and 1%, respectively (Frolov et al., 1985; Pasechnik and Kuznetsova, 1987; Shevelev et al., 1989a). These three components of the TES signal should have various latent periods in their maximum amplitude from the beginning of cortical activation, thus opening the possibility of their temporal separation. The task may be complicated by diversely directed changes of local blood flow in the brain and metabolism in the cortex, as well as by requirements for a wide dynamic range of the measurement system.

One may imagine the following picture of the origin of TES signals, for example in sensory stimulation of an animal. Activation of the sensory representation of the cortex leads to an increase of its heat emission as a result of generation of action potentials (Ritchie and Keynes, 1985) and PSPs (Kusano and Tasaki, 1990), as well as an increase in neuroglial metabolic activity (Ivanov, 1990). These phenomena, along with the accumulation of vasoactive catabolites (Roy and Sherrington, 1890) or nervous regulation (Lou et al., 1987), change blood flow in the brain. In experiments on direct registration of cortical blood flow with the aid of the videomicroscopic method, it was shown that in somatosensory stimulation of the vibrissae in rats a local increase of blood flow occurs, but only in the activated portions of the cortex (Cox et al., 1991). This could explain the local TES responses that we obtained in similar experiments.

Because of the thermoconductive properties of brain tissues and vessel walls, the venous blood is heated and carries heat from the heating zones (convection mechanism). The removal of heat is not instantaneous and the period of time required depends both on the intensity of the heat emission and the rate and volume of blood flow. We suggest that the retarded neutralization of temperature effects by the inflowing colder arterial blood enables the heat to reach the upper membranes of the brain and head. Their local heating causes emission in the form of radiation, whose intensity is also recorded by the thermovisor.

TES is compatible with simultaneous EEG, autoradiography, and voltage-dependent fluorescence dyes. A number of studies were dedicated to IR spectroscopy of nervous tissue, in particular to changes in the spectrum of a nerve during excitation (Sherebrin, 1972). One may hope that a combination of the spectral resolution of the spectrophotometer with the spatial resolution of the multiwavelength IR camera (Hejazi, 1990) would enable in vivo identification of spatial patterns of couplings in the cortex.

TABLE 1. A brief comparison between modern methods of the cortical mapping. The data used to categorize these methods was obtained from the literature (Toga, 1990; Nakashima et al., 1992; Boppart et al., 1992; Ratzlaff and Grinvald, 1991).

	Temporal Resolution	Spatial Resolution	Repeated Measures	Brain Damage
GRID UNIT RECORDING	μsecs	200 μm	possible	possible
14C-IODOANTIPYRINE AUTORADIOGRAPHY	15-30 secs	100-200 μm	no	yes
14C-DEOXIGLUCOSE AUTORADIOGRAPHY	30-50 mins	50 μm	no	yes
VOLTAGE-SENSITIVE (dyes) OPTICAL IMAGING	msecs	2-100 μm	possible	yes
INTRINSIC SIGNAL OPTICAL (visible) IMAGING	msecs	50-100 μm	possible	possible
TES (infrared)	secs	50-100 μm	possible	no

COMPARATIVE ANALYSIS OF BRAIN THERMOMAPPING METHODS

Methods of measuring the brain temperature of humans and animals began to be applied in experimental neurophysiology as early as the last century (Mosso, 1892). In those days mercury thermometers or thermocouples were sunk into the brain. The measurements carried out by investigators under various physical stress and conditions enabled assessment of the cerebral circulation and metabolism in a number of areas of the brain. The creation of new sensitive temperature detectors of the point-contact type and their miniaturization sharply expanded the use of the methodology of implanting thermistors and thermocouples into living objects in the study of functional brain activity as well as the mechanisms of thermoregulation in the central nervous system and the thermophysical properties of brain tissues and fluids.

In the meantime, the behavioral and thermoregulatory functions of the body can be easily disturbed by even insignificant external effects. Transplantation of detectors into the brain that mechanically destroy brain tissues and membranes distorts the natural functional relationships between the structures and, inasmuch as the measuring instrument has its own heat capacity, redistributes the temperature gradients. Thus, for example, a heat sensor constructed with a thin film of polyvinylidene fluoride and used for studying heat generation on preparations of nervous tissue has a good temperature sensitivity (1 microdegree/mV) and time resolution (response time up to 2.3 ms) (Tasaki et al., 1987; 1989). However, the specific heat volume of up to 0.6 cal deg^{-1} cm^{-3} and the physical dimensions of such a detector (2 x 3 mm^2) do not allow using it for mapping of temperature fields. The same arguments apply to both thermistors and thermocouples.

If the spatial resolution of a temperature field of the brain is determined by the size of cortical functional modules, then by virtue of the small number of measuring points the patchiness of cortical activity cannot be assessed with the aid of single point-contact sensors (possibly with the exception of superposed matrices of them).

In recent years it has become possible to construct thermomaps of living objects using well-known relationships between molecular diffusion, which can be measured and imaged with MRI, and temperature. It has been demonstrated that echo-planar imaging (EPI) could greatly enhance the usefulness of MRI by providing mapping of changes in temperature in addition to high-resolution morphological information (Le Bihan et al., 1989). In experiments it was shown that temperature imaging in phantoms could be obtained with an accuracy of better than 0.5 °C (Delannoy et al., 1990). Inasmuch as EPI in principle allows for reduction of the scanning time to 30-100 ms, in the future the method may allow investigations of a three-dimensional temperature dynamic of the human brain. The spatial resolution of EPI--1.5 mm x 1.5 mm x 5 mm--apparently is the limiting factor for the use of this method for investigation of animals.

The possibility of measuring temperature in deep structures of the brain also allows passive and active methods of microwave radiometry (Mizushina, 1989). Study of dynamic temperature maps of the human brain with the aid of a radiothermograph with 9 antennas showed different heat pictures in general excitation of the brain, and a differentiation of the hemispheres during sleep (Gulyaev et al., 1988). Measurements on a phantom head with the aid of an active microwave computed brain tomography have shown that this technique could be used to visualize dielectric changes within the brain due to the variations in temperature, blood content, or other causes (Almirall et al., 1991). The most modern instruments of these types enable measurement of temperatures with an accuracy of 0.5 °C in depths of up to 5 cm with time resolution of 3-10 sec. Here the spatial resolution reaches 2-3 cm, which in our view is insufficient for constructing thermomaps that reflect the functional activity of the brain cortex.

It seems that for dynamic functional mapping of the brain cortex of animals according to the temperature indicator, the TES method that we have developed, in comparison to the others listed above, exhibits a successful combination of spatial, temporal, and temperature resolutions. A clear shortcoming of the method may be the limitation in principle of measuring temperatures in deep structures, for example, in the depths of the cortical sulci. Inasmuch as TES requires scalping the animal, another shortcoming is the fact that one can only conditionally call the method non-contact and noninvasive. Although the problem has not been studied, one may suggest that such a procedure changes the heat exchange of the brain and body as a whole, as well as the behavior of animals.

There also exist complexities in calibration of the method by temperature, inasmuch as for accurate surface temperature measurement by a thermography system, the emissivity of the bone surface should be known (there are no data of this topic in the literature).

The TES method may be used together with other methods of studying the brain with the aid of thermistors or thermocouples (Neill et al., 1990; LaManna et al., 1989), if the latter do not overlap the optic field for the thermoimager. Using the TES method for focusing, this allows implantation of heat detectors in the area of greatest interest and subsequently to monitor them. It is also possible to use TES together with other optical methods, for example, the method of potential-sensitive dyes. This allows for monitoring the functional activity of the brain cortex according to its potentials and temperature activity simultaneously. Modifications of TES enabled recording of temperature fields in the study of nervous tissue preparations (Khizhnyak et al., 1986) as well as solutions (Kondrashova et al., 1988), despite the evaporation processes.

In conclusion we list a number of ideas linked to the use of TES for studying the neurophysiologic correlates of behavior as well as the plasticity of the brain cortex of animals:

1. The high instrumental spatial resolution is limited only by the wavelength of the IR radiation, and this resolution may be increased to 20-30 μm/pixel by an IR microscope lens (Inframetrics, 1991).
2. The actual recording speed of IR scans currently possible (up to 25 frames/s) may be increased, and this will equalize the frequency bands recorded by IR imaging and electroencephalography (Amber, 1991).
3. The use of controlled focusing of the lens enables determination of the relationship of the global functional mapping of the cortex (regional level of study) with the functional mapping of local neural networks (modal level of cortical study).
4. In chronic experiments with animals it is possible to create a cranial window or an artificial skull with a given characteristic of thermoconductivity and emissivity, which replaces the upper part of the skull and allows IR rays to pass through. Then the spatial, temperature, and time resolution of TES may be increased.
5. There are possibilities for real-time recording of functional activity of the cerebral cortex of animals in the process of their moving about in a confined space. For this it is essential to use a multi-lens IR optical system with overlapping fields of vision. Computer processing would enable recreation of the animal's head, taking into account its spatial position.
6. The use of optically restructured filters enables measurement of cortical activity in different wavelength ranges, including the visible range, which enables differentiation of functional maps linked to the blood flow and metabolism of the cerebral cortex.

The non-invasive and dynamic nature of TES in combination with its excellent temporal and spatial resolution and sensitivity will be useful for physiological analysis. One may regard TES as one of the few methods which is potentially capable of being used directly in the operating room for functional neurovisualization. This would potentially provide neurosurgeons a rapid comprehensive image of localized functional zones, distinguishing normal from abnormal (e.g., tumor and epileptic) cortex.

ACKNOWLEDGMENT: The author greatly appreciates the interest and assistance of Dr. Bradley Roth from National Center for Research Resources, National Institutes of Health USA.

REFERENCES

Almirall H., Broquetas A., Jofre L. (1991) Active microwave computed brain tomography: the response to a challenge. J Neurosci Methods 36:239-251.

Amber Engineering Inc. (1991) AE-4256 InSb IR Camera System. Catalog 1991.

Baker M., Frye F., Millet V. (1973) Origin of temperature changes evoked in the brain by sensory stimulation. Exper Neurol 38:502-519.

Baker M. (1982) Brain cooling in endotherms in heat and exercise. Ann Rev Physiol 44:85-96.

Boppart S., Wheeler B., Wallace C. (1992) A flexible perforated microelectrode array for extended neural recordings. IEEE Trans Biomed Eng 39:37-40.

Budko K., Godik E., Gorbach A., Gulyaev Yu., Petrov A., Taratorin A., Tsicalov E., Shevelev I. (1985) Thermal responses of the brain to sensory stimulation. Doklady Biological Sciences March:616-619.

Cabanac M., and Brinnel H. (1985) Blood flow in emissary veins of the human head during hyperthermia. Eur J Appl Physiol 54:172-176.

Cooper L., and Trezek G. (1972) A probe technique for determining the thermal conductivity of tissue. J Heat Transfer:133-140.

Cox S., Woolsey T., Rovainen C. (1991) Blood flow increases in rat barrel cortex during whisker stimulation. Soc Neurosci Abstr 17:27.

Delannoy J., Le Bihan D., Hoult D., Levin R. (1990) Hyperthermia system combined with a magnetic resonance imaging unit. Med Phys 17:855-860.

Duckrow R. (1991) Regional cerebral blood flow during spreading cortical depression in conscious rats. J Cereb Blood Flow Metab 11:150-154.

Falk D. (1990) Brain evolution in Homo: the "radiator" theory. Behav Brain Sci 13:333-381.

Frolov A., Tsicalov E., Kiseleva N. (1985) Assessment of the characteristics of heat sources in the brain according to temperature distribution on the bone surface. in: The use of mathematical and computational methods in biology. Puschino, (in Russian):132-133.

George J., Lewine J., Moore S., Flynn E. (1991) IR thermal imaging of a monkey's head: local temperature changes in response to somatosensory stimulation. Soc Neurosci Abstr 17:475.

Glossary of terms for thermal physiology (1987) Pflugers Arch 410:567-587.

Gorbach A., Tsicalov E., Kyznetsova G., Shevelev I., Budko K., Sharaev G. (1989a) Infrared mapping of the cerebral cortex, Thermology 3:108-111.

Gorbach A., Tsicalov E., Kyznetsova G., Shevelev I., Budko K., Sharaev G. (1989b) Infrared visualization of spreading wave processes in brain cortex. Abstr XVII Annual Meeting of the American Academy of Thermology:2.20.

Gorbach A., and Tsicalov E. (1990a) Thermoencephaloscopy - novel technology for research activation of cortex. The Physiologist 33:A-129.
Gorbach A., and Tsicalov E. (1990b) Visualization of processes in the brain cortex: a new method. Proc XII Ann Conf IEEE Engin Med Biol Society 3/5:1245-1246.
Gorbach A., and Tsicalov E. (1991) Cortex activation during investigation of the associative learning. J Cereb Blood Flow Metab 11:S378.
Grinvald A., Frostig R., Lieke E., Hildesheim R. (1988) Optical imaging of neuronal activity. Physiol Rev 68:1285-1365.
Grayson J. (1952) Internal calorimetry in the determination of thermal conductivity and blood flow. J Physiol 118:54-72.
Gulyaev Yu., Godik E., Petrov A., Taratorin A. (1985) Remote functional diagnosis of biological subjects by their intrinsic infrared radiation. Doklady Biophysics 277:159-162.
Gulyaev Yu., Godik E., Dementienko V., Kalashnikov I., Krasyuk N., Kuznetsov I. (1988) Radiothermal dynamic mapping of biological entities. Doklady Biophysics 299:60-62.
Hejazi S. (1990) A thermal imaging system based on multiwavelength infrared detection. Ph.D. Dissertation, SUNY at Buffalo, NY.
Inframetrics, Inc. (1991) Inframetrics auxiliary optics for 500 & 600 series imaging radiometers. Catalog 1991.
Ivanov K. (1990) "The principles of energetics in organism: theoretical and practical aspects," Nauka, Leningrad (in Russian).
Khizhnyak Ye., Bragin A., Ivanitskii G., Krinskii V., Tyazhelov V. (1986) Thermal activity of an isolated brain fragment. Biophysics 31:971-983.
Kondrashova M., Khizhniak E., Tiazhelov V. (1988) Heat production in mitochondria during oxidation of various substrates. Biofizika (in Russian) 33:727-728.
Koval'zon B., and Mukhametov L. (1983) Temperature fluctuations of the dolphin brain corresponding to unihemisperic slow-wave sleep. J Evol Biochem Physiol 18:222-224.
Kusano K., and Tasaki I. (1990) Heat generation associated with synaptic transmission in the mammalian superior cervical ganglion. J Neurosci Res 25:249-255.
Kuznetsova G., Nezlina N., Mutuskina E., Podolets A., Gurvich A. (1989a) Changes in the spreading depression wave in rats in the postresuscitation period. Fiziol J USSR (in Russian) 75:1032-1037.
Kuznetsova G., Nezlina N., Petrova E. (1989b) Thermovisualization of the hemisphere asymmetry in the animal brain. Doklady Biol Sci March:623-627.
LaManna J., McCracken K., Patil M., Prohaska O. (1989) Stimulus-activated changes in brain tissue temperature in anesthetized rat. Metab Brain Dis 4:225-237.
Le Bihan D., Delannoy J., Levin R. (1989) Non-invasive temperature mapping using magnetic resonance imaging of molecular diffusion: application to hyperthermia. Radiology 171:853-857.
Lemons D., Chien S., Crawshaw L., Weinbaum S., Jiji L. (1987) Significance of vessel size and type in vascular heat transfer. Am J Physiol 253:R128-R135.
Lou H., Edvinsson L., MacKenzie E. (1987) The concept of coupling blood flow to brain function: revision required? Ann Neurol 22:289-297.
Madsen P., and Vorstrup S. (1991) Cerebral blood flow and metabolism during sleep. Cereb Brain Metabol Rev 3:281-296.
Maquet P., Dive D., Salmon E., Sadzot B., Franco G., Poirrier R., von Frenckel R., Franck G. (1990) Cerebral glucose utilization during sleep-wake cycle in man determined by positron emission tomography and [F]2-fluoro-2-deoxy-D-glucose method. Brain Res 513:136-143.

Melzack R., and Casey K. (1967) Localized temperature changes evoked in the brain by somatic stimulation. Exp Neurol 17:293-312.

McGinty D., and Szymusiak R. (1990) Keeping cool: a hypothesis about mechanisms and functions of slow-wave sleep. Trends Neurosci Meth 13:480-487.

Meyer J., Amano T., Karacan I. (1981) Changes in LCBF measured by CT scan during REM and non-REM human sleep. J Cereb Blood Flow Metab Suppl 1:S465-466.

Meyer J., Ishikawa Y., Hata T., Karacan I. (1987) Cerebral blood flow in normal and abnormal sleep and dreaming. Brain Cogn 6:266-294.

Mosso A. (1892) The temperature of the brain, especially in relation to psychical activity. Proc R Soc Lond 51:83-85.

Mizushina S. (Ed.) (1989) "Non-invasive temperature measurement," Gordon & Breach Sci Publ, New York.

Nakashima M., Yamada S., Shinono S., Maeda M., Satoh F. (1992) 448-Detector optical recording system: development and application to Aplysia gill-withdrawal reflex. IEEE Trans Biomed Eng 39:26-36.

Neill K., Crain B., Nadler V. (1990) A simple, inexpensive method of monitoring brain temperature in conscious rodents. J of Neurosci Methods 33:179-183.

Pasechnik V., and Kuznetsova G. (1987) Evolution of heat production induced by a wave of spreading depression. Biophysics 32:155-160.

Petrova E., and Markov A. (1989) Thermal asymmetry of the rat brain in catalepsy. Zhurnal Vysshei Nervnoi Deyatel'nosti imeni I. Pavlova (in Russian) 39:1160-1162.

Ramm P., and Frost B. (1983) Regional metabolic activity in the rat brain during sleep-wake activity. Sleep 6:196-216.

Ratzlaff E., and Grinvald A. (1991) A tandem-lens epifluorescence macroscope: hundred-fold brightness advantage for wide-field imaging. J Neurosci Methods 36:127-137.

Ritchie J., and Keynes R. (1985) The production and absorption of heat associated with electrical activity in nerve and electric organ. Q Rev Biophys 18:451-476.

Roy C., and Sherrington C. (1890) On the regulation of the blood supply of the brain. J Physiol (London) 11:85-108.

Rusinov V., Tsykalov E., Kuznetsova G. (1990) Temperature reactions of the cerebral cortex in the rat and the distant effect of an experimentally evoked excitatory focus. Neurosci Behav Physiol 20:167-173.

Sawaya R., and Ingvar D. (1989) Cerebral blood flow and metabolism in sleep. Acta Neurol Scand 80:481-491.

Serota H., and Gerard R. (1938) Localized thermal changes in the cat's brain. J Neurophysiol 1:115-124.

Sherebrin M. (1972) Changes in infrared spectrum of nerve during excitation. Nature New Biol 235:122-124.

Shevelev I., Gorbach A., Tsicalov E., Budko K., Sharaev G. (1986a) Thermovisual study of reaction to light in the human brain. Doklady Biological Sciences 1-6:644-647.

Shevelev I., Kotlyar B., Volovik M., Martinson Yu., Budko K., Tsykalov E., Sharaev G., Gorbach A. (1986b) Detection of activation of the cat motor cortex during limb movement by thermomapping of the brain. Neirofiziologiya (in Russian) 18:266-269.

Shevelev I., Kuznetsova G., Gulyaev Yu., Godik E., Petrov A., Taratorin A., Tsicalov E., Koroleva V., Budko K., Gorbach A. (1986c) Dynamic thermal mapping of rat brain during sensory stimulation and spreading depression. Neurophysiology September:18-25.

Shevelev I., Tsicalov E., Gorbach A., Budko K., Sharaev G. (1986d) Thermovision parameters of brain responses to visual stimulation. Hum Physiol May:227-232.
Shevelev I., Tsicalov E., Gorbach A., Budko K., Sharaev G. (1986e) Thermoencephaloscopy. Human Physiology November:1-8.
Shevelev I., Tsicalov E., Budko K. Gorbach A., Sharaev G. (1986f) Spatial characteristics and dynamics of activation foci in the cortex during elaboration of a conditioned reflex (Thermographic Studies). Zhurnal Vysshei Nervnoi Deyatel'nosti imeni I.P.Pavlova (in Russian) 36:74-83.
Shevelev I., Tsicalov E., Volovik M., Budko K., Sharaev G., Gorbach A. (1986g) Sensory mapping of the cerebral cortex by thermovision methods in rats. J of Evol Biochem and Physiol 21:348-355.
Shevelev I., Tsykalov E., Budko K., Gorbach A., Sharaev G. (1987a) Movement of thermowaves over the cerebral cortex in the white rat. Neurophysiology January:250-255.
Shevelev I., Volovik M., Sharaev G., Tsykalov E., Budko K., Gorbach A. (1987b) Parameters of photic stimulation and characteristics of thermal response in the white rat cerebral cortex. Neurophysiology January:243-249.
Shevelev I., Tsicalov E., Gorbach A., Budko K., Sharaev G. (1988) Dynamic infrared images of the brain. IBRO NEWS 16:8-9.
Shevelev I., Kuznetsova G., Tsicalov E., Gorbach A., Budko K., Sharaev G. (1989a) "Thermoencephaloscopy" Nauka, Moscow (in Russian).
Shevelev I., Tsicalov E., Gorbach A., Budko K., Sharaev G. (1989b) Spread of fast thermowaves through the cerebral cortex produced by visual stimulation. Neurophysiology March:331-338.
Shevelev I., Tsicalov E., Gorbach A., Budko K., Sharaev G. (1991a) Dynamic infrared functional mapping of the brain cortex. Biomed Sci 1:571-577.
Shevelev I., Tsicalov E., Gorbach A., Budko K., Sharaev G. (1991b) Dynamic functional mapping of the brain cortex by its infrared radiation, in: "Machinery of the Mind. Data, Theory, and Speculations About Higher Brain Function," R. John, ed., Boston-Basel-Berlin, Birkhauser/Springer:512-531.
Shevelev I., Tsicalov E., Gorbach A., Budko K., Sharaev G. (1991c) Cortical processing: dynamic functional imaging by infrared radiation. Third IBRO World Congr Neurosci:397.
Shevelev I., Tsicalov E., Gorbach A., Budko K., Sharaev G. (1992) Thermoimaging of the brain. J Neurosci Methods, in press.
Shevelev I., Kuznetsova G., Tsicalov E., Gorbach A., Sharaev G. (1993) Dynamic thermoimaging of the brain. (in preparation)
Stolwijk J. (1970) Physiological and behavior temperature regulation. Heat Transfer in Medicine and Biology. New York:703-745.
Tasaki I., Byrne P., Masumura M. (1987) Detection of thermal responses of the retina by use of polyvinylidene fluoride multilayer detector. Jpn J Physiol 37:609-619.
Tasaki I., Kusano K., Byrne P. (1989) Rapid mechanical and thermal changes in the garfish olfactory nerve associated with a propagated impulse. Biophys J 55:1033-1040.
Tasaki I., and Byrne P. (1991) Demonstration of heat production associated with spreading depression in the amphibian retina. Biochemical and Biophys Research Communication 174:293-297.
Toga A., ed. (1991) "Three-Dimensional Neuroimaging," Raven Press, New York.
Tsicalov E., Kuznetsova G., Shevelev I., Budko K., Gorbach A., Sharaev G. (1987) Thermal distribution over the cortical surface of the rat brain under direct electrical stimulation. Neurophysiology, November:169-174.

Tsicalov E., Petrov A., Taratorin A., Kuznetsova G., Koroleva V. (1985) Intrinsic temperature fields related to excitation of the rat cortex. Doclady Biological Sciences March:613-616.

Volovik M., Shevelev I., Vinogradova O., Bragin A., Podachin V., Lushchekina E., Ermakova I., Tsykalov E., Budko K., Gorbach A., Sharaev G. (1988) Thermovision control of neurotransplant in rats. Neurosci Behav Physiol 18:492-497.

Volovik M., Shevelev I., Tsykalov E., Sharaev G., Budko K., Gorbach A. (1990) Change in the configuration and localization of thermal foci in the cerebral cortex of the white rat during visual stimulation. Sensory Systems 3:118-123.

IR THERMAL IMAGING OF A MONKEY'S HEAD: LOCAL TEMPERATURE CHANGES IN RESPONSE TO SOMATOSENSORY STIMULATION

J. S. George, J. D. Lewine, A. S. Goggin, R.B. Dyer, and E.R. Flynn

Los Alamos National Laboratory
Los Alamos, New Mexico, USA

INTRODUCTION

The past two decades have produced dramatic advances in the diagnostic imaging of the structure of the human brain, but there is a relative lack of techniques for the noninvasive characterization of neural function. Much progress has been made with techniques such as PET (Positron Emission Tomography, Phelps and Maziotta 1985) and SPECT (Single Photon Emission Computed Tomography) which measure blood flow or glucose uptake associated with tissue metabolism, but these techniques require internalization of radioactive tracers, a factor which compromises their overall utility. In addition, these techniques require complex and expensive instrumentation (e.g., a PET scanner must be located in close proximity to a cyclotron) which limits their availability.

A series of reports from the laboratory of Shevelev and colleagues in the (former) Soviet Union (Shevelev et al. 1988, Tsicalov et al. 1987, Gorbach et al. 1989, Shevelev et al. 1990) suggests that the mapping of thermal emissions from the head or brain surface can provide a safe and inexpensive alternative for mapping brain function. This is because physiological activation of small populations of cortical neurons is accompanied by heat associated with metabolism, as well as compensatory changes in regional blood flow which also produce measurable changes in local temperature (Melzack and Casey 1967, McElligott and Melzack 1967, Zeschke and Krasilnikov 1976).

Thermal mapping is accomplished using specialized camera systems sensitive to infrared (IR) radiation in the range of wavelengths from 3-20 microns. Available data indicate that thermoencephalographic (TEG) mapping requires systems that must meet the additional criterion of being sensitive to changes of less that 0.01 degree C. This poses stringent requirements for the imager, image acquisition system, and image processing capabilities. Another important factor in TEG experiments is that the IR photons visible to the camera arise primarily from the head surface; heat transients originating in deep structures are visible only as a spatio-temporal pattern of

Optical Imaging of Brain Function and Metabolism, Edited by
U. Dirnagl *et al.*, Plenum Press, New York, 1993

temperature changes from heat diffusing up to the scalp. By making appropriate assumptions about the configuration of heat sources and the physical properties of intervening tissues, it may be possible to provide detailed characterization of deep sources on the basis of surface measurements.

Described herein are the results of a series of computational, phantom, and *in vivo* experimental studies which examined the feasibility of IR thermal imaging for mapping areas of brain activity in the intact head. On the one hand, the computational and phantom studies have helped to clarify some physical factors which constrain the utility of thermal imaging techniques. On the other hand, these investigations have allowed us to identify several technical improvements that should significantly extend the sensitivity and range of application of the technique. *In vivo* experiments focused on noninvasive characterization of the TEG response elicited by electrical stimulation of the median and tibial nerves of lightly anesthetized monkeys. In these cases we had strong *a priori* knowledge (derived from neuroanatomy and physiology) as to where the focus of elicited thermal activity should have been, and indeed, our expectations matched the reality. Furthermore, TEG data were cross-validated with respect to electroencephalographic and magnetoencephalographic data obtained from the same monkeys under identical stimulation conditions.

METHODS

Phantom Experiments and Computer Modeling

The experimental work of Shevelev and colleagues which motivated the current project, mostly involved imaging of rats and other small mammals, and in many instances, the skull had been removed to improve the quality of recordings. These investigators have reported detectable thermal responses with latencies on the order of 100 milliseconds to several seconds depending on the experimental preparation. In order to better understand the relationship between surface recorded thermal patterns and internal thermal transients, a series of phantom experiments was initiated to empirically characterize the physical processes of heat diffusion, infrared emission and detection. These phantom experiments employed a simple physical phantom consisting of a resistor located within a spherical flask (diameter approx. 8 cm) filled with water. The resistor was positioned at various locations near the surface of the flask viewed by the IR camera (details of the imaging system are described below). After a sequence of baseline images was collected, a voltage pulse was applied to the resistor, and a series of images was collected to map the evolution of the thermal transient on the surface of the flask. Using a 10 Kohm resistor, we found that a 150 V pulse applied for 1 second, at locations from 5 mm to 1 cm from the surface of the flask produced a robust heat transient that could be seen in single pass experiments.

Most experiments were conducted using an evoked response paradigm. In each experimental trial, we collected (in digitizer memory) a series of images that bracketed the delivery of a current pulse, time-locked to the vertical sync of the video signal. During a typical phantom experiment, sixteen video frames, acquired at a rate of 1/sec were obtained prior to delivery of a current pulse. Another sixteen frames (rate 1/sec) were then acquired immediately post-stimulus; and these were followed by 32 frames sampled at a rate of 0.5/sec. A total of 64 images spanning a time course of 16 seconds pre-stimulus and 80 seconds post-stimulus were thereby obtained. In each experiment, the current pulse was presented on 16 distinct occasions, and the data were averaged across presentations. This resulted in a data set of 64 sequential images, each reflecting an average over 16 events.

In thermal imaging as in other approaches to functional neural imaging, the signal of interest is a relatively small change on top of a large background signal. Our strategy was therefore to calculate the average of the pre-stimulus frames, and to subtract this baseline image from each of the images in the temporal sequence. That is, to better reveal the time-course of thermal transients, a composite baseline image (derived from the average of all 16 pre-stimulus average frames) was subtracted from each of the 64 average images. These difference images were then averaged across every four video frames, that is, the data set of 64 images was reduced to 16 (64/4) sequential difference images. This yielded a sequence of 16 averaged difference images, the first four of which spanned the pre-stimulus interval (1 image every 4 seconds), the next four spanning the first 16 seconds post-stimulus (1 image every 4 seconds), and the final eight images spanning from 16-80 seconds post-stimulus (1 image every 8 seconds).

As an initial step toward solving the inverse problem required for localizing internal sources on the basis of surface thermal measurements, two computational approaches towards the analysis of phantom data were explored. The first approach utilized an analytical formula for diffusion of heat from a point source in an infinite solid. This formula allowed for derivation of an expression for the temperature at the surface of an eccentric sphere, as a function of position and time. From such calculations it was possible to plot the predicted time-course of temperature changes at a given location, and to generate a movie sequence illustrating both the spatial and temporal course of the thermal transient.

Because of concerns about potential shortcomings of the analytical model (due to invalid assumptions) a finite element calculation using a system of differential equations was developed to explicitly model the phantom situation. To produce an efficient calculation, the cylindrical symmetry of the phantom was exploited, with the system being modeled as a stack of slices, each consisting of a set of concentric cylindrical shells. In addition to the outward diffusion of heat from the point source (reflected in the analytical formulation), the finite element approach models the lateral diffusion at the air-glass interface. In principle, the point source assumption can be relaxed; although we have not yet explored the consequences of this modification. Imaging experiments of the resistor in air suggest that the initial heating is focused at the center of the resistor, though presumably the ends of the resistor heat more quickly than the surrounding medium. Our existing computer code does not model radiative or convective dissipation of heat from the surface of the sphere, nor does it consider differences in thermal conductivity of water and glass (approx. 30%). Nevertheless, it provides a viable first-order model of the situation.

Physiological Preparation

Animal experiments employed anesthetized pigtail monkeys (Macaca nemestrina). A light anesthetic state was achieved via intramuscular injection of a Ketamine & Rompin 'cocktail', administered in combination with atropine to control salivation. Supplementary injections were provided whenever movements or other responses indicated the animal was awakening. Typical experiments lasted a total of 4 hours. Somatosensory stimulation was applied to the tibial or median nerves, through electrodes located at the ankle or wrist. Each stimulus was a short pulse (typically 10 milliamps for 0.2 ms) generated by an isolated constant current source (Grass Model SIU8-S88). Prior to the actual experiment, pulse current was adjusted until a slight flexion of the digits was consistently observed. Similar stimulation protocols are routinely used with awake human volunteers in our laboratory, and are not considered painful. The monkeys' heads were shaved for imaging experiments since hair obscures thermal emission from the scalp surface. The monkeys used for this experiment had previously served as subjects in similar experiments conducted with neuromagnetic and scalp electric recording techniques. Volumetric magnetic resonance images which allow identification of anatomical regions associated with specific somatosensory responses, were available for each animal.

During testing, an animal was laid on plastic tray and placed in a wooden enclosure with a front curtain which darkened the chamber and limited drafts from laboratory ventilation. The principal features of the experimental setup are illustrated in figure 1. The IR camera was mounted on top of the box, a front-surface mirror being

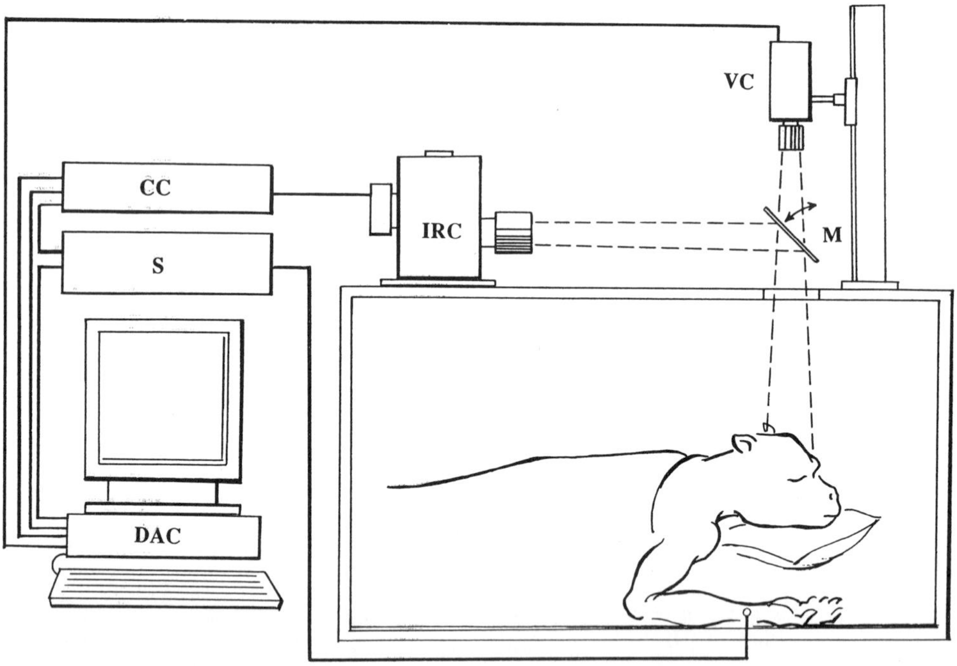

FIGURE 1. Experimental setup for IR thermal evoked response studies. A monkey was placed on a plastic tray inside a wooden enclosure. An IR Camera (IRC) viewed the dorsal surface of the monkey's head through an opening in the box, using a front surface mirror (M). The mirror could be shifted aside to provide a corresponding view of the preparation with a CCD-based video camera (VC). Images were acquired with a PC/AT compatible data acquisition computer (DAC), which also controlled the stimulator subsystem (S) and IR camera control unit (CC). During the acquisition of a sequence of images, a stimulus was delivered at a known time, synchronized to the imager frame interval.
Transient response data were obtained and processed in a fashion identical to that described above in the phantom experiments.

used to visualize the top of the monkey's head. Because the IR detector array we used was housed in an open dewar, it could not be tilted to extreme angles; other models now available are housed in a pressurized dewar that can be operated in any position. The reflecting mirror could be moved aside to allow the experimental field to be viewed with a downward looking video camera. A warm calibration disk with several holes near the perimeter produced images that were detectable with both IR and video cameras, and this was used to establish fiducial markers for alignment of images.

Imagers, Image Acquisition and Processing

Two different imagers were evaluated in our experimental work. An initial series of steady-state physiological experiments (see Results section) employed an Inframetrics model 600 Thermovision camera. This device is based on a single, liquid nitrogen cooled, bolometric (heat) detector. In this imager, the measurement site is scanned over the field of view using a system of galvanometrically driven mirrors. The camera support electronics generate a continuous RS-170 compatible video signal (30 interlaced frames per second), although the frame rate of the underlying device is ca. 8 frames per second.

The primary experiments reported herein used a solid state imager, the Amber Engineering 4128. This is a nitrogen cooled focal plane array (FPA) consisting of 128x128 indium arsenide quantum detectors. The maximum frame rate of the camera system is 217 Hz although the FPA supports somewhat higher readout rates. The full-well charge capacity of photosites is 40×10^6, implying that the detector shot noise limit is better than 15 bits. However, the FPA output is internally digitized at 12 bits. This digital signal steam is subsequently corrected for detector response characteristics in order to produce a flat video field, and the resulting (12 bit) digital data is provided at an RS-422 (differential digital video) port. Analog and digital images are also provided in RS-170 compatible timing and pixel formats.

Focal plane arrays appear to have a number of theoretical advantages, particularly from the standpoint of noise reduction. However, FPA technology is less mature, and such cameras suffer from nonuniformity of response across the array. In the Amber camera, this problem is addressed through digital signal processing in the camera controller. Proper calibration produces a satisfactory image, though at the high gains and contrast employed for this work, some fixed pattern noise is noticeable.

Calibration of detector response was performed as suggested by the camera manufacturer, using a black painted aluminum disk at room temperature and another maintained at elevated temperature in a water bath. The disks were placed near the lens of the imager, out of focus. Temperatures were selected to span 60-80 percent of the range of the transimpedence amplifier output of the detector array, typically a range of 5-7 degrees C. Based on this data, the controller calculates a two point correction (gain and offset) that is applied to the digital data stream on a pixel by pixel basis. For absolute temperature calibration of images, a pair of in-focus images were collected of the calibration sources at known temperatures. Given the multiple stages of gain and contrast enhancement in the image the acceptable temperature range for this calibration was much tighter: < 2 degrees C.

In these experiments we captured the analog video signals using an Imaging Technology Inc. VS-100 frame grabber in an IBM AT-bus compatible, Intel 286-based microcomputer. Amplification and offset controls in the camera controller and in the frame-grabber were used to produce a high contrast video signal that was acquired at 8 bits. Images were summed into 16 bit integer arrays in extended memory of the host computer. By summing thermal images across multiple frames, it was possible to achieve thermal sensitivity of <.005 degrees Celsius. Data acquisition software consisted of minor extensions of a system previously developed in our laboratory for general purpose scientific image acquisition (Rich et al. 1989). Initial analyses were performed on the data acquisition computer, but more extensive analyses and visualization were performed on a SUN Sparcstation, using IDL and/or Precision Visuals Wave.

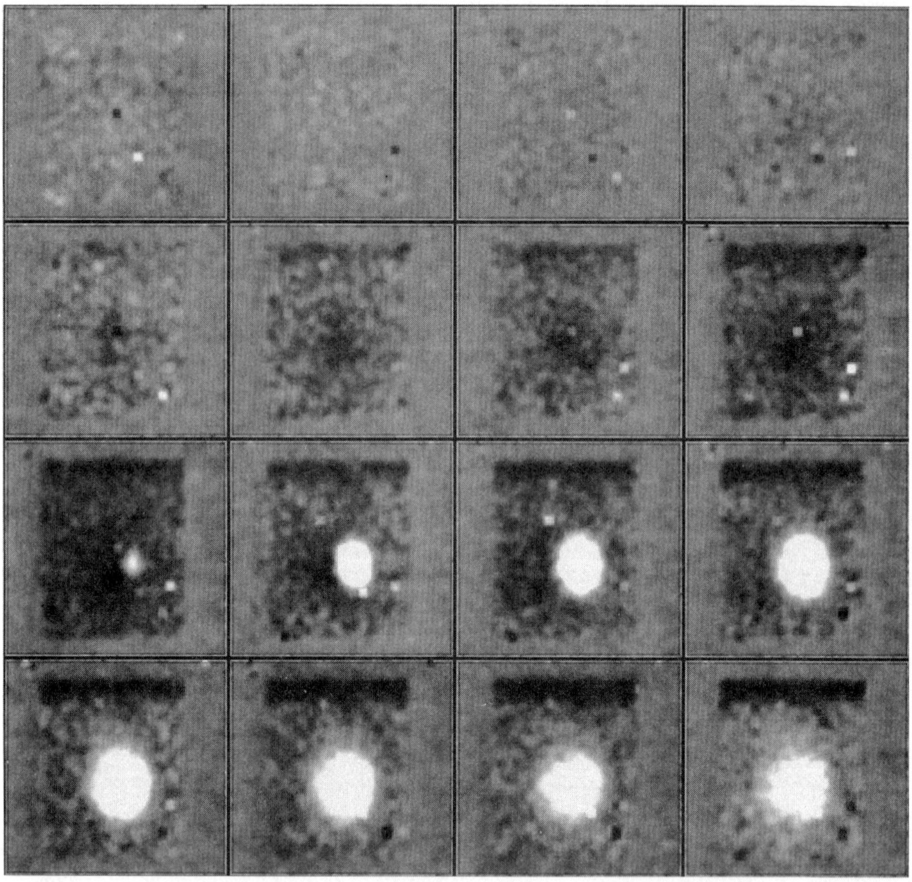

FIGURE 2. A sequence of images from a phantom experiment. The heat source was a resistor located at a nominal position 5 mm from the surface of a water filled spherical flask; the heat transient generated by application of a 150V pulse for 1 second. A total of 64 images (16 pre and 48 post-stimulus) were collected with respect to each of 16 stimulus presentations. The first 32 images were collected at a rate of 1/sec, whereas the last 32 were collected at a rate of 0.5/sec. Corresponding data frames were averaged across stimulus presentations. A baseline image calculated by averaging the first 16 pre-stimulus frames was then subtracted from each of the 64 sequential images, to produce 64 difference images. Sets of 4 difference images were then averaged to produce the 16 difference images shown here.

RESULTS

Phantom and Computer Modeling Experiments:

Figure 2 shows a sequence of 16 difference images derived from an experiment where the resistor was located 0.5 cm below the surface of the flask. The data show that focal heating was not clearly apparent until around 20 seconds post-stimulus. This suggests that physiologically induced thermal changes will display a similar, slow time course. The transient began as a small, well-defined spot which grew in size and intensity. In later frames the spot became larger, more diffuse and flatter in its intensity profile. The apparent cooling observed during the pre-stimulus period (the first 4 images) is believed to be an artifact related to our failure to allow the system to return to a complete thermal equilibrium between stimulus pulses. When trials are repeated before the temperature distribution returns to its steady state, residual heat is averaged into the pre-stimulus baseline, and the cooling of subsequent frames is apparent in the difference images.

Determining the distribution of heat sources and sinks from measurements at the head surface is an ill-posed inverse problem. Given the variable latency and amplitude of the thermal response as a function of source distance from the head surface, it is possible to postulate multiple source configurations to account for the observed data. By making appropriate assumptions concerning the source configuration and time-course, it should be possible to find a set of model parameters that best account for the observed data in some numerical/statistical sense. Initial simulations modeled a point source of heat embedded in a homogeneous medium. Figure 3 illustrates the time-course of the temperature transient at the central (hottest) point on the surface of the spherical volume (radius = 5 cm). Results are illustrated for both the analytical model and the finite element model described above. In the finite element computation, the point heat source was located along the axis of the cylindrical computational geometry, corresponding to a radial axis in the spherical volume. Even given the differences in model assumptions, agreement between these computed amplitudes and time-courses is very reasonable.

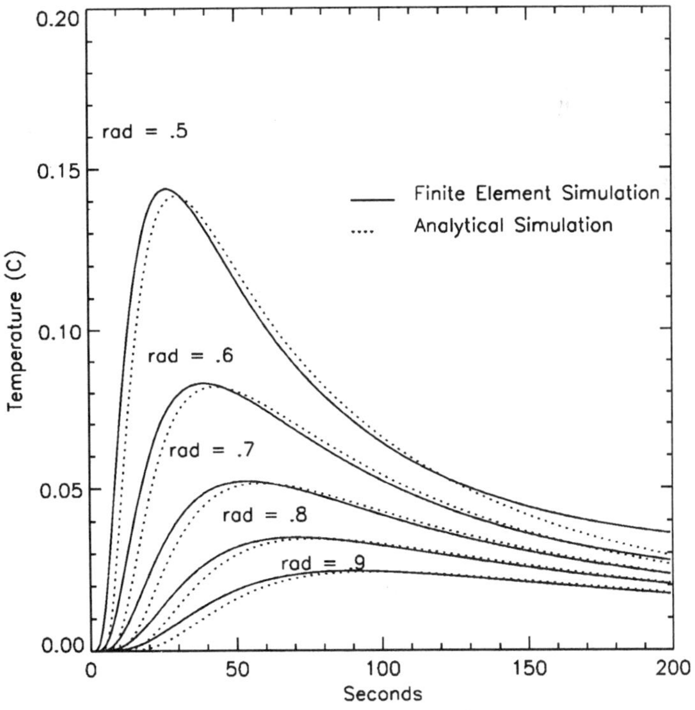

FIGURE 3. Computational simulations of the time-course of surface heating. Geometry was a simplified spherical case comparable to the phantom experiment described in figure 2. A point heat source was assumed, located a varying distances from the surface of the sphere along the axis of the calculation (0.5 to 0.9 cm are plotted). Calculations were based on an analytical model (a point heat source in an infinite solid) or a cylindrically symmetrical finite element calculation.

In initial phantom experiments the response latency was longer and the subsequent temperature rise was sharper than predicted by the models. As suggested by figure 2, some of this may have been due to the apparent cooling process obscuring the early phase of the temperature rise. Also, in recent experiments we have removed a plastic and silicon sleeve originally placed around the resistor to electrically insulate it from the water in the flask. This change, in conjunction with a longer delay between trials has produced better agreement between empirical phantom data and computational simulations.

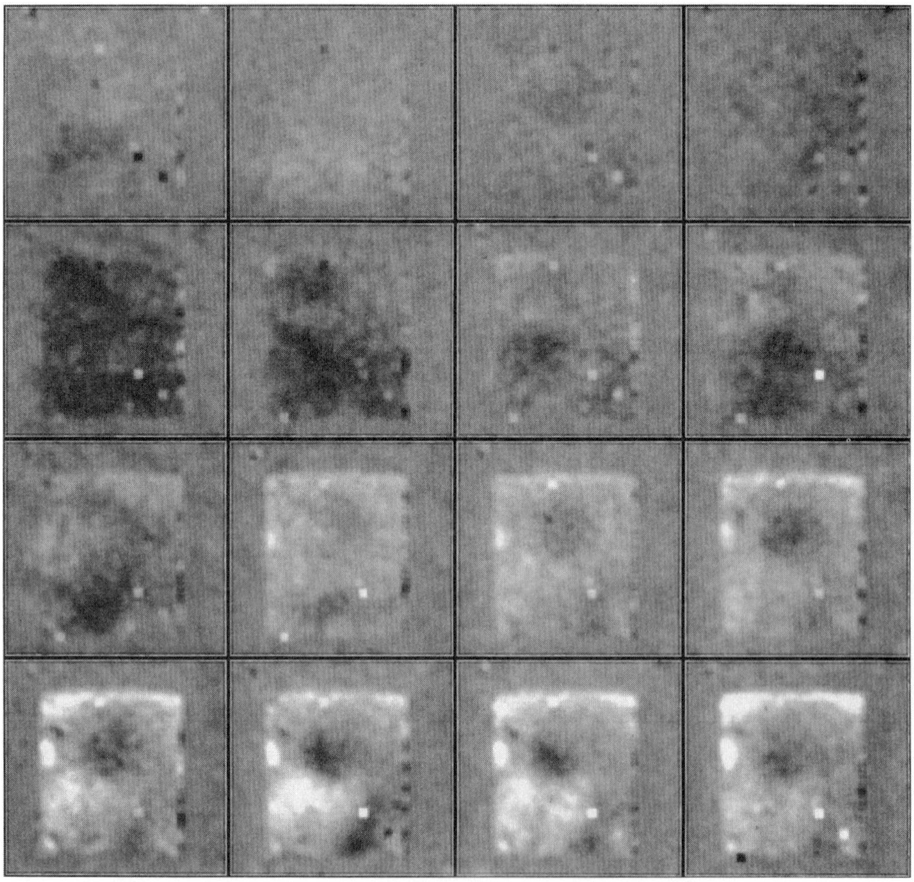

FIGURE 4. Difference-image sequence derived from monkey data recorded in response to electrical stimulation of the left median nerve. The data were processed as described in figure 2.

Physiological Experiments

Most of the experiments described here were conducted with an Amber Engineering IR imager. A typical transient-response experiment attempted to resolve the time-course and spatial distribution of the thermal transient associated with individual electrical stimuli delivered to the median or tibial nerve. Figure 4 illustrates a sequence of difference images collected over the region of somatosensory cortex during stimulation of the median nerve.

In this sequence, a focal thermal transient (apparent as a bright spot) was observed near the lower left corner of the image frame beginning at approximately 20 seconds post-stimulus, and peaking around 40-44 sec post-stimulus. There is also a dark region suggesting surface cooling in the same region during the early pre-stimulus frames. As discussed for the phantom data, this appears to be an artifact due to inadequate inter-stimulus interval -- two minutes in this example!

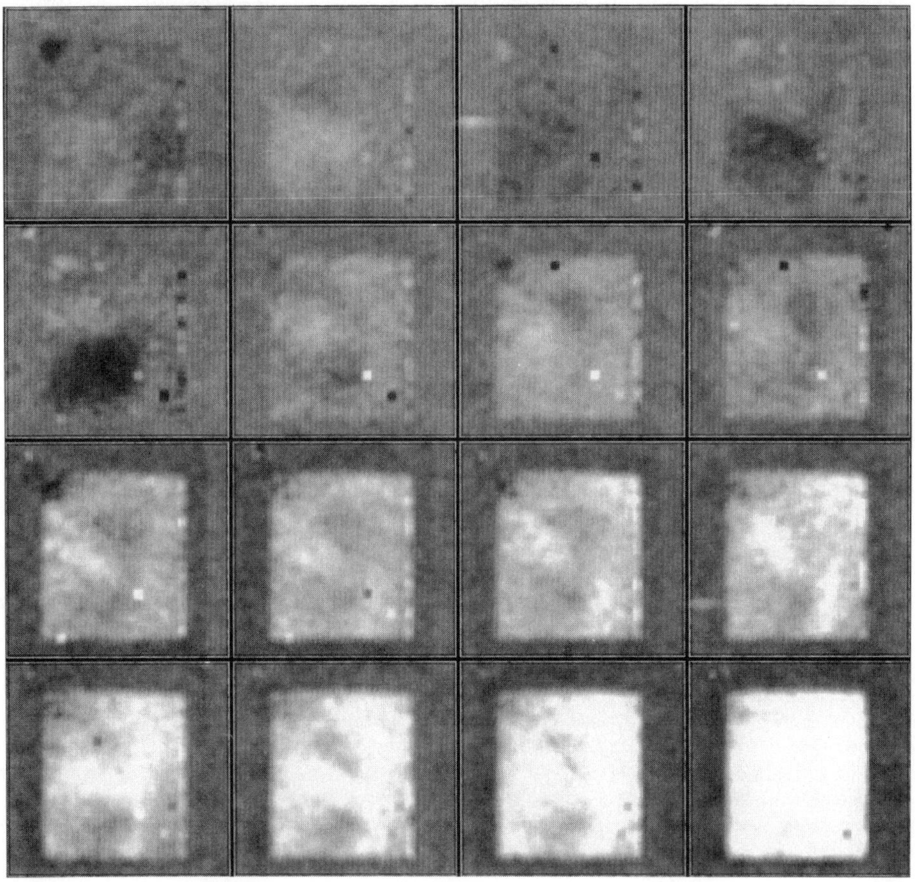

FIGURE 5. A series of difference-images comparable to those of figure 4, but with stimulation of the left tibial nerve, near the ankle.

Figure 5 displays a comparable sequence of difference images collected during stimulation of the tibial nerve. In this case, the heat transient appears slightly earlier, and is higher and more central than the corresponding image feature observed during median nerve stimulation. The peak of the transient is difficult to identify with certainty because of the gradual increase in basal temperature in the sequence.

Figure 6 compares the locations of heat transients associated with median or tibial nerve stimulation. Characteristic IR difference images were selected near the peak of the response and were superimposed on a line drawing of the monkey's head. This outline was constructed by aligning IR and video images using the fiducial techniques described above, and then using interactive computational drawing tools to trace the relevant contours. The markers near the centers of the hot spots reflect the predicted locations of the peaks of the heat distribution as predicted from neuromagnetic and neuroelectric experiments. Neuromagnetic and neuroelectric techniques had been used previously to locate sources of somatosensory evoked activity in MRI volumetric images of the experimental animals, and the closest approach to the head surface was identified. The predicted loci of the thermal transients estimated in this manner agree reasonably well with the observed patterns of response. Certainly, the general pattern of these locations and their hemispheric asymmetry is consistent with known patterns of functional anatomical organization. In particular, note the shift posterior and toward the midline as stimulation switched from median to tibial nerve.

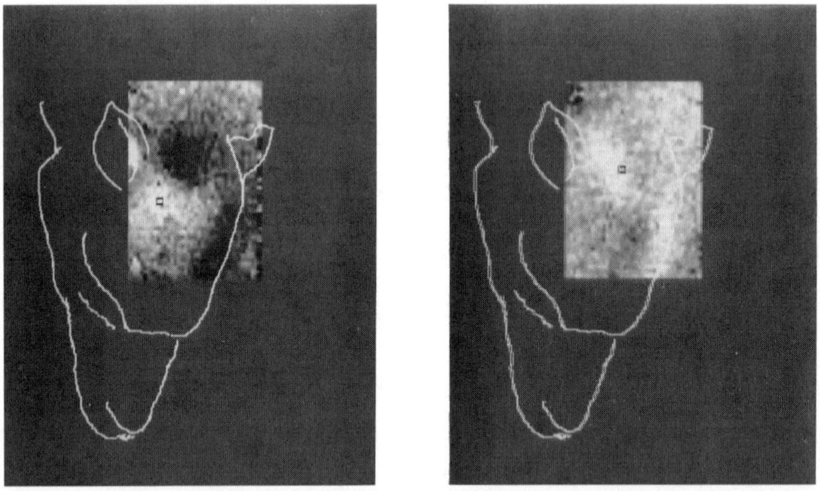

FIGURE 6. A comparison of median (left panel) or tibial nerve (right panel) stimulation. For this illustration, characteristic images of the focal transient were chosen.. The median nerve stimulation image is an average of 4 images from 57-64 sec post-stimulus. The tibial stimulation example is an average of images from 41-48 sec post-stimulus. The gray levels represent a range of temperature changes from +.030 degrees C (white) to -.030 degrees C (black) with the background gray equivalent to no change. The line drawing is based on a video image covering the field of view of the IR camera. The box represents the location of the corresponding source identified by electromagnetic methods, projected onto the head surface.

In several early experiments we attempted to image temperature changes associated with steady-state repetitive stimulation, and it is informative to compare the results of these experiments with those of the transient experiments. In these experiments continuous stimulation was applied at a rate of 3/second during stimulus-on phases. Figure 7 illustrates the result of one such experiment utilizing the Inframetrics Thermovision camera. A sequence of 64 frames (approximately 2 seconds of continuous video) was collected at each of the following time points: (a) 2 minutes before the onset of repetitive stimulation; (b) 1 minute after the onset of stimulation; (c) 3 minutes after the onset of stimulation; and (d) 3 minutes after the offset of stimulation. All images in each sequence were averaged, and difference images were calculated as (b+c)-(a+d). These images are illustrated for median nerve stimulation (left panel) and tibial nerve stimulation (right panel). More features are apparent in these images than in the transient stimulation trials; in particular, a hot spot near the vertex of the monkey's head that is seen in both images. However, the other dominant features in the images are consistent with the pattern of tibial nerve projection near the midline and a more lateral, lower projection of the median nerve.

DISCUSSION

Phantom experiments and computational simulations suggest that thermal responses of cortical origin detectable at the scalp surface of man or the larger mammals will have latencies on the order of seconds to tens of seconds, and our physiological experiments are consistent with this. This factor, coupled with the long duration of such responses makes the technique undesirable for evoked response paradigms in which the stimulus is repeated and image sequences are averaged across

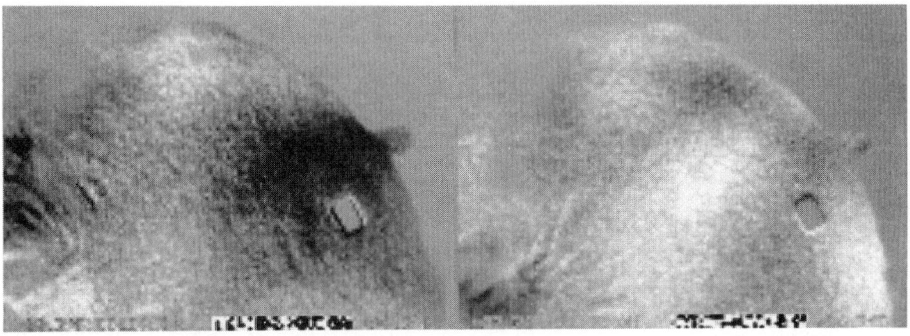

FIGURE 7. Steady state heat distributions associated with median (left panel) or tibial nerve (right panel) stimulation. Control or stimulation conditions were established as described in the text, and a series of 64 images were acquired, using the Inframetrics model 600 imager. All images in a sequence were averaged. Figures above illustrate differences between averages of stimulated and pre + post stimulus control sequences.

multiple trials. However, the low temporal frequencies anticipated in the thermal response suggest that for source depths of even a few millimeters it should be sufficient to sample frames at intervals of a second or more. For a modest investment in equipment and development time, it should be possible to exploit the high frame rate of the FPA imager to achieve very good signal/noise performance in single pass experiments by averaging over 100 frames for each sampled image. This capability could be exploited in cognitive experiments or by sensory paradigms using steady state stimuli. Such experiments could be designed to modulate the relative activation of different brain regions over relatively long time scales. Although the transient experiments described here are desirable for describing the response function of the system, it is likely that sustained stimulation would produce larger thermal responses.

Even if it is possible to adequately describe the system response function for a simple heat source, it may not be possible to produce a tomographic reconstruction of heat sources within the head volume. In addition to the computational difficulties posed by such an ill-posed, under-determined problem, there are biological unknowns associated with the patterns of neural activation and cerebral circulation. For example, what are the implications of the centrifugal pattern of blood flow in cortex? Might deeper activation produce compensatory changes in blood flow that would affect overlying cortex? Clearly the biological origin of the thermal response and factors affecting its spread need to be characterized much more thoroughly.

Requirements for sensitivity and stability make thermal imaging experiments difficult to conduct, control and interpret. In particular, the small size, long latency and long duration of a thermal transient associated with a deep source poses serious problems for reliable detection. Obvious technical improvements can address some but not all of these problems. Nevertheless, we have obtained results which suggest that somatosensory evoked thermal responses can be observed using IR sensitive imaging equipment. These techniques may prove most useful for imaging metabolic activation over large regions of exposed cortex, minimizing problems of response latency and amplitude. However,the prospect of functional neural imaging through the intact skull is a powerful incentive. Given suitable experimental designs, with consideration of the caveats imposed by the physics of the system, thermal imaging may prove to be a useful tool for noninvasive exploration of brain function.

REFERENCES

Gorbach A.M., E.N. Tsicalov, G. D. Kyznetsova et al., Infrared Mapping of the Cerebral Cortex, Thermology, 3:108-111 (1989).

McElligott J.G. and R. Melzack. Localized thermal changes in the brain evoked by visual and auditory stimulation. Exp. Neurol. 17:293-312, (1967).

Melzack R. and K. L. Casey. Localized temperature changes in the brain evoked by somatic stimulation. Exp. Neurol. 17:276-292, (1967).

Phelps M.E., and J. S. Maziotta, Positron emission tomography: Human brain function and biochemistry, Science 228:799-809 (1985).

Rich P.M., D. M. Ranken, and J. S. George. A Manual for Microcomputer Image Analysis. Los Alamos. LA-11732-M (manual), (1989)

Shevelev I.A., E. N. Tsicalov, A. M. Gorbach, et al. , Dynamic Infrared Images of the Brain, IBRO News, 16:8-9 (1988).

Shevelev I.A., E. N. Tsicalov, A. M. Gorbach, K. P. Budko and G.A. Sharaev. Dynamic Functional Mapping of the Brain by its Infrared Radiation. in: Machinery of the Mind: Data, Theory and Speculations About Higher Brain Function. E. Roy John ed. , Birkhauser, Berlin (1990).

Tsicalov E.N., G. D. Kuznetsova, I.A. Shevelev et al. Thermal Distribution over the Cortical Surface of the Rat Brain under Direct Electrical Stimulation, Neurophysiology, 19:169-174 (1987).

Zeschke G. and V. G. Krasilnikov. Decreases of local brain temperature due to convection (local brain blood flow) and increases of local temperature due to activity. Acta Biol. Med. Germ. 35:935-941, (1976).

SPATIOTEMPORAL INHOMOGENEITY OF $[Ca]_i$ IN NEURONS

S.L. Mironov

Department of Neurophysiology,
Max-Planck Institute for Psychiatry
Planegg-Martinsried, Germany

INTRODUCTION

The concept that Ca might serve as an intracellular messenger to couple stimulus to response in excitable tissues is well established. An understanding of the precise mechanisms by which second messengers act, requires information on the temporal nature of changes in second messenger concentration and their spatial distibution within the cell. By optical imaging with fluorescent Ca-indicators, it was found that the change in intracellular free Ca-concentration is often transient and spatially inhomogeneous (for a recent review see Tsien and Tsien, 1990). Appreciation of this particular paradox leads to the recognition that Ca-cycling in the target cell has an important, spatially and temporally distinct messenger function. The purpose of this communication is to consider the intracellular messenger role of Ca in generation of specific transients and gradients in the cytoplasm of nerve cells.

Ca-RECOVERY AND Ca-RELEASE MEDIATED BY INTERNAL Ca-STORES

The mechanisms of Ca-regulation have been studied using snail neurons and rat sensory neurons. The results obtained were qualitatively similar. Intracellular Ca was calculated from fura-2 signal measured with a photomultiplier in epi-fluorescence mode, giving either the average, $<Ca>_i$, or local, $[Ca]_i$, concentration of free Ca. Fig. 1A shows that pulse trains of various frequencies established a new steady level of $<Ca>_i$ and after the end of stimulation it recovered with a time course, corresponding to a single exponent. The Ca-recovery is mainly determined by Ca-pump (CaP) of internal stores (CaS).

This can be argued (Kostyuk et al. 1990; Miller 1991) first by excluding all other mechanisms known from biochemical studies, namely mitochondria, Ca-pump or Na/Ca-exchange in the surface membrane. Second, the temperature coefficient was close to that of CaP and injection of vanadate, an inhibitor of ATPases, counteracted with Ca-recovery (Fig. 1).

Optical Imaging of Brain Function and Metabolism, Edited by
U. Dirnagl *et al.*, Plenum Press, New York, 1993

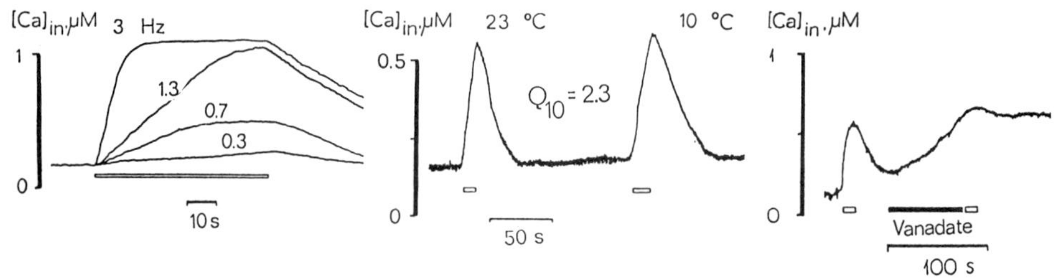

FIGURE 1. Ca-pump of internal Ca-stores determines the recovery of snail neurons from the Ca-load. Left - Ca-transients evoked by a train of 100-ms pulses of different frequency (indicated near each trace) taking the membrane potential to 0 mV. Middle - the temperature dependence of Ca-recovery. Right - the vanadate injection inhibits Ca-recovery. In each trace the period of membrane stimulation is shown by a horizontal bar.

Ca pumped into CaS could be subsequently released via a pathway, corresponding to the Ca-induced Ca-release. When cell stimulation raised $<Ca>_i$ to a certain threshold, which in neurons is low enough only in the presence of caffeine, it triggered a further spontaneous $<Ca>_i$ increase (Mironov and Usachev, 1990). Then $<Ca>_i$ recovered with the time course, similar to that observed after Ca-load by electric stimulation. Periodic Ca-transients were also observed. They occured either spontaneously or could be more consistently triggered by Ca-influx as discussed below.

Ca-INDUCED Ca-RELEASE: MECHANISM, MODEL AND IMPLICATIONS

The model of Ca-release (CaR, Mironov and Tupikin, 1988) was based on the concept proposed earlier by Fabiato (1983) which is shown schematically in Fig. 2A. The functioning of Ca-dependent activation and inactivation gates of CaR were represented by variables m and h, respectively, in line with the model of nerve excitability (Hodgkin and Huxley, 1952). The basic model consisted of two differential equations

$$dC/dt = P\, m^2_{\infty}(1 - h)^2 - gC/(K' + C) + j \qquad (1a)$$

$$dh/dt = k_+C(1 - h) - k_-h \qquad (1b)$$

For CaP the Michaelis - Menten equation was used with the maximal rate g = 10 μM/s and the dissociation constant $K' = 0.5$ μM. CaR activation was assumed to proceed instantaneously with the maximal rate P = 50 μM/s and it was described by a steady variable $m_{\infty} = C/(K + C)$ with K = 0.2 μM. On- and off-rate constants $k_+ = 4\ \mu M^{-1}s^{-1}$ and $k_- = 1\ s^{-1}$ were estimated from the experimental data (Fabiato 1983; Bezprozvanny et al.; Finch et al. 1991).

When Ca-influx j into the cytoplasm was large enough, the model started to generate periodic Ca-transients (Fig. 2B). Further increase in influx decreased the peak [Ca] and the period of oscillations due to accumulating inactivation after each spike, until only damped oscillations could be observed. The frequency of spikes was dependent on different model parameters, but most dramatically it can be changed, varying the rate constants for inactivation (Fig. 2C). It produced a 100-fold change in the oscillation frequency while the spike amplitude remained constant. The model predicts that $[Ca]_i$ oscillations should be initiated and modulated by a steady Ca-influx, an effect, which was observed experimentally (Lipscombe et al. 1988; Mironov and Usachev, 1991).

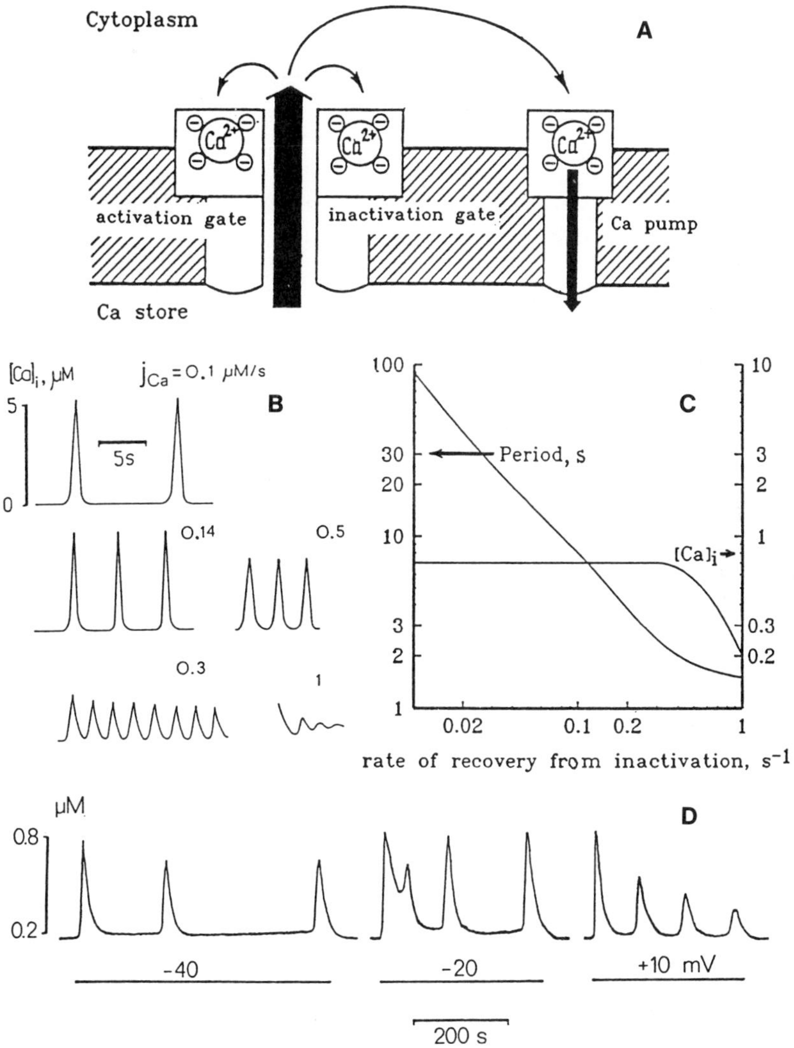

FIGURE 2. $[Ca]_i$ oscillations produced by Ca-induced Ca-release. A - the model. B - the modulation of [Ca] oscillations by Ca-influx (in μM/s, as indicated near each trace). C - the period and peak [Ca] are dependent on the inactivation characteristics of Ca release channel. D - Ca-influx induces $[Ca]_i$ spiking in the presence of 10 mM caffeine. The cell membrane was voltage-clamped and depolarized to different voltages as indicated by a horizontal bar.

$[Ca]_i$-oscillations in nerve cells could underlie slow waves of membrane potential intrinsic for bursting neurons (Gorman and Thomas 1978; Meech 1979; Mironov 1983). Both phenomena have the same period. Periodic Ca-release can control the membrane potential via the modulation of Ca-activated current(s).

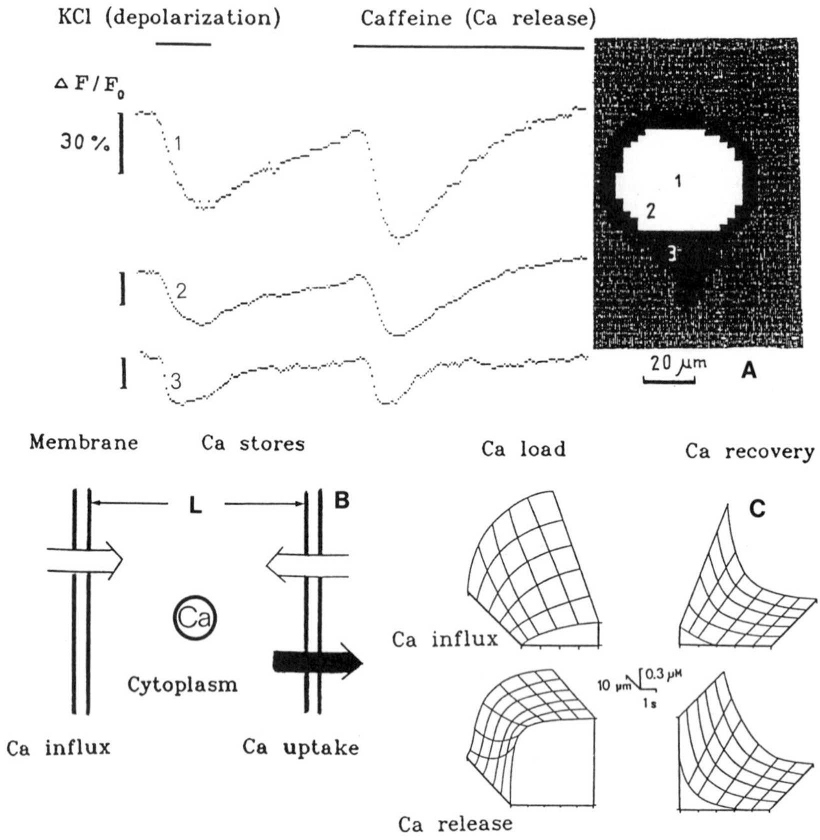

FIGURE 3. The role of diffusional volume in the time-course of Ca-transients. A - The distribution of fura-2 fluorescence (excitation wavelength 387 nm) in the isolated snail neuron and the measurement points is shown in the right part. Left - the changes of relative fluorescence. The bathing solution contained (in mM): 100 NaCl, 5 KCl, 10 $CaCl_2$, 2 $MgCl_2$, 20 Tris-HCl (pH 7.3) and 50 mM KCl or 10 mM of caffeine were applied as indicated. B - the model. C - Ca-transients calculated for Ca-load by Ca influx and Ca-release and subsequent Ca-recovery.

SPATIAL INTERRELATIONSHIP BETWEEN CA-INFLUX, UPTAKE AND RELEASE: THE DIFFUSION VOLUME CAN DETERMINE THE TIME-COURSE OF CA-TRANSIENTS

The local concentration of Ca was measured by scanning the fluorescence signal with a dissector (a sort of photomultiplier) as described elsewhere (Kirischuk et al. 1989). The full frame acquisition was relatively slow in comparison with CCD cameras, but this technique allowed to measure the fluorescence in the given predetermined point(s) with a time resolution of 10-30 ms. Being interested mainly in the kinetics of

Ca-transients, we measured them as a decrease in fura-2 fluorescence excited at 387 nm. No spatial difference in resting $[Ca]_i$ were usually found, but the time course of Ca-transients was faster in the soma than in the axon (Fig. 3A).

Ca-stores are often attributed to sheet-like structures located near the surface membrane (O'Sullivan et al. 1989), sometimes being almost parallel to it. This forms the geometrical basis for the model which consists of two planes, representing the surface membrane and the membrane of CaS (Fig. 3B). Calculated [Ca] changes had a complicated spatial kinetics (Fig. 3C), but the the time-course of average [Ca] was described by a single exponent. The apparent time constant τ was the same for Ca-load and Ca-release which could explain the exponential kinetics of $\langle Ca \rangle_i$ recovery. It was dependent on the Ca-diffusion coefficient D and the rate of Ca-uptake g according to the following empirical formula

$$\tau = L/g + L^2/\pi D \tag{2}$$

In this expression numerical values of τ, L, D and g are given in ms, μm, 100 $\mu m^2/s$ and 100 μm/s, respectively. This result provides some measure of correlation between the kinetics of $\langle Ca \rangle_i$ transients and the cytosolic volume where Ca is redistributed.

Some other data can be considered as a support of this suggestion. Lipscombe et al. (1988) observed faster Ca-transients in growth cones in comparison with a cell body. In barnacle neurones the time constant of Ca-recovery increased in a sequence: dendrite < axon < soma (Ross, Stockbridge and Stockbridge, 1989). Ca-transients in muscle cells have a characteristic time constant, ranging from 50 to 300 ms, whereas in neurons they are much slower. This can be due to a closer proximity of the sarcoplasmic reticulum to the surface membrane.

Ca-STORES AS A PATHWAY FOR Ca-DELIVERY TO A GIVEN POINT IN CYTOPLASM

From the general analysis (Neher, 1986; Mironov, 1990) follows that a single Ca-channel in the surface membrane in the presence of a cytoplasmic buffer is able to raise Ca-concentration only within $\approx$1 μm from the channel mouth. The association of CaS to the surface membrane and their abundance in a cell interior allow to suggest that they form a 'grid' whereby a local [Ca] increase can propagate in a cytoplasm. These Ca-waves have been described for various cell types (Tsien and Tsien, 1990). Via such mechanisms, Ca can be delivered to a given point in a cell interior without increasing significantly the Ca-concentration in other cell regions. The calculated velocity of Ca-wave ranged from 1 to 300 μm/s, depending on model parameters (Mironov, 1990). It was only weakly dependent on a distance between two neighbour Ca-release channels, a feature, that is essential to produce a non-dissipating Ca-wave.

CYTOPLASM AS THE 'Ca-EXCITABLE MEDIUM'

In extension to previous formulations which considered the fate of Ca in the cell mainly due a passive redistribution via diffusion, the recent data (cf. Tsien and Tsien, 1990) unequivocally demonstrated that active processes of Ca-uptake and Ca-release associated with CaS plays a very important role. Under some conditions a crosstalk between two mechanisms results in such properties of the cytoplasm that it can be considered as the Ca-excitable medium. Intracellular Ca-spikes possess many features, similar to the excitability of nerve membranes. It includes the activation threshold, the refractory period, repetitive firing and the propagation in the cytoplasm. By the analogy, kinetic properties of Ca-release mechanism must be similar to that of Na channels, initiating the action potential in nerve. Two molecular devices of this sort have been described: IP3- and Ca-dependent Ca-release channels (cf. Bezprozvanny et al. 1991). They are found in distinct intracellular organelles and have different pharmacology. Both are activated by μM-ranges of Ca (the first one only after IP3 binding) and are inactivated by Ca, but in different [Ca] range, μM and mM, respectively. Like for Na channels in the Hodgkin - Huxley formalism the inactivation of Ca-release channels is of crucial importance. Without it neither single not repetitive Ca-spikes could be generated. Therefore the second type of channels inactivated in a non-physiological range of [Ca] seems to be of secondary importance in the events considered. The other argument is that IP3-sensitive channels are much more abundant (Bezprozvanny et al. 1991).

External stimuli activate various membrane receptors that stimulate IP3 production. Considering it together with IP3 metabolism, it can be shown that within several seconds a steady state is reached for which IP3 concentration ranges from 1 to 10 μM, depending on the cell volume (Mironov, unpublished data). IP3 stimulation of Ca-release corresponds to EC50 $\approx$ 0.6 μM (Bezprozvanny et al. 1991; Finch et al. 1991). This means that under these conditions IP3-sensitive Ca-release mechanisms are almost fully activated. Then spatiotemporal kinetics of free Ca is determined by Ca-sinks and Ca-sources located within a cell and in the surface membrane.

Keeping in mind possible future developments, it seems very important to have indicators for internal Ca-stores able to monitor their Ca-content and visualize their spatial distribution. The corresponding measurements can be made via optical imaging which is ideally suited for this purpose.

REFERENCES

Bezprozvanny I. Watras J. and Ehrlich B. E. (1991) Bell-shaped Ca-response curves of IP3- and Ca-gated channels from endoplasmic reticulum of cerebellum. Nature 351:751-753.

Fabiato A. (1983) Ca-induced release of Ca from the cardiac sarcoplasmic reticulum. Amer. J. Physiol. 245: 1-14.

Finch E. A. Turner T. J. and Goldin S. M. (1991) Ca as a coagonist of IP3-induced Ca release. Science 252: 443-445.

Gorman A. and Thomas M. (1978) Changes in the intracellular Ca concentration in a pacemaker neuron. J. Physiol. 275: 357-373.

Hodgkin A. and Huxley A. (1952) A quantitative description of membrane current and its application to conduction and excitation in nerve. J. Physiol. 117: 500-542.

Kirischuk S. I. Zhukarev V. A. and Mironov S. L. (1990) 2D-analysis of biological microobject images by means of digital fluorescence microscopy. Biol. memb. 7: 399-405.

Neher E. (1986) Concentration profile of intracellular Ca in the presence of diffusible chelators. Exp. Brain Res. S14: 80-96.

Kostyuk P. Mironov S. Tupikin A. and Belan P. (1990) Cytoplasmic free $[Ca]_i$ in isolated snail neurones as revealed by fura-2. J. Membrane Biol. 110: 11-20.

Lipscombe D. Madison D. Poenie M. Reuter H. Tsien R.Y. and Tsien R. W. (1988) Spatial distribution of Ca channels and cytosolic Ca transients in growth cones and cell bodies of sympathetic neurones. Proc. Natl. Acad.Sci. USA 85: 2398-2402.

Meech R. (1979) Membrane potential oscillations in molluscan bursting neurones. J. Exp. Biol. 81: 93-115.

Miller R. J. (1991) The control of neuronal Ca homeostasis. Progr. Neurobiol. 37: 255.

Mironov S. L. 1983 Slow wave membrane potential oscillations in mollusc neurones. Neuroscience 10:899-907.

Mironov S. L. and Tupikin A. V. (1988) On the nature of periodic changes in the intracellular Ca concentration. Biol. Memb. 5: 528-534.

Mironov S. L. (1990) Theoretical analysis of propagation of $[Ca]_i$ wave along the surface of intracellular stores. J.Theor. Biol. 146: 87-98.

Mironov S. L. and Usachev J. M. (1990) Sr and Ba transients in isolated snail neurones studied with fura-2. Neurosci. Lett. 112: 184-189.

Mironov S. L. and Usachev J. M. (1991) Caffeine affects Ca uptake and Ca release from intracellular stores. Neurosci.Lett. 123: 200-204.

O'Sullivan A. Check T. Moreton R. Berridge M. and Burgoyne R. (1989) Localization and heterogeneity of agonist-induced changes in cytosolic Ca concentration in single bovine adrenal chromaffine cells. EMBO J. 8: 401-411.

Ross W. N. Stockbridge L. L. and Stockbridge N. L. (1986) Regional properties of Ca entry in barnacle neurons determined with Arsenazo III and a photodiode array. J.Neurosci. 6: 1148-1156.

Tsien R. W. and Tsien R. Y. (1990) Ca channels stores and oscillations. Ann. Rev. Cell. Biol. 6: 715-745.

INTRACELLULAR ION CONCENTRATIONS IN THE BRAIN: APPROACHES TOWARDS IN SITU CONFOCAL IMAGING

A. Them

Department of Neurology, University of Munich
Munich, Germany

INTRODUCTION

In the central nervous system, regulation of intracellular ion concentrations is of paramount importance for cell function. Intracellular ion concentrations control fundamental cellular processes like gene expression, metabolism, or excitability. Accordingly, disturbances of ion concentrations play a key role in many diseases, particularly in cerebral ischemia. This article focusses on:

1. a brief review of physiological and pathophysiological mechanisms of intracellular pH ($[H^+]_i$) and Calcium ($[Ca^{2+}]_i$) regulation,
2. the description of available techniques to reveal intracellular ion concentrations with particular emphasis on optical approaches, and
3. the presentation of first results from a new approach to measure intracellular ion concentrations in acute rat brain slices and in the intact brain with an optical method.

PHYSIOLOGICAL IMPORTANCE OF $[H^+]_i$ AND $[CA^{2+}]_i$ IN THE BRAIN

$[H^+]_i$

Heterotrophic cell metabolism is associated with acidification through carbon dioxide and acid production. Buffering and rapid energy-dependent transport of H^+ are important mechanisms of intracellular pH-regulation. With development of improved measurement techniques like double-barreled-electrodes, $[H^+]_i$ of brain cells has been investigated in an increasing number of studies, demonstrating the involvement of H^+ in many important cell functions such as enzyme-activities (Busa, 1986), neurotransmitter action (Kaila, Voipio, 1987) or neuronal activity (Chesler, Kraig, 1987; Moody, 1984). $[H^+]_i$ is kept in almost all cell types of the CNS around a value of pH 7.0 under normal steady state conditions, whereas the extracellular hydrogen ion concentration $[H^+]_e$ in brain was determined around pH 7.3 (Kraig et al.1983).

Optical Imaging of Brain Function and Metabolism, Edited by
U. Dirnagl *et al.*, Plenum Press, New York, 1993

Although brain-pH-homeostasis has been emphasized, it becomes clear that there is an array of pH-alterations among neurons, glia, and the interstitial space in the active tissue. Functional changes cause heterogeneity in $[H^+]_i$ among different types of cells because of different abilities in regulation mechanisms like Na^+/H^+-exchange (Grinstein, Rothstein, 1986). It is not clear to which extend pH-shifts influence cellular processes. Most enzymes do not show a significant sensitivity to pH-changes in the physiological range, but there might be local pH-shifts large enough to interact with local enzyme activity in subcellular domains. In some preparations, an initial alcaline-shift has been reported in glial cells (Chesler, Kraig, 1989), and an acidification in neurons. Therefore, a hypothetical flux of H^+ among neurons, glia and interstitial space has been postulated upon neuronal activity (Chesler, 1990). However, little is known about heterogeneous pH regulation in neighbouring compartments, because of the technical difficulties of simultaneous recordings with pH sensitive electrodes.

$[Ca^{2+}]_i$

Intracellular Ca^{2+} is involved in important signaling-pathways inside the cell. Ca^{2+} is identified as a fundamental second messenger in neurons regulating protein kinases and phospholipases activities, neurotransmitter release and influencing the excitatory patterns of cells. The complex functional significance of the calcium ion in cell physiology has been extensively reviewed (Smith, Augustine, 1988; Tsien, Tsien, 1991; Miller, 1988; Blaustein, 1988; Tsien et al.1988; Meldolesi et al.1988; Meyer, 1989).

A summary of Ca^{2+} mediated pathways in an hypothetical neuron is shown in Fig 1. In short, there is a gradient in $[Ca^{2+}]$ between extra and intracellular space in the magnitude of 10^4 under physiological conditions. Because of the low resting $[Ca^{2+}]_i$ (< 10^{-7}M), even a small flux of ions is capable of changing the concentration significantly. $[Ca^{2+}]_i$ is regulated through plasmalemma-channels (Tsien et al.1988), buffering mechanisms by cytosolic proteins (calmodulin, parvalbumine; Blaustein, 1988), sequestration into organelles like mitochondrias or the smooth endoplasmatic reticulum (SER; Meldolesi et al.1988) and extrusion mechanisms like the $3Na^+/Ca^{2+}$ exchanger or the ATP-driven Ca^{2+} pump (Miller, 1988).

Ca^{2+} influx through the plasmalemma is provided by different classes of Ca^{2+}-channels (Spedding, Paoletti, 1992):

1). *Voltage dependent calcium channels* (VDCC's). VDCC's are distinguished according to their functional characteristics into transient (=T-; Carbone, Lux, 1987), long lasting (=L-), (dihydropyridine sensitive, Campbell et al.1988), neither L nor T (=N-; Miller 1987), and P-type (Hillman et al.1991).

2). *Agonist dependent calcium channels* (ADCC's), allowing either direct Ca^{2+} influx (=ionotropic channels), or acting via second messenger mediated steps (=metabotropic channels).

Fast, "ionotropic" ADCC's are directly gated by receptors sensitive to the excitatory amino acid L-glutamate. Those receptors can be distinguished by different glutamate analogues which act as selective agonists to an *NMDA*- (N-methyl-D-aspartate) receptor operated Ca^{2+} channel (Segal, Manor, 1992; MacDermott et al.1986; Finkbeiner, Stevens, 1988; Mayer, Westbrook, 1987; Cotman et al.1992; Watkins, Olverman, 1987) and a Ca^{2+} channel with a *AMPA*- (A-amino-3-hydroxy-5-methyl-isoxalole-4-propionic acid) sensitive receptor (Olsen et al.1987; Collingridge, Lester, 1989).

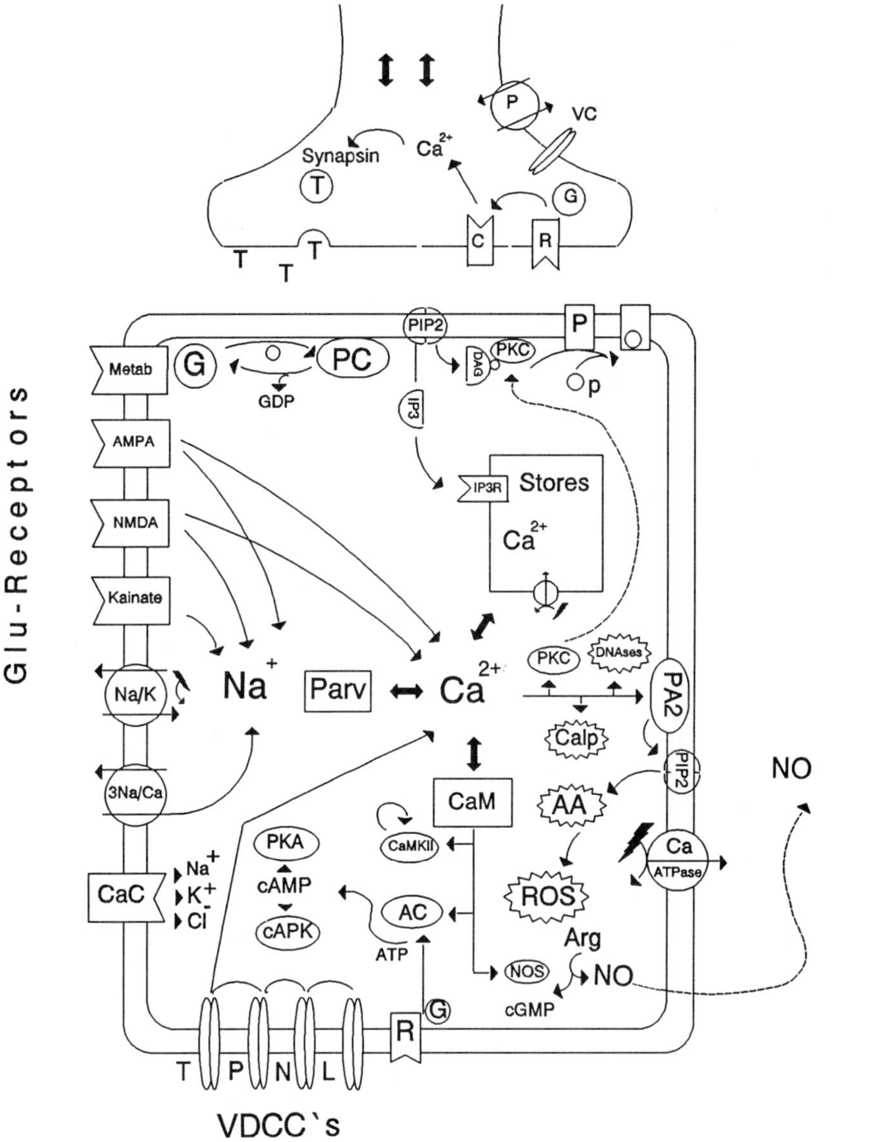

FIGURE 1. Important pathways of cellular metabolism are mediated by Ca^{2+}. Calcium is involved in ion-concentration homeostasis, various enzyme-activation steps, the release of excitatory neurotransmitters and catabolic cascades. **Abbreviations**: Glutamate-receptors: kainate, NMDA, AMPA, metabotropic. Voltage dependent Ca-channels = VDCC: T, P, N, L-types. CaC = Ca^{2+} sensitive ion channel; G = G-protein; PLC = phospholipase-C; PKC = protein kinase-C; Parv = parvalbumine; CaM = calmoduline; Calp = calpaine; AA = arachidonic acid; ROS = reactive oxygene species; CaMKII = Ca^{2+}-calmoduline dependent proteine kinase II; AC = adenylyl-cyclase; cAPK = cAMP dependent proteine kinase; PKA = proteine kinase-A; PA2 = phospholipase-A2; NOS = nitric oxide synthase; Arg = arginine. For details see text.

The NMDA receptor has a slower response kinetic and is blocked by Mg^{2+} ions unless the membrane is depolarized. A kainate sensitive glutamate-receptor (Sommer et al.1992a) is gating water and monovalent ion influx leading to depolarization. This can cause Ca^{2+} influx through VDCC's and activation of NMDA-receptors. There is recent evidence for direct Ca^{2+} influx through certain subtypes of kainate and AMPA-gated channels (Brorson et al.1992; Sommer et al. 1992b; Hollmann et al. 1991). Not only neurons but also glial cells express different glutamate receptors, exept the NMDA-sensitive receptor (Cull-Candy, Wyllie, 1991; Teichberg, 1991). Glia may therefore be implemented in the current view of synaptic currents.

Another type of Ca-signaling is mediated by inositol phospholipids. They play a key role in transmitting signals from cell surface receptors to intracellular compartments:

A tAPCD (trans-1-amino-cyclopentyl-1,3-dicarboxylate) sensitive metabotropic receptor (Baskys, 1992) is activating Ca^{2+}-dependent phospholipase C (PLC) through a G-protein mediated step. PLC can liberate inositol (1,4,5)trisphosphate (IP_3) and diacylglycerol (DAG) from phospholipid phosphatidyl 4,5-bisphosphate (PIP_2). DAG activates protein kinases (Nahorski, 1988) to phosphorylate an array of enzymes and IP_3 mobilizes Ca^{2+} from intracellular stores (Ross et al.1992; Hirano et al.1991; Kao et al.1989; Meyer et al.1988), a cascade which can phosphorylate Ca-channels (Meyer, 1989). A small Ca^{2+} influx may be amplified by additional release from intracellular stores, hence representing a means of intracellular signal amplification.

Another important function of $[Ca^{2+}]_i$ is the modulation of enzyme systems (Morgan, Curran, 1989), such as the Calmodulin dependent activities of Ca-calmodulin kinaseII (CaMKII), adenylyl-cyclase, and cytoplasmic NO-synthase. NO represents a new class of neurotransmitter, is easily diffusible to neighbouring cells (Izumi et al.1992) and may be involved in modifying presynaptic cell function through induction of cGMP dependent pathways.

This modification of presynaptic function may stimulate long term potentiation (LTP), a mechanism of learning and memory (Connor et al.1988; Collingridge, Bliss, 1987; Ben-Ari, Represa, 1990).

Possible direct interactions between $[H^+]_i$ and $[Ca^{2+}]_i$ are still obscure, although attempts have been made to monitor them simultaneously (Martinez-Zaguilan et al.1991; Hayashi et al.1992). An example of interactions between $[Ca^{2+}]_i$ and $[H^+]_i$ is the finding of improved neuronal survival in anoxic and acidic conditions because of a suppressed (NMDA receptor mediated) Ca^{2+}-current under low pH conditions (Tang et al.1991).

$[CA^{2+}]_I$ AND $[PH]_I$ DURING CEREBRAL ISCHEMIA

Cerebral ischemia is among the leading causes of death and adult disability in the industrialized countries. Cerebral ischemia manifests in severe changes of ion fluxes, which have been extensively reviewed (Zivin, Choi, 1991; Campbell, 1987; Choi, 1988; Meyer, 1989; Siesjö, Bengtsson, 1989; Siesjö, 1988).

The hypothesis that increased release of excitatory neurotransmitters is neurotoxic was initiated by early findings that glutamate causes neuronal loss (Lucas, Newhouse, 1957; Olney, 1969). Further support came from the finding that the extracellular Ca^{2+} concentration declines during ischemia (Nicholson et al.1977) as a consequence of increased influx into cells. The investigation of glutamate toxicity revealed the NMDA-receptor operated Ca^{2+}-channel (Miller et al.1990a; Michaels, Rothman, 1990; Cline, Tsien, 1991; Katayama et al.1991; Buchan et al.1991; Nayler,

Britnell, 1991; Marcoux, Probert, 1988; Miller et al.1990b) as the major site of direct Ca^{2+} influx.

Accordingly, the "excitotoxic hypothesis" (Olney, 1969; Choi, 1988) states that the failure of energy dependent pumps causes not only influx of monovalent ions followed by depolarisation and an immediate form of cell death through osmolysis, but also an elevation of $[Ca^{2+}]_i$.and sustained release of excitatory neurotransmitters. This initiates catabolic cascades which finally destroy the cell entity.

These pathophysiological events have been briefly classified in three stages: induction, amplification and expression (Choi, 1990):

Induction: lack of oxygen leads to failure of ATP dependent enzymes, like the Na^+/K^+-pump causing Na^+-influx with subsequent depolarisation. VDCC's may open and the large electrochemical gradient promotes Ca^{2+} influx (Benveniste, Diemer, 1988; Chen et al.1987; Kogure et al.1985; Benveniste, 1991), and release of excitatory amino acids (EAA's) like L-glutamate. This recruits additional neurons by ADCC's like NMDA receptor gated channels to rise their intracellular concentrations of Ca^{2+}, Na^+ and Cl^-. Subsequent H_2O influx and cell swelling preceede an immediate form of cell death through osmolysis.

Amplification: The increased Na^+ -concentration can reverse the $3Na^+/Ca^{2+}$ exchanger. Furthermore, an elevated $[Ca^{2+}]_i$ activates several enzyme families: *Phospholipase-A_2* mediates degradation of membrane phospholipids, thus generating arachidonic acid, which can be further metabolized to highly reactive *oxygen free radicals* (Pazdernik et al.1992; Pellegrini-Giampietro et al.1990).

A Quisqualate *metabotropic receptor* triggers *phospholipase c* mediated production of IP_3 and additional release of Ca^{2+} from internal stores (Ahmed et al.1990). Recently, some investigators however denied a neurotoxic effect of metabotropic receptor stimulation in cortical neurons and cerebellar granule cells (Koh et al.1991b; Rodriguez et al.1991; Koh et al.1991a).

Ca^{2+} -Calmodulin dependent enzymes like *CaMKII* (Aronowski et al.1992) and *NO-synthase* mediated pathways are activated (Frandsen et al.1992).

Mitochondrial Ca^{2+} accumulation uncouples oxydative phosphorylation and further increases the energy failure.

Expression: High $[Ca^{2+}]_i$ will further amplify the free radical production and activate highly proteolytic enzymes, like calpain I and endonucleases, inducing DNA fragmentation.

The rise of $[Ca^{2+}]_i$ may therefore be a "final common pathway" leading to the delayed form of ischemic cell death (at least in neurons) (Choi, 1988), as well as irreversible damage after hypoglycemia or status epilepticus (Meyer, 1989; Meldrum, 1985). The hypothesis is supported by the fact that the distribution of areas with low ischemic tolerance corresponds well with the one for the NMDA-type of L-glu-receptors, namely the hippocampal regions CA1 and CA4 (Crépel et al.1992), as well as neocortical neurons in layer 3,5 and 6. The fact that dentate granule cells tolerate ischemia (Peng et al.1991; Schmidt-Kastner et al.1990; Crain et al.1990) despite their NMDA-channels, demonstrates the potential involvement of other entry mechanisms like AMPA-channels (Siesjo et al.1991).

The impact of $[H^+]$-accumulation during cerebral ischemia is still contradictory. It is recognized that intra- and extracellular pH fall with the accumulation of lactate and failure of the Na^+/H^+-antiport driving Na^+/K^+-pump (Fujiwara et al.1992). However, several groups (Kraig et al.1985; Chesler, Kraig, 1987; Kraig et al.1983; Chesler, Kraig, 1989; Silver, Erecinska, 1990) reported short alkaline transients mainly in glia and proposed a compartimentalisation of $[H^+]$ between neurons and glia, while others reported homogeneous behavior (Katsura et al.1991).The initial alkalinisation can be explained by a H^+ consuming process of the creatine phoshate/creatine system.

Because of potentially pH dependent Ca^{2+} channel receptors, such mechanisms may influence the viability of brain cells under hypoxic conditions.

TECHNIQUES

To investigate ion concentrations with cellular resolution in the CNS, very different types of tissue preparations and measurement techniques have been applied.

Principally, three different model systems are used to study the physiology and pathophysiology of intracellular ion concentrations of living cells in the CNS. *Cell cultures* or *cell compartment preparations* can be used where intact synaptical connections between cells are not necessary. However, it is difficult to transpose the results gathered in cell cultures into the highly complex context of the intact organ, particularly where blood flow is involved. Due to the disaggregation and in vitro culture process, artefacts ensue. *Brain slice* methods show intact cell interrelations at least in a tissue fragment and easy access to any anatomical region of interest, but this preparation represents a subsystem under in vitro conditions (Dingledyne, 1984; Schurr et al.1991)). The *in vivo and in situ* preparation offers access to the intact organ, although it depends on the approach whether the sourrounding physiological entity is more or less unaffected (Uematsu et al.1988b; Uematsu et al.1988a). Complex physiological interactions involving blood-flow can only be examined in vivo and in situ.

Table 1. Intracellular ion concentrations measurement techniques for the CNS.

Method	penetration	temporal resolution	spatial resolution	characteristics
Ion selective electrodes	++	+/-	++	single cells, invasive
NMR-spectroscopy	++	-	-	large volume
Flow-cytometry	-	++	-	statistics of cell populations
Fluorescence microscopy	-	++	-	
CLSM	+/-	++	++	

++ = high; - = low. CLSM: confocal laser scanning microscopy

Ion selective micro- and patch clamp-electrodes are powerful tools with high temporal resolution (Alford, Collingridge, 1990; Alford, Collingridge, 1991; Hansen, 1985), but are principally restricted to the investigation of single cells or the status in parts of it at any one time. Furthermore, these approaches interfere with the plasma-membrane. Nuclear magnetic resonance spectroscopy (NMRS) (Pirttilä, Kauppinen, 1992; Rijen van et al.1991; O'Shhaughnessy et al.1991; Knight et al.1991) offers poor temporal resolution in relation to the dynamics of ion-concentration changes and does not reach cellular resolution, but offers excellent penetration-depth. Flow cytometry (Sei, Arora, 1991; Rijkers et al.1990; Vandenberghe, Ceuppens, 1990) allows use of ion binding fluorescent dyes and statistics of large cell populations but no spatial resolution.

Optical methods with cellular resolution

Traditional fluorescence microscopy used in vitro or in vivo has been used to measure ion concentrations with cellular resolution (Spring, 1991a; Spring, 1991b). Digital image processing and deconvolution techniques (Trump, Berezesky, 1990) can improve the results to some extend. Confocal laser scanning microscopy (CLSM) (Villringer et al.1991; Takamatsu et al.1991; McNaughton et al.1990; Hernandez-Cruz et al.1990) leads to a further increase in resolution. CLSM provides better penetration into the tissue and very high spatial resolution. Depending on the detection system temporal resolution is somewhat lower than in conventional fluorescence microscopy, which offers real time imaging. The introduction of CLSM into neurobiology has helped to overcome many difficulties in microscopical visualisation within thick objects and particularly in CNS tissues (Fine et al.1988).

CLSM is characterized by increased spatial resolution through the use of a point light source, illumination of a single point through a pinhole and exclusion of out of focus information by a second pinhole. A potent illumination source is needed because of the high proportion of light rejected by the pinholes. Advances in laser technologies provided a major breakthrough in the development of confocal microscopy. Now arc-lamps can also reach comparable intensities (Hell et al.1991). The principle of this promising and relatively new technique has been reviewed in great detail elsewhere (Stelzer et al.1991; Dixon et al.1991; Fine et al.1988; Brakenhoff et al.1989).

In combination with fluorescent dyes, CLSM offers the possibilty to localize substances with high resolution in ex vivo stained preparations (Vincent et al.1991). Monitoring of free ion concentrations has been performed at every level of preparation from subcellular compartments (Smith, 1990), cell culture (Jensen, Chui, 1990; Jensen, Chiu, 1991; Daniell, 1991; Cline, Tsien, 1991; Connor, 1986) and brain slices (Yuste et al.1992; Alford et al.1991; Alford, Collingridge, 1991; Dingledyne, 1984; Schurr et al.1991; MacVicar, Tse, 1989; Tank et al.1991), to the in vivo imaging approach presented here. Combinations of electrophysiological and fluorescence measurements have been used in voltage clamped neurons (Barish, 1991).

Fluorescent dyes

Because fluorescent probes interfere chemically with the system under investigation, it is important to evaluate the chemical properties of the fluorescent dyes and the concentrations involved in the experimental setup.

In contrast to chemiluminescence (Hashimoto et al.1992), fluorescence provides high photon-quantum yield and reexcitability. Fluorescence offers high spatial and temporal resolution in the millisecond domain. Fluorescent dyes have been extensively reviewed (Tsien, 1989a; Tsien, 1988; Moore et al.1990; Tsien, 1989b). The chemically engineered dye-molecule changes its fluorescence upon binding the target molecule. Ideally, the dyes binding site is highly specific for the ion under study and its dissociation constant (K_D) is in the middle of the ion concentration range. Indicator molecules have been developed for detecting the ion-concentrations of sodium, calcium, chloride, hydrogen, or magnesium. For a comprehensive list of the latest commercially available dyes, see (Haugland, 1992).

Because of their polar structure, dyes have been loaded into cells by microinjection or by temporarily permeabilizing the membrane (Cohan et al.1987). Tsien et al. introduced the elegant approach of labile ester masking of dyes, which makes them membrane permeant (Tsien, 1989b). Subsequent hydrolysis by cytosolic esterase activity releases the active lipophobe dye and traps it in the cell. With these dyes, it has become possible to load very different types of cells (Read et al.1992) just by incubating them with the AM-ester, which itself is not responding to free Ca^{2+} until cleaved intracellularly. Ester-masking is the most commonly used method for dye laoding in animal cells.

There are three possible types of fluorescence responses of dyes upon binding the appropriate ion: the "single wavelength" (1.) response shows decrease or increase of fluorescence across the whole emission spectrum. Besides the measured ion concentration itself, the fluorescence intensity depends on the absolute amount of dye and its detectability at exactly the same position in the specimen. This can be affected by factors such as leakage and photobleaching, as well as motion artifacts and non uniform distribution in the cell. Therefore quantification is more reliable with so called "ratioable" or "dual wavelength"-dyes. These types of dyes exhibit a shift either in their emission (2.) or excitation (3.) spectrum upon binding the specific ion. The ratio between the two wavelengths provides the association equilibrium, which is directly proportional to the free dye concentration, and thereby circumvents many quantification problems. The combined use of a sensitive and an unsensitive dye for the ion under investigation and monitoring of both dyes fluorescence, can provide a similar approach, supposing equal distribution of both dyes in the cell (Rijkers et al.1990). Another criterion for the suitability of a dye is the magnitude of its Stoke shift, which is the difference between the excitation and emission wavelength. Large Stokes' shifts facilitate the separation of the excitation light from the emitted fluorescence with optical filters. The development of dyes, excitable in the visible range of the spectrum, circumvents possible damage by UV irradiation. Furthermore, visible emission wavelengths have the advantage to be less affected by intrinsic background fluorescence which is high in the UV-range, while the simultaneous use of UV-activated caged compounds to alter ion concentration noninvasively is made possible (Adams et al.1989; Peres et al.1991).

In a living cell population, a large number of factors, such as solubility, selectivity, hydrolysis, photodecomposition, quantum yield, fluorescence intensity, retention, sequestration, etc. influence the loading and imaging characteristics of a dye. Furthermore, there is an increasing number of newly developed dyes (Haugland, 1992; Hodgkiss et al.1991). Therefore we wanted to establish a system which allows the characterisation of different fluorescent probes and the selection of the most suitable dye for use in brain tissue. As a model, we implemented the acute rat brain slice in confocal fluorescence microscopy (Them et al.1992) to select dyes for the intact brain preparation. Different stimuli were applied in both preparations to evoke ion concentration shifts and to check whether they can be monitored optically.

EXPERIMENTS

Material

The confocal imaging system used in this study was a Biorad MRC-600 CLSM system (Biorad Microscience Division, Watford, UK), attached to a Nikon Optiphot microscope (Nikon Optics, Japan). We used a x40 water immersion objective (Zeiss, Oberkochen, Germany) with a numerical aperture of 0.75 providing a working distance of 1.6 mm, corrected for a cover slip. The laser source was a Krypton Laser with principal lines at 488 nm (blue), 568 nm (yellow) 647 nm (red) and an Argon Laser with 488 nm and 514 nm (green) respectively (Ion Laser Technologies). The laser power is 4.8 mW per line at the source, reduced by neutral density filters and energy loss along the beam path (mirrors, filters, lenses, cover slip e.g.). The standard neutral density filter used was the ND 3 filter (1% transmission). For a standard setup the laser power per pixel reaching the specimen needed to produce a sufficient signal to noise ratio has been measured to be approximately 0.1 mW (Brad Amos, pers.comm.). For excitation and emission imaging of fluorochromes, appropriate filter sets (Bio Rad Microscience Division, UK) were used. In the case of double emission labelling, the emitted fluorescence can be splitted by a second dichroic mirror to a second photomultiplier, allowing two channel imaging of the fluorescent signal. An increase in temporal resolution is achieved by fixing the laser beam at a single line instead of scanning the whole xy-image. Thereby resolution can be increased to 2ms, but spatial resolution is reduced to a single line.

All dyes were obtained from Molecular Probes, Eugene/OR, USA. We only used the membrane permeable acetoxymethyl (AM-) esters. Different concentrations of dimethylsulfoxid (DMSO, Sigma Chemicals, USA) and the non-ionic detergent Pluronic F-127 (Pluronic, Hoechst, USA) altered the intensity and speed of cell loading. We found the following optimal loading procedure. The dye is dissolved in dry DMSO, added up to 25 v/w % of Pluronic for better dispersion and loading into brain cells (Haugland, 1992). After stirring at 40 °C for about 30 min, artificial cerbrospinal fluid (ACSF, composition see below) is added up to a final concentration of 0.1 - 0.05 v/v % DMSO. The dyes in solution are unstable to hydrolysis in aqueous solutions and have to be prepared freshly for each experiment.

Table 2. The following fluoroprobes were used in concentrations from 1 - 50 μM. ex: excitation, em: emission

Dye	AM-ester	Affinity	ex [nm]	em [nm]	ratio-able	fluor-escence upon binding	references
BCECF	y	$[H^+]$	490	525	n	↓	Haugland, 1992
Ca-Crimson	y	$[Ca^{2+}]$	589	607	n	↑	Haugland, 1992
Fura-Red	y	$[Ca^{2+}]$	488	660	+Fluo-3	↓	Eberhard, Erne, 1991
Fluo-3	y	$[Ca^{2+}]$	490	525	n	↑	Minta et al.1989
Ca-Green	y	$[Ca^{2+}]$	507	530	n	↑	Eberhard, Erne, 1991

Slice preparation

The *small volume slice imaging chamber* for submerged rat brain slices is closed by a coverslip to allow optical imaging with the CLSM (Them et al.1992). Rapid media exchange is provided by switching the inflow tubes of the < 0.5 ml volume chamber. For pharmacological studies, different compounds can be added to the ACSF (T: 22 °C; O2: >550mmHg; concentrations [mM]: Na^+:150; K^+: 5; Ca^{++}: 2; Mg^{++}: 2; Cl^-: 133; $HCO3^-$: 26; Glc: 10; $H2PO4^-$: 1,25). For oxygenation and hypoxia we used carbogen gas (95% O_2, 5% CO_2) or N_2, providing oxygen tensions of >650 and <50 pO_2 [mmHg], respectively.

Wistar rats (Savo, Germany) with 1 to 30 days of age were anesthetized with halothane and decapitated. The brain was removed rapidly in cold (3 °C), oxygenated

ACSF and the neocortex and hippocampus were tangentialy sliced (500μm) using a vibratome. The procedure lasted 5 min approx., from decapitation to placement of the slice in the imaging chamber, while slices were continously covered by the ACSF.

Loading

The slices were incubated in the oxygenated dye solution (1 to 50 μM) for 5 to 180 min and transferred into the imaging chamber. Using fixed laser and photomultiplier settings, the fluorescence signal was compared between various loading times, concentrations and different dyes in order to optimize loading characteristics for best signal and stimulation intensity. The fluorescence increase was determined relatively to the respective baseline.

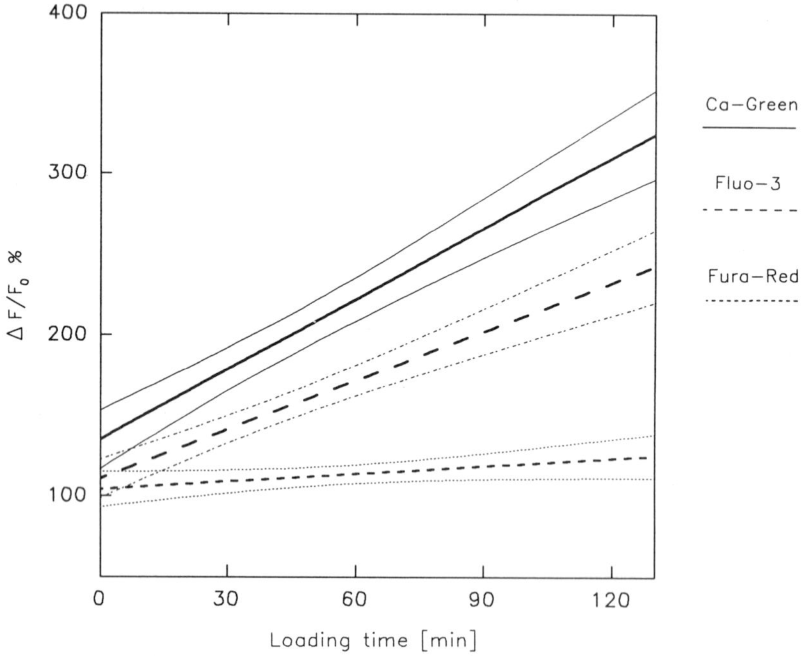

FIGURE 2. Fluorescence increase against time for three $[Ca^{2+}]_i$-sensitive dyes at 1μM concentration. Video noise subtracted. Signal expressed as percentage of increase from baseline (100%). First order regression with 95% confidence intervalls. Note that Fura-Red showed a much slower increase than Fluo-3, while Ca-green showed the fastest increase.

We compared the *fluorescent signal* after loading the slices with different Ca^{2+} sensitive dyes. All dyes showed a relative increase in fluorescence with loading time (Fig.2). The Ca^{2+} sensitive dyes Fluo-3 and Ca-Green showed the strongest signal increase, whereas Ca-Crimson and Fura-Red showed much slower increases and did not reach comparable brightness. With the strong fluorescence provided by Fluo-3 and Ca-Green, we were able to image single cells down to > 50μm below the surface. The increase of fluorescence was linear during the loading periods, indicating no saturation of the intracellularly cleaved dye. The slope of fluorescence-increase was steeper at high dye concentrations, resulting in higher signal intensity at shorter loading intervals (Fig.3).

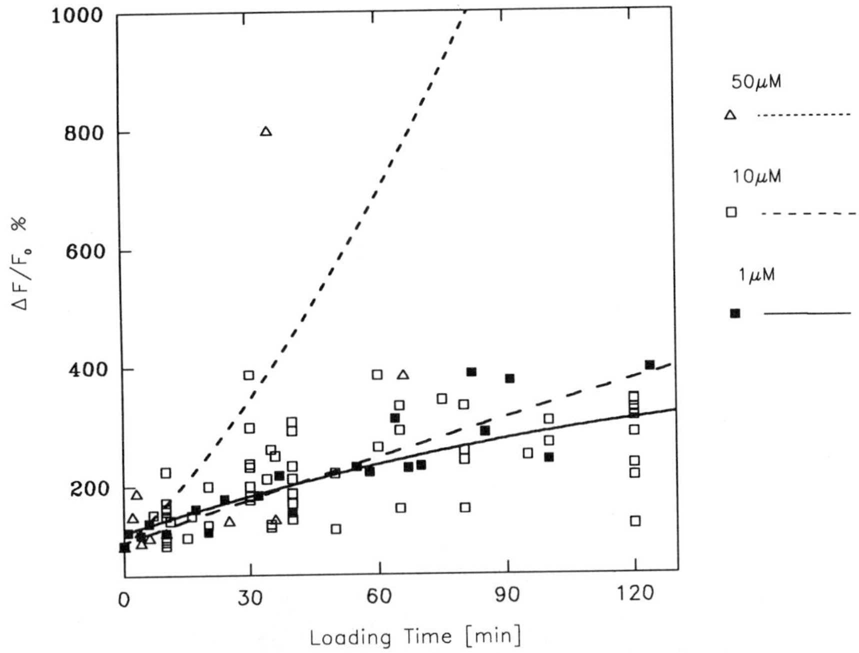

FIGURE 3. Fluorescence increase of different Calcium-Green concentrations against loading time. Video noise subtracted. Signal expressed as percentage of increase from baseline (100%). Second order regression. Note that the 50μM solution produced a very steep increase, whereas the difference between 1 and 10μM was minor.

With increasing *age* of the rats (approx. > 7 days), cell labelling became more and more sporadic, while neonatal to approx. 4 day old preparations showed a dense, homogeneous packing of labelled cell bodies throughout the cortex. This property was seen with variations of temperature from 22° to 38° C, different concentrations of pluronic detergent from 0,1 - 0,005%, different concentrations of DMSO (0,1 - 0,003%), and various loading times and dye concentrations. Very low dye concentrations (1μM) in the superfusing medium loaded readily into the cells, providing good signal to noise ratio and excitability after loading times of 15 min to 2 hours.

Stimulation

Induction of ion concentration changes was used to test whether a stimulus evoked alterations can be detected with the fluoroprobes and served as criterion for the viability of the slices. Slices which did not respond to the excitatory stimuli, were discarded. Repetitive stimulation of $[Ca^{2+}]_i$-rises were performed with either hypoxia, EAA, or K^+. Use of such stimuli was performed by switching the superfusion to L-glu (500μM), K^+ (50mM), or low O_2 pressure (< 50 mmHg) containing ACSF. The signal intensity was imaged before, during, and after repetitive stimulation at different time intervalls.

Weakly stained neurons never responded to stimulation and slices from older rats had a lower number of stained cells. However, the labelled cells did react similar to the ones in the young slices.

Different *excitatory stimuli* caused an increase (with Fura-Red decrease) in fluorescence signal. In every case, a subsequent recovery of signal intensity was taken as proof for reversibility of the stimulus triggered Ca^{2+}-influx and to exclude artifacts due to Ca^{2+} overload (overstimulation). Furthermore, in the neonatal preparation spontaneous transients in $[Ca^{2+}]_i$ were observed. No significant difference in the amount of loaded cells was found between cortical and hippocampal layers. The responses observed upon stimulation were nonhomogeneous between different cells. Differences in duration and intensity of stimulation were monitored. Successful stimulations were evoked even after several hours in the observation chamber.

Superfusion with 500μM L-glu containing ACSF caused a significant change in fluorescence intensity, indicating a rapid elevation of $[Ca^{2+}]_i$. Subsequent outwash decreased signal intensity slowly. Fig.(4) shows examples of the stimulation time-course. Images were taken before, during, and after end of stimulation.

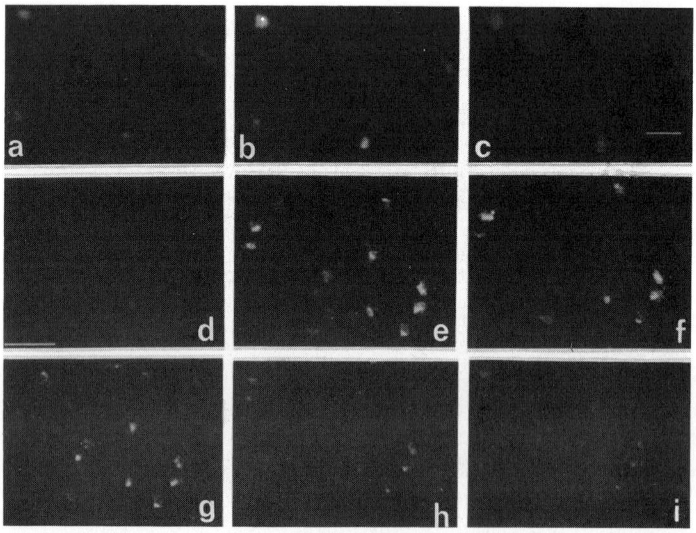

FIGURE 4. Stimulation with L-glutamate in brain slice and in situ preparation after loading with 1μM Ca-green for 30 min (slice) and 120min (in situ). a).- c): Before (a), during (b) and after (c) 30 sec stimulation with 500 μM L-glutamate in the brain slice preparation. Scalebar is 25μm. d). - i): Before (d), during (1 (e);2 (f) and 3 (g) min) and after (3 (h) and 5 (i) min) superfusion of 100μM L-glutamate containing ACSF in the in situ preparation. Scalebar is 40μm.

We imaged fluorescence increases of up to 400% upon L-glu (500μM)-stimulation of $[Ca^{2+}]_i$ in the acute slices. Immediately after switching the superfusion to hypoxic ACSF, a change in the fluorescence signal was obtained. Many repetitive stimulations were possible and signal returned to baseline upon resuperfusion with oxygenated ACSF. As with the peak response, the time course of the return to baseline was inhomogeneous between different cells.

Imaging of *dose response* patterns was performed by different duration of hypoxic ACSF superfusion (Fig. 5).

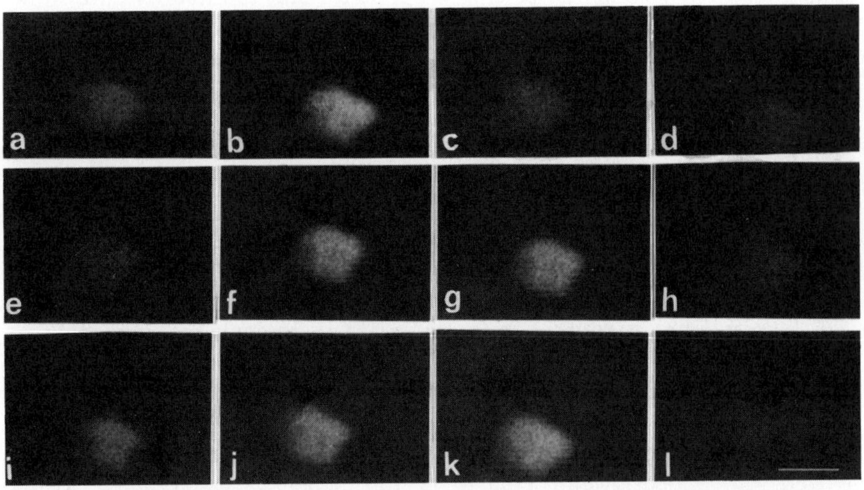

FIGURE 5. Hypoxic dose response in the acute brain slice. Loading with 1μM Ca-green for 15 min. **a) - d):** Before hypoxia (a), 5 sec after onset of a 15 sec period of hypoxic ACSF superfusion (b), 15 sec post hypoxia (c) and 60 sec post hypoxia (d). **e) - h):** Same procedure as in a-d). but with a hypoxic period of 30 sec. Before (e), 5 sec after onset of a 30 sec period of hypoxic ACSF superfusion (f), 15 sec post hypoxia (g) and 60 sec post hypoxia (h). **i) - l):** Same procedure as in a-d), but with a hypoxic period of 60 sec. Before (i), 5 sec after onset of a 30 sec period of hypoxic ACSF superfusion (j), 15 sec post hypoxia (k) and 60 sec post hypoxia (l).

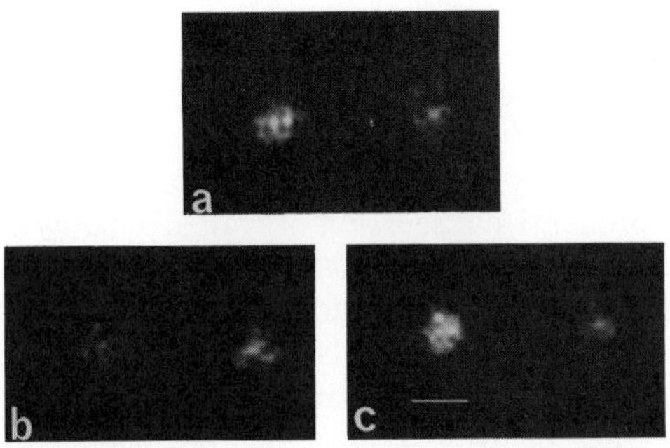

FIGURE 6. Double loading with the $[Ca^{2+}]_i$ sensitive dye Fura-Red and $[H^+]_i$ sensitive dye BCECF, both 1μM. Since the two dyes are excited with the same wavelength (488 nm) and emit at different wavelenths simultaneous imaging is feasable. The emission wavelength are separated with filters and dichroic mirrors and detected on two separate photomultipliers. Left image: red light emitted from Fura-Red, right image green light emitted from BCECF (pseudocoloring). a) Fluorescence before stimulation. Left (Fura-Red) image with stronger fluorescence because of 1 hour loading time, right image shows weaker fluorescence because of shorter loading time (5 min). b) During Hypoxia of one minute. Left image decreases fluorescence, indicating rising $[Ca^{2+}]_i$, whereas the right one shows an increase in fluorescence, indicating an alcaline shift. c) 120 sec post Hypoxia. Ca^{2+} and pH dependent fluorescence has returned to baseline. Scalebar = 10 μm.

In one experiment, we *double-loaded* the pH and Ca^{2+} sensitive dyes BCECF and Fura-Red. Although the BCECF derived signal is much stronger, a signal separation was achieved by dichroic mirrors and image aquisition from two photomultipliers. The channel for $[Ca^{2+}]_i$-dependent Fura-Red fluorescence showed a decrease in fluorescence, while the channel for $[H^+]_i$-dependent BCECF signal from exactly the same area showed an increase in fluorescence, thus indicating an alcaline shift during the first minute of hypoxia, while $[Ca^{2+}]_i$ is still elevated (Fig. 6).

Since the cell membrane is slightly permeable to K^+, elevated $[K^+]_e$ also serves as a depolarizing stimulus, triggering elevated $[Ca^{2+}]_i$. This was imaged in a preliminary experiment with 50μM K^+.

In an attempt to block the rise of $[Ca^{2+}]_i$ upon stimulation, we added different concentrations of the selective glutamate receptor antagonists MK-801 (NMDA-antagonist), BNQX (AMPA-antagonist) and the dihydropyridine antagonist Nimodipine to the superfusing ACSF. No change in the response to hypoxic stimulation was observed. This preliminary result is an example for pharmacological protocolls.

IN SITU PREPARATION

The *intact brain preparation* used male Wistar rats (250 gr., Savo, Germany), anesthetized with thiobarbiturate (Inactin™, 100 mg/kg/bw, Byk Gulden, Konstanz, Germany) and lidocain. The surgical procedure consisted of tracheotomy for artificial ventilation (Harvard App, USA) and continuous endexpiratory CO_2-monitoring (Heyer-EGM-I, Bad Ems, Germany); cannulation of the femoral artery and vein to allow serial monitoring of arterial blood pressure (SAP), pO_2, pCO_2 and hematocrit at the beginning and the end of each experiment, and finally implantation of a closed cranial window over the right parietal cortex (modified after Morii et al. 1986). All physiological parameters were controlled in the normal range. The brain was superfused with ACSF (CO_2: 35 mmHg; concentrations [mM]: Na^+:150; K^+: 3; Ca^{++}: 1.5; Mg^{++}: 1; Cl^-: 150; $HCO3^-$: 26; Glc: 3). Microsurgical removement of the dura mater was performed avoiding any bleeding; even minute erythrocyte - brain parenchym contact led to exclusion from the protocoll. Our experience in former studies (Dirnagl et al.1992) indicates that various markers for the viability of the preparation (blood brain barrier, responsiveness of pial vasculature) indicate massive alterations after brain-erythrocyte contact. Intracranial pressure (ICP) was controlled at 2-5 mmHg. For comprehensive discussion of cranial window approaches see Haberl in this issue. Continuous recording of relevant physiological parameters (temperature, ICP, SAP) was performed after a/d conversion of the signal on a microcomputer (IBM-PC-AT; Asyst, MacMillan Software, NYC/NY, USA). The animal was restrained in a stereotactic frame, the body temperature was measured by a rectal temperature probe, and the body temperature controlled by a heating pad. The cranial window was positioned under the microscope-objective, which was corrected for water immersion. The water immersion acts as a thermodynamic connection to the microscope. To avoid any alteration of the brain temperature, we therefore heated the lens constantly with an auto-heated waterflow to control the temperature of the lens immersion fluid at 37°C. A software controlled x-y-z-stepper (BioRad) allowed exact positioning, z-scanning and memorizing of regions of interest. Furthermore imaging was triggered to respiration to reduce motion artifacts.

Use of the dyes in situ

BCECF, the hydrogen ion sensitive dye was added to the ACSF (10μM) and superfused for 60 min loading time, followed by 30 min of washout. Signal was retrieved from the outer 100-200μm of the cortex with a cellular resolution down to >50μm below the surface of the brain.

To determine the rate of bleaching, we exposed a certain brain area loaded with BCECF to continuous laser illumination. Imaging up to 2000 frames (x1 frame/sec) caused less than 20% signal loss. Because the rate of signal loss was low, continous monitoring of signal seemed possible for at least several minutes. Although BCECF showed the strongest signal intensity, similar signal stability was observed with all Ca^{2+}-indicators used in this study (Dirnagl et al.1991).

Leakage was assessed by intermittent sampling of the outwash-ACSF and fluorometrical determination of dye content. The results indicate a good retention inside the cells. We calculated less than 10% fluorescence loss per hour. Another indication for good retention of fluorescent free dye inside the cell was the low background fluorescence.

The effect of BCECF on cerebral blood flow (CBF) was studied with laser doppler flowmetry (LDF Laserflo 403 A, Vasamedics, St Paul/MN, USA). CBF showed less than 10 % change upon start of superfusion with subsequent return to baseline during outwash.

In an attempt to load Ca^{2+} sensitive dyes into the intact brain, we superfused Ca-Crimson (10μM), Fura-Red (10μM) or Fluo-3 (10μM) for 30min or 60min; and Ca-Green (1μM; 2,5μM) for 2 hours, followed by a 30 min outwash time.

After loading, outwash and subtraction of background fluorescence, signal was retrieved from the outer 100-200μm of the cortex. Individual loaded cells were resolved down to a depth of approx. 50 μm.

To induce ion concentration changes, we used spreading depression (SD), EAA's and global ischemia:

Spreading Depression

SD is a slowly propagating cessation of electrical activity, accompanied by a large negative DC-potential in the extracellular space and an increase in $[Ca^{2+}]_i$ (Nedergaard, Hansen, 1988; Scheller et al.1990; Lauritzen, Hansen, 1992). This event triggers a phasic rise in CBF and can be elicited by various chemical or mechanical stimuli. Monitoring of propagating waves of SD was performed by LDF-measurement of regional CBF through the imaging window in an area including the cells under study.

SD was induced by application of 50 μM K^+ in a second small craniotomy 5 mm apart from the imaging window. An example for such a $[Ca^{2+}]_i$-stimulation by induction of SD is given by Villringer et al. in this issue.

As a consequence of the superior imaging characteristics of Ca-Green (1μM) in our slice experiments (see above), we loaded Ca-Green in the same concentration for two hours in the intact cortex. Stimulation of Ca^{2+}-inflow was then induced by superfusing 100μM L-glu. The signal rised immediately upon arrival of L-glu, and dropped after washout of the EAA within minutes (Fig. 4 d-i).

Ischemia

Global cerebral ischemia was induced by temporary occlusion of both common carotid arteries. SAP was lowered by applying lower body negative pressure with a suction device, as described elsewhere (Dirnagl et al.1993). Verification of cortical cerebral blood flow (CBF) reduction was performed with LDF, allowing continuous monitoring and recording of CBF before, during, and after ischemia at hypo- and hyperperfusion. The pH sensitive dye BCECF was loaded as in all other experiments and after one hour of post ischemic hyperperfusion, calibration of the signal was performed by superfusion of nigericin (10μM), valinomycin (10μM), 130mM $K^+$and

30mM buffer titrated to pH 6 (MES) or 8 (TRIS) respectively, in order to clamp intracellular to extracellular pH.

Upon induction of ischemia, the BCECF derived signal dropped immediately, indicating a decrease in pH. After 1 hour of hyperperfusion, pH-calibration by the nigericin-valinomycin-high potassium method was performed at pH=6 and pH=8. The estimated pH values from the calibration (corrected for leakage) were 6.9±0.2 for the normal brain and 5.9±0.5 for the ischemic period (Fig. 7).

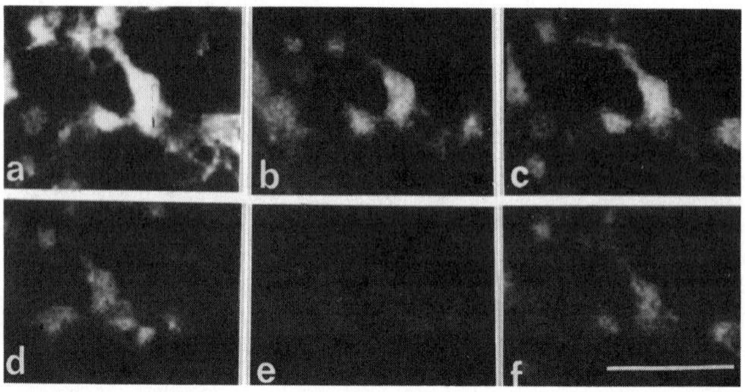

FIGURE 7. In vivo imaging of $[H^+]_i$-concentration in the intact rat neocortex. Brain cells loaded for 30 min with the pH sensitive dye BCECF and imaged through a closed cranial window preparation after 30 min outwash period. a) Signal before ischemia. b) During 5 min of ischemia induced by bilateral common carotid occlusion and hypobaric hypotension. Drop of signal intensity indicating acidification. c) 1 min after onset of reperfusion. Immediate recovery of fluorescence. d) 30 min after onset of reperfusion. Drop of signal indicating leakage or increasing acidity. e) Calibration with clamping at pH=6. f) Subsequent calibration with clamping at pH=8. Scalebar= 50μm.

DISCUSSION

This study demonstrates the feasibility of imaging intracellular ion concentrations and their alterations in acute rat brain slices and in the intact brain with confocal laser scanning fluorescence microscopy. Various dyes were successfully loaded into nervous tissue and signal changes were obtained upon stimulation with physiological and pathophysiological stimuli.

Use of fluoroprobes

The application of fluorescence CLSM with its improved imaging capability allows deeper penetration into the object and higher spatial resolution if compared to conventional fluorescence microscopy (Murray, 1992). Temporal resolution was 1/16 sec - >1 sec, depending on the scanned areas size. The fact that photobleaching did not reduce fluorescence intensity significantly in most cases demonstrates that relatively long imaging times are feasible. With a scan rate of 1 image/sec several minutes of observation time are possible. The AM-esters loaded in various cell types, including neural tissue. For a comprehensive list of the increasing number of publications, see (Haugland, 1992). The homogeneous loading of almost all cells in young slices confirmed the potentially good loading capability.

Problems

The use of fluorescent dyes may have several intrinsic restrictions (Tsien, 1989b): low fluorescence yield; fluorescence quenching upon conjugation; photobleaching and phototoxicity; spectral overlap (Carlsson, Mossberg, 1992); incomplete deesterfication (Scanlon et al.1987); small Stoke shifts and low solubility in water. Interaction with the living cell, e.g. leakage, binding, buffering and compartimentalisation or sequestration properties have to be taken into account (Williams et al.1991; Williams et al.1987; Przywara et al.1991; Tsien, 1988). The principle of binding an ion with high selectivity is inherently coupled with the problem of buffering the ions' concentration, particularly if excessive amounts of ligand are introduced. Therefore the lowest possible dye concentration has to be used. If relevant cell functions are normal, it is reasonable to assume that buffering has not been significant (Tsien, 1988).

We used only dyes with a excitation wavelength greater or equal 488 nm. Development of fluoroprobes for the visible spectral range overcomes the problems of the photodamage and autofluorescence of tissue and equipment encountered in the UV spectral range (Grynkiewicz et al.1985).

Signal calibration

In most instances, a qualitative view of the temporal and spatial dynamics of $[Ca^{2+}]_i$-transients is the first concern. In such an approach, many problems intrinsic to exact quantification are avoided.

If a quantitative measurement of the ion concentration is desired, ratioable dyes would lead to best results, because simultaneous detection of two wavelength (concentration-dependent/concentration-independent) reduces errors due to heterogeneity in fluorescence. However, there are many parameters to be evaluated before a calibration can be achieved. Examples are the pH dependent binding constants of Ca-Green, Ca-Crimson (Eberhard, Erne, 1991) and Fluo-3 (Eberhard, Erne, 1989). They are reported to be small enough to allow independent use for Ca^{2+}-imaging in the physiological pH range, but differ at more extreme values. In addition, the dye properties may be affected by pH due to Ca^{2+}/Mg^{2+} selectivity upon acidification as reported for indo-1 and fura-2 (Lattanzio, Bartschat, 1991). CLSM minimizes the problem of out of focus blur, although unequal disribution of indicator in the cell may still affect the image intensity (Moore et al.1990). We did not detect unequal signal distribution, thus we propose the signals to be of cytosolic origin.

Since the dyes in this study can not be used in our microscope with ratioing of emission or excitation, correction for leakage and photobleaching as well as clamping to defined concentrations at the end of an experiment is necessary for calibration.

To clamp the intra- to the extracellular concentration of the ion to be calculated, one uses membrane permeabilizing ionophores like A23187 or ionomycin (Liu, Herman, 1978) for Ca^{2+} and valinomycin/ nigericin/ high potassium for pH clamping respectively. This can be done at the end of each experiment. The new dye Fura-Red offers some interesting possibilities for ratioable double labelling in combination with BCECF or Fluo-3. Since Fluo-3 increases and Fura-Red decreases fluorescence upon ion binding, the ratio should reflect absolute $[Ca^{2+}]_i$-changes. The combination with BCECF may allow simultaneous $[Ca^{2+}]_i$ and pH excitation ratio measurements (for all details and spectra, see the Bioprobes series from Molecular Probes, Eugene, OR, USA). Because there are many pitfalls in the calibration procedure, current $[Ca^{2+}]_i$ calibrations should be regarded as approximate rather than absolute.

Loading and imaging

All dye-esters, BCECF, Fluo-3, Ca-Crimson, Ca-Green and Fura-Red were loaded into the brain cells in vitro and in vivo. The weaker signal from Fura-Red may be explained by lower photon yield or slower loading, resulting in lower intracellular fluorescence. In contrast to earlier studies, we reduced the concentration of dyes to 1μM, even in the intact brain preparation.

During the loading time of 30 min to 2 hours, no significant saturation of dye uptake was seen even at high concentrations in the slice, suggesting that the intracellular dye concentrations did not buffer $Ca^{2+}{}_i$ significantly.

The loading time in the intact brain appeared longer than in vitro. This is not surprising, considering the buffering capacity of the intact extracellular space by constant supply of cerebrospinal fluid (CSF). Use of the detergent Pluronic F-127 was reported to enhance loading, while having no adverse effects as shown in lymphocytes (Vandenberghe, Ceuppens, 1990). In our series, when Pluronic was used, loading times were shorter, although no statistical significance was achieved. Beneath the brain surface, scattering, absorption, self quenching and shadowing of the dyes decrease the image quality. With increasing depth of penetration there is an increase in spherical aberration due to the mismatch between the refractive indices of tissue and immersion medium (Benson, Knopp, 1984). Long distance objectives are needed for microscopical imaging in vivo and in the sclice chamber. Such objectives have rather low numerical apertures (usually < 0.7). Since the thickness of the focal plane is directly related to the size of the numerical aperture, such lenses affect the confocal effect. New lenses with higher numerical apertures and use of long wavelength excitation dyes will improve penetration depth.

Some reports (Tsien, 1989b) claim that high temperatures in cell culture are responsible for sequestration processes, and report improved cytosolic stability of the dye by working at room temperature. We did not see compartimentalisation in the 22°C slices, nor the 37°C ones, while slices treated unphysiologically showed an increasing number of membrane bound fluorescence.

We can only speculate, why young animal-slices and the intact preparation showed better labelling than the slices from adult rats. A newborn rat is faster to dissect and the brain cooling is faster in the critical slicing phase; additionaly, there is an increased ability of young animals to resist hypoglycemia (Crépel et al.1992). This could have a synergistic effect, providing better viability in the loading procedure. In addition, the developing brain shows patterns of enzyme activity, stimulus sensitivity, and viability much different from the adult brain (Wahl et al.1989; Balduini et al.1991), possibly facilitating the uptake or cleaving of AM-ester-dyes.

Injury and toxicity

Since light in the presence of fluorescent dyes generates oxygen free radicals, the issue of "light and dye injury" has to be considered (Rosenblum, El-Sabban, 1977). In a previous study the parameters of our approach were compared with reports of the "light and dye model" (Chertok et al.1989; Herrmann, 1983; Miller et al.1992). In contrast to the mercury lamp of the light and dye model, where light damage is purposly induced to functionally destroy vascular endothelium, we use laser illumination with a smaller intensity/time integral. Only a 10-100 fold increase in intensity produced an alteration of capillary blood flow (Dirnagl et al.1991) in our setup.

In cell culture systems there are anecdotal reports of AM-ester dye toxicity. In isolated hepatocytes treated with 2,5μM Quin-2 (Carpenter-Deyo et al.1991) had toxic effects, whereas the same dye at 50 μM protected cultured motoneurons against glutamate induced toxicity, possibly because of Ca^{2+} buffering (Kudo et al.1990).

Because of the relative novelty of most of the dyes used here, reports of toxicity are not available at the moment.

Slice preparation viability

Some groups of loaded cells reacted homogeneously to a stimulus while others did not. This might be due to altered cell function due to the acute slice preparation process. It is well known that in vivo slices contain dead cells, especially on the surface. On the other hand physiologically altered cells may accumulate Ca^{2+} prior to cell death, therefore demonstrating high fluorescence when loaded with Ca^{2+}-sensitive dyes. Such highly fluorescent cells were observed only sporadically. In purposely altered slices the amount of highly fluorescent cells which did not react to stimulation was higher than in normal slices. When brains were kept in situ after decapitation for different amounts of time, as described by Leonard et al. (1991), the amount of highly fluorescent cells increased. BCECF and similar AM-ester dyes can be regarded as viability-test-dyes, since they are only cleaved and retained in cells with intact membranes.

Problems inherent to the in vivo preparation

In the in vivo preparation, respiration, heart beat, intrinsic rhythmic phenomena and also experimental manipulations cause motion artifacts. Because imaging can be triggered to respiration, this major source of motion artifacts can be controlled, however subsequent loss of temporal resolution due to the respiration rate has to be taken into account. Experimantal manipulations like ischemia or SD cause large alterations in CBF and brain volume, thus making it necessary to keep track of the imaged plane by continuous scanning and tracking of the focal plane. Brain geometry changes not equally in all three dimensions, and neighboring cells located on the same focal plane during baseline may not be in exactly the same relation to each other during ischemia. This is a severe problem in quantifying the signal.

Since small alterations in brain temperature are reported to determine the extend of neuronal outcome (Busto et al.1987), we used a heating system for the objective to provide temperature control.

Stimulation in different cell types

It was the aim of our preliminary experiments to determine conditions of loading ion sensitive dyes into acute slices and intact brain. Furthermore, we used different excitatory stimuli, to rise $[Ca^{2+}]_i$ and varied the age of the animals from 1 day to 1 month. Others reported large differences between neonatal and adult rat brain preparations regarding loading of fluorescent AM-ester dyes (Yuste, Katz, 1991). In slice cultures, different formation of cortical layers between neonatal and embryonic derived slices has been shown (Götz, Bolz, 1992).

As physiological excitatory stimulus to increase intracellular calcium we used glutamate. Toxic glutamate concentrations have been determined in different preparations. Toxic concentrations are reported for cultured murine cortical neurons (Choi, 1988; Koh, Choi, 1988) to be 50μM, in cultured hippocampal neurons (Finkbeiner, Stevens, 1988) 1-100μM, or 500μM - 1mM (Michaels, Rothman, 1990). Others reported only a small or moderate increase in $[Ca^{2+}]_i$ after exposure to 100μM (Wahl et al.1989; Jensen, Chui, 1990) in cultured cortical neurons, and Tsien noticed irreversible destruction at concentrations of 100μM in frog tectal cells (Cline, Tsien, 1991), which showed robust $[Ca^{2+}]_i$ transients at 5-20μM. The large variety of cell types used in those studies and the heterogeneity of Ca^{2+}-increases obtained hampers the comparison to our results.

When imaging cells in heterogeneous tissues, the question arises as to the type of brain cell under study. To adress this question, I will briefly review some differences among glial (Orkand, 1992; Paley, Chan-Paley, 1992), neural or other brain cells (Tilton, 1991; Revest et al.1991). As early as 1886, neuroglia was advanced as the possible substrate "of intelligence" (Galambos, 1961). Those studies remained without impact on the traditional definition of neuroglia as a non excitable cell population or "nerve glue". During the last decade there has accumulated evidence of a role for glia in signal transduction, since it has been shown that glial cells possess a variety of ion channels (Glaum et al.1990; Inagaki et al.1991a; Inagaki et al.1991b; Gehrmann et al.1992), and Cornell-Bell et al. (Cornell-Bell et al.1990) reported propagating Ca^{2+}-transients in cultured astrocytes.

Spontaneous $[Ca^{2+}]_i$ transients have been reported in neurons of the developing somatosensory cortex (Katz, Yuste, 1991), in spinal neurons (Holliday, Spitzer, 1990), in neonatal hippocampal neurons, and in astrocytes of organotypic slices (Dani et al.1991). In our in vitro approach we confirmed (Them et al.1992) observations of spontaneous activity reported by others (Katz, Yuste, 1991). In cultured glia release from internal stores has been shown after glutamate stimulation of 10, 100μM and 1mM without presence of $[Ca^{2+}]_e$ (Ahmed et al.1990), thus suggesting the existence of metabotropic signaling. Reaction of glial $[Ca^{2+}]_i$ upon ischemia (Jensen et al.1990), EAA stimulation, neuronal activation, (Dani et al.1992) and mechanical stimulation (Charles et al.1991) has been reported. Subtle differences in the stimulus response behaviour to glutamate (Jensen, Chiu, 1991; McNaughton et al.1990) between the neuroglial subtypes of astroglia, oligodendroglia, and microglia (Kettenmann et al.1990) have been elucidated. Because of the large variety of receptor subtypes in neurons and glia a cells' $[Ca^{2+}]_i$-reaction to different stimuli is probably not suitable for the discrimination of differnt cell types.

According to their general morphology (Paley, Chan-Paley, 1992; Orkand, 1992), at least large neurons can be distinguished from astroglia. Neurons are discriminated morphologically by their large cell bodies and apical dendrites in some reports (Katz, Yuste, 1991; Yuste, Katz, 1991). In our experiments, many of the excitable cells had no typical neuron like anatomy. Further studies have to positively identify the type of those cells. Besides electrophysiological identification (Turner et al.1991; Smith et al.1991), which is not feasible in a closed slice chamber and the cranial window preparation, the most specific identification of glia is immunohistochemistry. Microglia has been identified with specific antibodies (Gehrmann et al.1992), and astroglia with glial-fibrillary acid protein (GFAP) - immunohistochemistry, specific to an antigen solely expressed on glial cells (Faddis, Vijayan, 1988). This labelling procedure was used in organotypically cultured slices, were fixation facilitates recognition of the tissue area imaged before the immuno-counterstaining process (Dani et al.1992). Currrently we are performing GFAP-counterstaining in the acute slice and in vivo.

OUTLOOK

The optical approach provides a fascinating *insight* in the activity of the CNS, opening a window to the function of the brain. Recent advances in both dye chemistry and microscope technique will improve the impact of the imaging approach. The temporal resolution will be increased by developments like acusto-optical mirrors for fast beam deflection. The penetration depth will be improved by excitation wavelengths in the red or near infrared. Better compatibility and properties of dyes will extend their use towards real time and noninvasive imaging. It will be possible to measure second messenger levels like cAMP (Adams et al.1991) and brain cell electrical activities (Cinelli, Kauer, 1992). Immunohistochemistry will facilitate the identification of the cells under study.

ACKNOWLEDGEMENTS: I wish to thank Dr. U. Dirnagl for very helpful discussions in all stages of this work. Supported by the DFG (Di 454/4-1) and the Wilhelm Sander Stiftung.

REFERENCES

Adams SR, Harootunian AT, Buechler YJ, Taylor SS, Tsien RY (1991) Fluorescence ratio imaging of cyclic AMP in single cells. Nature 349: 694-697

Adams SR, Kao JPY, Tsien RY (1989) Biologically useful chelators that take up Ca^{2+} upon illumination. J Am Chem Soc 111: 7975-7986

Ahmed Z, Lewis CA, Faber DS (1990) Glutamate stimulates release of Ca^{2+} from internal stores in astroglia. Brain Res 516: 165-169

Alford S, Collingridge GL (1990) Simultaneous whole-cell patch-clamp recording and imaging of single pyramidal neurones in rat hippocampal slices. J Physiol 435: 5P

Alford S, Collingridge GL (1991) Somatic Ca^{2+} entry following repetitive synaptic stimulus in patch-clamped CA1 pyramidal neurones in rat hippocampal slices. J Physiol 435: 44P

Alford SA, Schofield G, Collingridge GL (1991) Dentritic Ca^{2+} transients associated with NMDA receptor-mediated synaptic currents in rat hippocampal slices. J Physiol 483: 255P

Aronowski J, Grotta JC, Waxham MN (1992) Ischemia-induced translocation of Ca^{2+}/Calmodulin-dependent protein kinaseII: potential role in neuronal damage. J Neurochem 58: 1743-1753

Balduini W, Candura SM, Costa LG (1991) Regional development of carbachol-, glutamate-, norepinephrine-, and serotonin-stimulated phosphoinositide metabolism in rat brain. Dev Brain Res 62: 115-120

Barish ME (1991) Increases in intracellular calcium ion concentration during depolarisation of cultured embryonic xenopus spinal neurones. J Physiol 444: 545-565

Baskys A (1992) Metabotropic receptors and 'slow' excitatory actions of glutamate agonists in the hippocampus. Trends Neurosci 15: 92-96

Ben-Ari Y, Represa A (1990) Brief seizure episodes induce long-term potentiation and mossy fibre sprouting in the hippocampus. Trends Neurosci 13: 312-318

Benson DM, Knopp JA (1984) Effect of tissue absorption and microscope optical parameters on the depth of penetration for fluorescence and reflectance measurements of tissue samples. Photochem Photobiol 39: 495-502

Benveniste H (1991) The excitotoxin hypothesis in relation to cerebral ischemia. Cereb Brain Metab Rev 3: 213-245

Benveniste H, Diemer NH (1988) Early postischemic 45Ca accumulation in rat dentate hilus. J Cereb Blood Flow Metab 8: 713-719

Blaustein MP (1988) Calcium transport and buffering in neurons. Trends Neurosci 11: 438-443

Brakenhoff GJ, van-der-Voort HT, van-Spronsen EA, Nanninga N (1989) Three-dimensional imaging in fluorescence by confocal scanning microscopy. J Microsc 153: 151-159

Brorson JR, Bleakman D, Chard PS, Miller RJ (1992) Calcium directly permeates kainate/alpha-amino-3-hydroxy-5- methyl-4- isoxazolepropionic acid receptors in cultured cerebellar Purkinje neurons. Mol Pharmacol 41: 603-608

Buchan A, Li H, Pulsinelli A (1991) The N-methyl-D-aspartate antagonist mk-801 fails to protect against neuronal damage caused by transient severe forebrain ischemia in adult rats. J Neurosci 11: 1049-1056

Busa WB (1986) Mechanisms and consequences of pH-mediated cell regulation. Annu Rev Physiol 48: 389-402

Busto R, Dietrich WD, Globus MY-T, Valdez I, Scheinberg P, Ginsberg MD (1987) Small differences in intraischemic brain temperature critically determine the extend of ischemic neuronal injury. J Cereb Blood Flow Metab 7: 729-738

Campbell AK (1987) Intracellular calcium: friend or foe? Clin Sci 72: 1-10

Campbell KP, Leung AT, Sharp AH (1988) The biochemistry and molecular biology of the dihydropyridine- sensitive calcium channel. Trends Neurosci 11: 425-430

Carbone E, Lux HD (1987) Single low-voltage-activated calcium channels in chick and rat sensory neurones. J Physiol Lond 386: 571-601

Carlsson K, Mossberg K (1992) Reduction of cross-talk between fluorescent labels in scanning laser microscopy. J Microsc 167: 23-37

Carpenter-Deyo L, Duimstra JR, Hedstrom O, Reed DJ (1991) I. Toxicity to isolated hepatocytes caused by intracellular calcium indicator, Quin 2. J Pharm Exp Ther 258: 739-746

Charles AC, Merrill JE, Dirksen ER, Sanderson MJ (1991) Intercellular signaling in glial cells: Calcium waves and oscillations in response to mechanical stimulation and glutamate. Neuron 6: 983-992

Chen ST, Hsu CY, Hogan EL, Juan HY, Banik NL, Balentine JD (1987) Brain calcium content in ischemic infarction. Neurology 37: 1227-1229

Chertok VM, Kotsiuba AE, Lariushkina AV (1989) [Histophysiology of tissue basophils of the cerebral dura mater after laser irradiation]. Biull Eksp Biol Med 108: 493-495

Chesler M (1990) The regulation and modulation of pH in the nervous system. Prog Neurobiol 34: 401-427

Chesler M, Kraig RP (1987) Intracellular pH of astrocytes increases rapidly with cortical stimulation. Am J Physiol 253: R666-R670

Chesler M, Kraig RP (1989) Intracellular pH transients of mammalian astrocytes. J Neurosci 9(6): 2011-2019

Choi DW (1988) Calcium-mediated neurotoxicity: relationship to specific channel types and role in ischemic damage. Trends Neurosci 11: 465-469

Choi DW (1990) Cerebral hypoxia: some new approaches and unanswered questions. J Neurosci 10: 2493-2501

Cinelli AR, Kauer JS (1992) Voltage-sensitive dyes and functional activity in the olfactory pathway. Annu Rev Neurosci 15: 321-351

Cline HT, Tsien RW (1991) Glutamate-induced increase in intracellular Ca^{2+} in cultured frog tectal cells mediated by direct activation of NMDA receptor channels. Neuron 6: 259-267

Cohan CS, Connor JA, Kater SB (1987) Electrically and chemically mediated increases in intracellular calcium in neuronal growth cones. J Neurosci 7: 3588-3599

Collingridge GL, Bliss TV (1987) NMDA receptors- their role in long-term potentiation. Trends Neurosci 10: 288-293

Collingridge GL, Lester RA (1989) Excitatory amino acid receptors in the vertebrate central nervous system. Pharmacol Rev 41: 143-210

Connor JA (1986) Digital imaging of free calcium changes and of spatial processes in single, mammalian central nervous system cells. Proc Natl Acad Sci USA 83: 6179-6183

Connor JA, Wadman WJ, Hockberger PE, Wong RK (1988) Sustained dendritic gradients of Ca^{2+} induced by excitatory amino acids in CA1 hippocampal neurons. Science 240: 649-653

Cornell-Bell AH, Finkbeiner SM, Cooper MS, Smith SJ (1990) Glutamate induces calcium waves in cultured astrocytes: long- range glial signaling. Science 247: 470-473

Cotman CW, Monaghan DT, Ottersen OP, Storm-Mathisen J (1992) Anatomical organisation of excitatory amino acid receptors and their pathways. Trends Neurosci 10: 273-279

Crain BJ, Evenson DA, Polsky K, Nadler JV (1990) Electron microscopic study of the gerbil dentate gyrus after transient forebrain ischemia. Acta Neuropathol 79: 409-417

Crépel V, Krnjevic K, Ben-Ari Y (1992) Developmental and regional differences in the vulnerability of rat hippocampal slices to lack of glucose. Neuroscience 47: 579-587

Cull-Candy SG, Wyllie DJ (1991) Glutamate-receptor channels in mammalian glial cells. Ann N Y Acad Sci 633: 458-474

Dani JW, Chernjavsky A, Buchanan J, Smith SJ (1991) Neuronal activity elicits astrocyte Ca^{2+} waves and oscillations within hippocampal slices. Soc Neur Abstr 17: 57(Abstract)

Dani JW, Chernjavsky A, Smith SJ (1992) Neuronal activity triggers calcium waves in hippocampal astrocyte networks. Neuron 8: 429-440

Daniell LC (1991) N-Methyl-D-Aspartate increases cytoplasmatic free calcium in mouse hippocampus. Neuropharmacology 30: 539-545

Dingledyne R (1984) Brain Slices. Plenum Press, New York

Dirnagl U, Thorén P, Villringer A, Sixt G, Them A, Einhäupl KM (1993) Global forebrain ischemia in the rat: controlled reduction of cerebral blood flow by hypobaric hypotension and two-vessel occlusion. Neurol Res (in press)

Dirnagl U, Villringer A, Einhäupl KM (1991) Imaging of intracellular pH in normal and ischemic rat brain neocortex using confocal laser scanning microscopy in vivo. J Cereb Blood Flow Metab 11,Suppl.2: S206(Abstract)

Dirnagl U, Villringer A, Einhäupl KM (1992) In-vivo confocal scanning laser microscopy of the cerebral microcirculation. J Microsc 165: 147-157

Dixon AE, Damaskinos S, Atkinson MR (1991) A scanning confocal microscope for transmission and reflection imaging. Nature 351: 551-553

Eberhard M, Erne P (1989) Kinetics of calcium binding to FLUO-3 determined by stopped-flow fluorescence. Biochem Bioph Res Commun 163: 309-314

Eberhard M, Erne P (1991) Calcium binding to fluorescent calcium indicators: calcium green, calcium orange and calcium crimson. Biochem Biophys Res Commun 180: 209-215

Faddis BT, Vijayan VK (1988) Application of glial fibrillary acidic protein immunohistochemistry in the quantification of astrocytes in the rat brain. Am J Anat 183: 316-322

Fine A, Amos B, Durbin RM, McNaughton PA (1988) Confocal microscopy: applications in neurobiology. Trends Neurosci 11: 346-351

Finkbeiner S, Stevens CF (1988) Applications of quantitative measurements for assessing glutamate neurotoxicity. Proc Natl Acad Sci USA 85: 4071-4074

Frandsen A, Andersen CF, Schousboe A (1992) Possible role of cGMP in excitatory amino acid induced cytotoxicity in cultured cerebral cortical neurons. Neurochem Res 17: 35-43

Fujiwara N, Takashi A, Endoh H, Warashina A, Shimoni K (1992) Changes in intracellular pH of mouse hippocampal slices responding to hypoxia and/or glucose depletion. Brain Res 572: 335-339

Galambos R (1961) A glial-neural theory of brain function. Proc Natl Acad Sci USA 47: 129-136

Gehrmann J, Bonnekoh P, Miyazawa T, Hossmann KA, Kreutzberg GW (1992) Immunocytochemical study of an early microglial activation in ischemia. J Cereb Blood Flow Metab 12: 257-296

Glaum SR, Holzwarth JA, Miller RJ (1990) Glutamate receptors activate Ca^{2+} mobilization and Ca^{2+} influx into astrocytes. Proc Natl Acad Sci USA 87: 3454-3458

Götz M, Bolz J (1992) Formation and preservation of cortical layers in slice cultures. J Neurobiol 23: 783-802

Grinstein S, Rothstein A (1986) Mechanisms of regulation of the Na^+/H^+ exchanger. J Membrane Biol 90: 1-12

Grynkiewicz G, Poenie M, Tsien RY (1985) A new generation of Ca^{2+} indicators with greatly improved fluorescence properties. J Biol Chem 260: 3440-3450

Hansen AJ (1985) Effect of anoxia on ion distribution in the brain. Physiol Rev 65: 101-148

Hashimoto K, Kikushi H, Ishikawa M, Kobayashi S (1992) Regional imaging of brain tissue calcium ions using Aequorin. J Cereb Blood Flow Metab 12: 306-310

Haugland RP (1992) Molecular Probes handbook of fluorescent probes and research chemicals. Molecular Probes, Eugene,OR,

Hayashi H, Miyata H, Noda N, Kobayashi A, Hirano M, Kawai T, Yamazaki N (1992) Intracellular Ca^{2+} concentration and pH_i during metabolic inhibition. Am J Physiol 262: C628-C634

Hell S, Witting S, Schickfus v. M, Winjendts van Resandt RW, Hunklinger S, Smolka E (1991) A confocal beam scanning white-light microscope. J Microsc 163: 179-187

Hernandez-Cruz A, Sala F, Adams PR (1990) Subcellular calcium transients visualized by confocal microscopy in a voltage-clamped vertebrate neuron. Science 247: 858-862

Herrmann KS (1983) Platelet aggregation induced in the hamster cheek pouch by a photochemical process with excited fluorescein isothiocyanate- dextran. Microvasc Res 26: 238-249

Hillman D, Chen S, Aung TT, Cherksey B, Sugimori M, Llinas RR (1991) Localization of P-type calcium channels in the central nervous system. Proc Natl Acad Sci USA 88: 7076-7080

Hirano Y, Okajima F, Tomura H, Majid MA, Takeuchi T, Kondo Y (1991) Change of intracellular calcium of neuronal cells induced by extracellular ATP. FEBS Letters 284: 235-237

Hodgkiss RJ, Jones GW, Long A, Middleton RW, Parrick J, Stratford MRL, Wardman P, Wilson GD (1991) Fluorescent markers for hypoxic cells: a study of nitroaromatic compounds, with fluorescent heterocyclic side chains, that undergo bioreductive binding. J Med Chem 34: 2268-2274

Holliday J, Spitzer NC (1990) Spontaneous calcium influx and its roles in differentiation of spinal neurons in culture. Dev Biol 141: 13-23

Hollmann M, Hartley M, Heinemann S (1991) Ca2+ permeability of KA-AMPA gated glutamate receptor channels depends on subunit composition. Science 252: 851-853

Inagaki N, Fukui H, Ito S, Wada H (1991a) Type-2 astrocytes show intracellular Ca^{2+} elevation in response to various neuroactive substances. Neurosci Lett 128: 257-260

Inagaki N, Fukui H, Ito S, Yamatodani A, Wada H (1991b) Single type-2 astrocytes show multiple independent sites of Ca^{2+} signaling in response to histamine. Proc Natl Acad Sci USA 88: 4215-4219

Izumi Y, Benz AM, Clifford DB, Zorumski CF (1992) Nitric oxide inhibitors attenuate N-methyl-D-aspartate excitotoxicity in rat hippocampal slices. Neurosci Lett 135: 227-230

Jensen AM, Chiu SY (1991) Differential intracellular calcium responses to glutamate in type 1 and type 2 cultured brain astrocytes. J Neurosci 11: 1674-1684

Jensen AM, Chui SY (1990) Fluorescence measurement of changes in intracellular calcium induced by excitatory amino acids in cultured cortical astrocytes. J Neurosci 10: 1165-1175

Jensen MB, Jorgensen MB, Finsen BR, Castellano B, Diemer NH (1990) Microglial and astroglial reactions to transient cerebral ischemia in the adult rat hippocampus. Eur J Neurosci 3: 220(Abstract)

Kaila K, Voipio J (1987) Postsynaptic fall in intracellular pH induced by GABA-activated bicarbonate conductance. Nature 330: 163-165

Kao JPY, Harootunian AT, Tsien RY (1989) Photochemically generated cytosolic calcium pulses and their detection by Fluo-3. J Biol Chem 264: 8179-8184

Katayama Y, Kawamata T, Tamura T, Hovda DA, Becker DP, Tsubokawa T (1991) Calcium-dependent glutamate release concomitant with massive potassium flux during cerebral ischemia in vivo. Brain Res 558: 136-140

Katsura K, Ekholm A, Asplund B, Siesjö BK (1991) Extracellular ph in the brain during ischemia: relationship to the severity of lactic acidosis. J Cereb Blood Flow Metab 11: 597-599

Katz LT, Yuste R (1991) Spontaneous calcium transients in developing neocortical neurons. Soc Neur Abstr Vol 17 Part 2:1470 (Abstract)

Kettenmann H, Hoppe D, Gottmann K, Banati R, Kreutzberg G (1990) Cultured microglial cells have a distinct pattern of membrane channels different from peritoneal macrophages. J Neurosci Res 26: 278-287

Knight RA, Ordidge RJ, Helpern JA, Chopp M, Rodolosi LC, Peck D (1991) Temporal evolution of ischemic damage in rat brain measured by proton nuclear magnetic resonance imaging. Stroke 22: 802-808

Kogure K, Ohtomo H, Hayashi T, Sakamoto N, Sato H (1985) Calcium Accumulation in ischemic cell death. J Cereb Blood Flow Metab 5 (Suppl.1): S331

Koh J, Choi DW (1988) Vulnerability of cultured cortical neurons to damage by excitotoxins: differential susceptibility of neurons containing NADPH-diaphorase. J Neurosci 8: 2153-2163

Koh JY, Palmer E, Cotman CW (1991a) Activation of the metabotropic glutamate receptor attenuates N- methyl-D-aspartate neurotoxicity in cortical cultures. Proc Natl Acad Sci USA 88: 9431-9435

Koh JY, Palmer E, Lin A, Cotman CW (1991b) A metabotropic glutamate receptor agonist does not mediate neuronal degeneration in cortical culture. Brain Res 561: 338-343

Kraig RP, Ferreira-Filho CR, Nicholson C (1983) Alkaline and acid transients in cerebellar microenvironment. J Neurophysiol 49: 831-850

Kraig RP, Pulsinelli WA, Plum F (1985) Heterogeneous distribution of hydrogen and bicarbonate ions during complete brain ischemia. In: Kogure K, Hossmann K-A, Siesjö BK, Welsh FA (eds) Progress in Brain Research. Elsevier, pp 155

Kudo Y, Takeda K, Yamazaki K (1990) Quin2 protects against neuronal cell death due to Ca^{2+} overload. Brain Res 528: 48-54

Lattanzio FA,Jr., Bartschat DK (1991) The effect of pH on rate constants, ion selectivity and thermodynamic properties of fluorescent calcium and magnesium indicators. Biochem Biophys Res Commun 177: 184-191

Lauritzen M, Hansen AJ (1992) The effect of glutamate receptor blockade on anoxic depolarisation and cortical spreading depression. J Cereb Blood Flow Metab 12: 223-229

Leonard BW, Barnes CA, Rao G, Heissenbuttel T, McNaughton BL (1991) The influence of postmortem delay on evoked hippocampal field potentials in the in vitro slice preparation. Exp Neurol 113: 373-377

Liu CM, Herman TE (1978) Characterisation of Ionomycin as a calcium ionophore. J Biol Chem 253: 5892-5894

Lucas DR, Newhouse DP (1957) The toxic effect of sodium L-glutamate on the inner layers of the retina. Arch Ophtalmol 58: 193-201

MacDermott AB, Mayer ML, Westbrook GL, Smith SJ, Barker JL (1986) NMDA-receptor activation increases cytoplasmic calcium concentration in cultured spinal cord neurones. Nature 321: 519-522

MacVicar BA, Tse FWY (1989) Localneuronal circuitry underlying cholinergic rhythmical slow activity in CA3 area of rat hippocampal slices. J Physiol 417: 197-212

Marcoux FW, Probert AW (1988) NMDA and calcium channel antagonists inhibit hypoxia induced calcium influx in cultured cortical neurons. Stroke 19: 23

Martinez-Zaguilan R, Martinez GM, Lattanzio F, Gillies RJ (1991) Simultaneous measurement of intracellular pH and Ca^{2+} using the fluorescence of snarf-1 and fura-2. J Cell Physiol 29: C297-C307

Mayer ML, Westbrook GL (1987) Permeation and block of n-methyl-d-aspertic acid receptor channels by divalent cations in mouse cultured central neurones. J Physiol 394: 501-527

McNaughton LA, Lagnado L, Socolovsky M, Hunt SP, McNaughton PA (1990) Use of the confocal microscope to measure changes in free (Ca^{2+})i in type 1 astrocytes cultured from rat cerebral cortex. J Physiol 424:5P (Abstract)

Meldolesi J, Volpe P, Pozzan T (1988) The intracellular distribution of calcium. Trends Neurosci 11: 449-452

Meldrum BS (1985) Intracellular accumulation of calcium in ischaemic and epileptic brain damage. J Cereb Blood Flow Metab 5 (Suppl.1) S333-S334
Meyer FB (1989) Calcium, neuronal hyperexcitability and ischemic injury. Brain Res Rev 14: 227-243
Meyer T, Holowka D, Stryer L (1988) Highly cooperative opening of calcium channels by inositol 1,4,5- trisphosphate. Science 240: 653-656
Michaels RL, Rothman SM (1990) Glutamate neurotoxicity in vitro: antagonist pharmacology and intracellular calcium concentrations. J Neurosci 10: 283-292
Miller FN, Sims DE, Schuschke DA, Abney DL (1992) Differentiation of light-dye effects in the microcirculation. Microvasc Res 44: 166-184
Miller RJ (1987) Multiple calcium channels and neuronal function. Science 235: 46-52
Miller RJ (1988) Calcium signaling in neurons. Trends Neurosci 11: 415-419
Miller RJ, Abele AE, Glaum SR, Scholz KP, Scholz WK (1990a) Pharmacological aspects of NMDA mediated neuronal death in vitro. In: Krieglstein J, Oberpichler H (eds) Pharmacology of cerebral ischemia. Wissenschaftliche Verlagsgesellschaft, Stuttgart, pp 227
Miller RJ, Abele AE, Glaum SR, Scholz KP, Scholz WK (1990b) Pharmacological aspects of NMDA mediated neuronal death in vitro. Am J Physiol
Minta A, Kao JPY, Tsien RY (1989) Fluorescent indicators for cytosolic calcium based on rhodamine and fluorescein chromophores. J Biol Chem 264: 8171-8178
Moody W,Jr. (1984) Effects of intracellular H^+ on the electrical properties of excitable cells. Annu Rev Neurosci 7: 257-278
Moore ED, Becker PL, Fogarty KE, Williams DA, Fay FS (1990) Ca^{2+} imaging in single living cells: theoretical and practical issues. Cell Calcium 11: 157-179
Morgan JI, Curran T (1989) Stimulus transcription coupling in neurons: role of cellular immediate-early genes. Trends Neurosci 12: 459
Morii S, Ngai AC, Winn HR (1986) Reactivity of rat pial arterioles and venules to adenosine and carbon dioxide: with detailed description of the closed cranial window technique in rats. J Cereb Blood Flow Metab 6: 34-41
Murray JM (1992) Neuropathology in Depth: The Role of Confocal Microscopy. J Neuropathol Neurol 51: 475-487
Nahorski SR (1988) Inositol polyphosphates and neuronal calcium homeostasis. Trends Neurosci 11: 444-448
Nayler WG, Britnell S (1991) Calcium antagonists and tissue protection. J Cardiovasc Pharm 18-suppl 1: s1-s5
Nedergaard M, Hansen AJ (1988) Spreading depression is not associated with neuronal injury in the normal brain. Brain Res 449: 395-398
Nicholson C, Bruggencate GT, Steinberg R, Stockle H (1977) Calium modulation in brain extracellular microenvironment demonstrated with ion selective micropipette. Proc Natl Acad Sci USA 74: 1287-1290
Olney JW (1969) Brain lesions, obesity and other disturbances in mice treated with monosodium glutamate. Science 164: 719-721
Olsen RW, Szamraj O, Houser CR (1987) [3H]AMPA binding to glutamate receptor subpopulations in rat brain. Brain Res 402: 243-254
Orkand RK (1992) Glial Cells. In: Handbook of physiology; Section I: The Nervous System, Volume 1. The American Physiological Sociey, Bethesda 1977, pp.855
O'Shaughnessy CT, Lythgoe DJ, Butcher SP, Kendall L, Wood B, Steward MC (1991) Effects of hypoxia on fetal rat brain metabolism studied in utero by 31p-nmr spectroscopy. Brain Res 551: 334-337

Paley SL, Chan-Paley V (1992) General morphology of neurons and neuroglia. In: Handbook of physiology; Section I: The Nervous System, Volume 1. The American Physiological Sociey, Bethesda 1977, pp.5

Pazdernik TL, Layton M, Nelson SR, Samson FE (1992) The osmotic/calcium stress theory of brain damage: are free radicals involved? Neurochem Res 17: 11-21

Pellegrini-Giampietro DE, Cherici G, Alesiani M, Carla V, Moroni F (1990) Excitatory amino acid release and free radical formation may cooperate in the genesis of ischemia induced neuronal damage. J Neurosci 10: 1035-1041

Peng LA, Juurlink BH, Hertz L (1991) Differences in transmitter release, morphology, and ischemia- induced cell injury between cerebellar granule cell cultures developing in the presence and in the absence of a depolarizing potassium concentration. Dev Brain Res 63: 1-12

Peres A, Bertollini L, Racca C (1991) Characterisation of Ca^{2+} transients induced by intracellular photorelease if InsP3 in mouse ovarian oocytes. Cell Calcium 12: 457-465

Pirttilä T-RM, Kauppinen RA (1992) Recovery of intracellular pH in cortical brain slices following anoxia studied by nuclear magnetic resonance spectroscopy: role of lactate removal, extracellular sodium and sodium/hydrogen exchange. Neuroscience 47: 155-164

Przywara DA, Bhave SV, Bhave A, Wakade TD, Wakade AR (1991) Stimulated rise in neuronal calcium is faster and greater in the nucleus than in the cytosol. FASEB J 5: 217-222

Read ND, Allan WTG, Knight H, Knight MR, Malhó R, Russel A, Shacklock PS, Trewavas AJ (1992) Imaging and measurement of cytosolic free calcium in plant and fungal cells. J Microsc 166: 57-86

Revest PA, Abbott NJ, Gillespie JI (1991) Receptor-mediated changes in intracellular $[Ca^{2+}]$ in cultured rat brain capillary endothelial cells. Brain Res 549: 159-161

Rijen van PC, Verheul HB, Echteld van CJA, Balasz R, Lewis P, Nasim MM, Tulleken CAF (1991) Effects of dextromethorphan on rat brain during ischemia and reperfusion assesed by magnetic resonance spectroscopy. Stroke 22: 343-350

Rijkers GT, Justement LB, Griffioen AW, Cambier JC (1990) Improved method for measuring intracellular Ca++ with fluo-3. Cytometry 11: 923-927

Rodriguez J, Jacques-Berg W, Patel AJ (1991) Differential regulation of cerebellar granule neurons by two types of quisqualate receptors. Neuroreport 2: 517-520

Rosenblum WI, El-Sabban (1977) Platelet aggregation in the cerebral microcirculation. Effect of aspirin and other agents. Circ Res 40: 320-327

Ross CA, Danoff SK, Schell MJ, Snyder SH, Ullrich A (1992) Three additional inositol 1,4,5-trisphosphate receptors: Molecular cloning and differential localization in brain and peripheral tissues. Proc Natl Acad Sci USA 89: 4265-4269

Scanlon M, Williams DA, Fay FS (1987) A Ca^{2+}-insensitive form of fura-2 associated with polymorphonuclear leucocytes. J Biol Chem 262: 6308-6312

Scheller D, Heister U, Dengler K, Peters T (1990) Do the excitatory amino acids aspartate and glutamate generate spreading depressions in vivo. In: Krieglstein J, Oberpichler H (eds) Pharmacology of cerebral ischemia. Wissenschaftliche Verlagsgesellschaft, Stuttgart,pp.205

Schmidt-Kastner R, Ophoff BG, Hossmann KA (1990) Pattern of neuronal vulnerability in the cat hippocampus after one hour of global cerebral ischemia. Acta Neuropathol 79: 444-455

Schurr A, Teyler TJ, Tseng MT (1991) Brain slices: fundamentals applications and implications. Karger, Basel

Segal M, Manor D (1992) Confocal microscopic imaging of $[Ca^{2+}]i$ in cultured rat hippocampal neurons following exposure to N-methyl-D-aspartate. J Physiol 448: 655-676

Sei Y, Arora PK (1991) Quantitative analysis of calcium (Ca^{2+}) mobilization after stimulation with mitogens or anti-CD3 antibodies. J Immunol Meth 137: 237-244

Siesjo BK, Memezawa H, Smith ML (1991) Neurocytotoxicity: pharmacological implications. Fundam Clin Pharmacol 5: 755-767

Siesjö BK (1988) Historical overview. Calcium, ischemia, and death of brain cells. Ann NY Acad Sci 522: 638-661

Siesjö BK, Bengtsson F (1989) Calcium fluxes, calcium antagonists, and calcium-related pathology in brain ischemia, hypoglycemia, and spreading depression: a unifying hypothesis. J Cereb Blood Flow Metab 9: 127-140

Silver IA, Erecinska M (1990) Intracellular and extracellular changes of $[Ca^{2+}]$ in hypoxia and ischemia in rat brain in vivo. J Gen Physiol 95: 837-866

Smith KL, Turner JN, Szarowski DH, Swann JW (1991) Three-dimensional imaging of neurophysiologically characterized hippocampal neurons by confocal scaning laser microscopy. Ann NY Acad Sci 627: 390-394

Smith SJ, Augustine GJ (1988) Calcium ions, active zones and synaptic transmitter release. Trends Neurosci 11: 458-464

Smith TL (1990) Regulation of intrasynaptosomal free calcium concentrations: studies with the fluorescent indicator, Fluo-3. Neurochem Int 16: 89-94

Sommer B, Burnashev N, Verdoorn TA, Keinanen K, Sakmann B, Seeburg PH (1992a) A glutamate receptor channel with high affinity for domoate and kainate. EMBO J 11: 1651-1656

Sommer B, Monyer H, Wisden W, Verdoorn TA, Burnashev N, Sprengel R, Sakmann B, Seeburg PH (1992b) Glutamate-gated ion channels in the brain. Genetic mechanism for generating molecular and functional diversity. Arzneimittelforschung 42: 209-210

Spedding M, Paoletti R (1992) Classification of calcium channels and the sites of action of drugs modifying channel function. Pharmacol Rev 44: 363-376

Spring KR (1991a) Illumination, wavelength selection, and detection in fluorescence microscopy. Kidney Int 33: S18-S22

Spring KR (1991b) Detectors for fluorescence microscopy. Scanning Microscopy 5: 63-69

Stelzer EH, Wacker I, De-Mey JR (1991) Confocal fluorescence microscopy in modern cell biology. Semin Cell Biol 2: 145-152

Takamatsu T, Minamikawa T, Kawachi H, Fujita S (1991) Imaging of calcium wave propagation in guinea-pig ventricular cell pairs by confocal laser scanning microscopy. Cell Struct Funct 16: 341-346

Tang C, Dichter M, Morad M (1991) Modulation of the NMDA channel by extracellular H^+. Proc Natl Acad Sci USA 87: 6445-6449

Tank DW, Sugimori M, Connor JA, Llinas RR (1991) Spatially resolved calcium dynamics of mammalian purkinje cells in cerebellar slice. Science 242:773-777

Teichberg VI (1991) Glial glutamate receptors: likely actors in brain signaling. FASEB J 5: 3086-3091

Them A, Dirnagl U, Villringer A (1992) Confocal fluorescence imaging of (Ca^{2+})i-transients in acute rat brain slices using Fluo-3-AM in a small volume submerged chamber system. Soc Neur Abstr Vol.18 Part 2:967 (Abstract)

Tilton RG (1991) Capillary pericytes: perspectives and future trends. J Electron Micr Techn 19: 327-344

Trump BF, Berezesky IK (1990) The importance of calcium regulation in toxic cell injury. Studies utilizing the technology of digital imaging fluorescence microscopy. Clin Lab Med 10: 531-547

Tsien RW, Ellinor PT, Horne WA (1991) Molecular diversity of voltage-dependent Ca^{2+} channels. Trends Pharmacol Sci 12: 349-354

Tsien RW, Lipscombe D, Madison DV, Bley KR, Fox AP (1988) Multiple types of neuronal calcium channels and their selective modulation. Trends Neurosci 11: 431-438

Tsien RW, Tsien RY (1991) Calcium channels,stores, and oscillations. Annu Rev Cell Biol 6: 715-760

Tsien RY (1988) Fluorescence measurement and photochemical manipulation of cytosolic free calcium. Trends Neurosci 11: 419-424

Tsien RY (1989a) Fluorescent indicators of ion concentrations. Methods Cell Biol 30: 127-156

Tsien RY (1989b) Fluorescent probes of cell signaling. Annu Rev Neurosci 12: 227-253

Turner JN, Szarowski DH, Smith KL, Marko M, Leith A, Swann JW (1991) Confocal microscopy and three-dimensional reconstruction of electrophysiologically identified neurons in thick brain slices. J Electron Micr Techn 18: 11-23

Uematsu D, Greenberg JH, Reivich M, Karp A (1988a) In vivo measurement of cytosolic free calcium during cerebral ischemia and reperfusion. Ann Neurol 24: 420-428

Uematsu D, Greenberg JH, Reivich M, Kobayashi S, Karp A (1988b) In vivo fluorometric measurement of changes in cytosolic free calcium from the cat cortex during anoxia. J Cereb Blood Flow Metab 8: 367-374

Vandenberghe PA, Ceuppens JL (1990) Flow cytometric measurement of cytoplasmic free calcium in human peripheral blood T lymphocytes with fluo-3, a new fluorescent calcium indicator. J Immunol Meth 127: 197-205

Villringer A, Dirnagl U, Them A, Schürer L, Krombach F, Einhäupl KM (1991) Imaging of leukocytes within the rat brain cortex in vivo. Microvasc Res 42: 305-315

Vincent SL, Sorensen I, Benes FM (1991) Localisation and high-resolution imaging of cortical neurotransmitter compartments using confocal laser scanning microscopy: GABA and glutamate interactions in rat cortex. Biotechniques 11: 628-634

Wahl P, Schousboe A, Honore T, Drejer J (1989) Glutamate induced increase in intracellular Ca^{2+} in cerebral cortex neurons is transient in immature cells but permanent in mature cells. J Neurochem 53: 1316-1319

Watkins JC, Olverman HJ (1987) Agonists and antagonists for excitatory amino acid receptors. Trends Neurosci 10: 265-272

Williams DA, Becker PL, Fay FS (1987) Regional changes in calcium underlying contraction of single smooth muscle cells. Science 235: 1644-1648

Williams DA, Fogarty KE, Tsien RY, Fay FS (1991) Calcium gradients in single smooth muscle cells revealed by the digital imaging microscope using fura-2. Nature 318: 558-561

Yuste R, Katz LC (1991) Control of postsynaptic Ca^{2+} influx in developing neocortex by excitatory and inhibitory neurotransmitters. Neuron 6: 333-344

Yuste R, Peinado A, Katz LC (1992) Neuronal groups revealed by optical recording of calcium transients in slices of developing neocortex. Soc Neur Abstr Vol.17 Part 2:1470 (Abstract)
Zivin JA, Choi DW (1991) Stroke therapy. Sci Am 7/91: 56-63

OPTICAL ACCESS TO THE BRAIN: HOW ARTIFICIAL ARE CRANIAL WINDOW TECHNIQUES?

R.L. Haberl

Department of Neurology, University of Munich
Munich, Germany

INTRODUCTION

Observation of brain surface vessels through a cranial window, implanted into the skull of anesthetized animals, is a standard technique for investigation of the cerebral circulation. It has been suggested that changes in pial vascular diameter in response to vasoactive stimuli are qualitatively similar to those of the cerebral circulation as a whole and that changes in cerebral blood flow may be estimated from the reaction of pial vessels (Rosenblum and Kontos, 1974). The assumption, however, that intraparenchymal vessels repond as pial vessels do, has not been proven. In addition, large cerebral arteries may dilate while pial arterioles constrict, thereby preventing a cerebral blood flow response (Faraci and Heistad, 1990). Pial arteriolar diameter reactivity is not predictive of CBF changes, therefore. The technique, however, is most valuable in the investigation of the mechanisms of arteriolar reactivity in the brain.

In situ investigations are considered to have advantages over in vitro experiments with vascular strips or rings, because the vessels are observed in their physiological milieu, continuous measurements of vessel diameter is possible and dilation can be elicited without prior generation of a pharmacologically induced constrictor tone (McCulloch and Edvinsson, 1984).

Different cranial window techniques have been described. In the "open cranial window" technique the brain is covered with mineral oil and test solutions may be administered into the periadventitial space by microapplication (Wahl et al., 1972a). In the "closed cranial window" the craniectomy is covered by a glass cover slip and the test solutions in artificial cerebrospinal fluid (CSF) are applied through inlet- and outlet-ports by superfusion over the brain surface (Levasseur et al., 1975). Although results with the two techniques are complementary in most instances, divergent results with some vasoactive stimuli may arise from methodological differences. Some of the factors, possibly influencing vascular reactivity in cranial window techniques, are discussed in the following paragraphs (Table 1).

Optical Imaging of Brain Function and Metabolism, Edited by
U. Dirnagl *et al.*, Plenum Press, New York, 1993

Table 1. Some factors affecting vascular reactivity in cranial window preparations.

Factor	alters reactivity to (examples)
CSF-pH	norepinephrine (NE), alterations in $paCO_2$
CSF-osmolarity	NE
CSF-ionic composition	NE, EDRF
CSF and brain temperature	?
sympathetic innervation	arterial hypertension
trigeminal innervation	arterial hypertension, reactive hyperemia, NE
primary vessel tone	5-HT, vasopressin, histamine
vessel size	alterations in $paCO_2$
endothelium	angiotensins, acetylcholine, bradykinin
metabolic-coactivation	NE, VIP, 5-HT
anesthesia (e.g. barbiturates, chloralose)	NE, induced hypertension, K^+
intracranial pressure	?

EFFECT OF CSF-PH ON VASCULAR REACTIVITY

Reactivity of cerebral arterioles depends on the pH of perivascular CSF. Navari et al. have shown that vasodilation in response to systemic hypercapnia is increased in the presence of elevated CSF-pH (Navari et al., 1978). Vasoconstriction in response to norepinephrine is increased at elevated CSF-pH values (Navari et al., 1978). The critique that constancy of CSF-pH within the physiological range is not garantueed in the open cranial window technique (Navari et al., 1978), refers to preparations covered with an inadequately thin layer of paraffin. Physiological CSF-pH in the open window preparation may be maintained by constant superfusion of the craniectomy site or by a superimposing paraffin layer of at least 1 cm thickness (Kuschinsky and Wahl, 1980).

EFFECT OF IONIC CONCENTRATIONS IN CSF ON VASCULAR REACTIVITY

Since early publications by Wahl and Kuschinsky, it is known that periarteriolar K^+ concentrations affect pial arteriolar diameter (Kuschinsky et al., 1972). In addition, the sensitivity to norepinephrine induced vasoconstriction is increased with rising CSF K^+-concentrations (Kuschinsky and Wahl, 1977). Similarily, CSF osmolarity both influences pial arteriolar diameter per se and sensitivity to vasoactive substances (Kuschinsky and Wahl, 1977; Wahl et al., 1972a).

EFFECT OF PREEXISTING TONE ON VASCULAR REACTIVITY

It is considered to be a principal advantage of in situ techniques that dilation of cerebral arterioles can be elicited without prior induction of a constrictor tone. This does not imply, however, that tone does not affect vascular reactivity in cranial window preparations. It has been demonstrated that pre-dilated pial arterioles in a piglet cranial window preparation constrict in response to topically applied vasopressin and do not have their usual dilator response to this stimulus (Armstead et al., 1989). The importance of initial tone as determinant of in vivo vascular responses is also shown by the effect of serotonin and histamin in the mouse open cranial window preparation. Preconstriction of pial arterioles with uridine-5'-triphosphate changed the usual constrictor response to serotonin and high dose histamin to vasodilation (Rosenblum and Nelson, 1990).

ALTERATIONS IN VASCULAR REACTIVITY BY ENDOTHELIAL DAMAGE

The strong impact of the endothelium on vascular responses is fully acknowledged since Furchgott's discovery of endothelium derived relaxing factor (EDRF) in 1980 (Furchgott and Zawadzki, 1980). Brain arterioles dilate in response to EDRF as peripheral vessels do. Mechanical damage of the brain surface during the preparation or minimal bleeding into the CSF prevents the endothelium dependent response of pial arterioles. Endothelial damage, with a reversible loss of responses to endothelial dependent substances like acetylcholine, also occurs with acute hypertension, free radical generation on the brain surface or during photochemical procedures like intravital fluorescence microscopy (Haberl et al., 1990; Marshall and Kontos, 1990; Rosenblum, 1986). Care has to be taken, therefore, when fluorescent techniques are applied to study the brain surface in vivo.

IMPACT OF PERIVASCULAR INNERVATION

Although there is an abundant sympathetic and parasympathetic innervation of cerebral vessels, its functional significance is not clear in many aspects. Sympathetic stimulation, with release of the transmitter norepinephrine, has little effect on the cerebral circulation under physiological conditions. Stimulation of sympathetic nerves during acute hypertension prevents dilation of large cerebral arteries, extends the upper limit of autoregulation and attenuates disruption of the blood-brain barrier (Gross et al., 1979). During parasympathetic activation, a number of neurotransmitters with known vasodilating potency (vasoactive intestinal polypeptide, peptide histidine isoleucin, neuropeptide Y) may be released, their functional significance, however, is speculative as yet (Edvinsson, 1991). Recently, a trigeminal, sensory innervation of cerebral vessels, with substance P and calcitonin gene-related peptide (CGRP) as neurotransmitters, has been described (Mayberg et al., 1981). Input from those sensory nerves has been shown

- to reduce the constrictor response of cerebral vessels to norepinephrine
- to enhance the vasodilation induced by acute severe hypertension
- to mediate postischemic cerebral hyperemia
- to enhance cerebral blood flow increases during seizures (MacFarlane et al., 1991).

Thus, trigeminal stimulation may markedly change cerebrovascular reactivity. Control experiments after trigeminalectomy or chemical denervation with capsaicin (MacFarlane et al., 1991) may be warranted.

BRAIN PARENCHYMAL INFLUENCES ON CEREBRAL VASCULAR REACTIVITY

Test solutions, which are superfused over the brain, or, which are filled into a closed cranial window chamber come into contact with both the pial vessels and the surface brain parenchyma. It is not possible to differentiate between direct vasomotor effects and indirect vascular effects through changes in the cerebral metabolic activity. This is a shortcoming of preparations, where a microapplicator cannot be used. Superfusion of the cortical surface with neurotransmitters such as norepinephrine, serotonin or vasoactive intestinal polypeptide has been shown to have actions on cerebral metabolism (Dora and Kovach, 1983; McCulloch and Edvinsson, 1984). Recently, it has been demonstrated that glutamate and N-methyl-D-aspartate (NMDA) dilate pial arterioles in a closed cranial window preparation (Busija and Leffler, 1989), although there do not seem to be NMDA-receptors on cerebral microvessels (Beart et al., 1988). It has been suggested that topically applied NMDA induces dilation through stimulation of neuronal generation of nitric oxide, a potent vasodilator (Faraci and Breese, 1992). These results emphasize the linkage of neuronal activity and local vascular reactivity.

In conclusion, a number of factors may alter cerebrovascular reactivity in cranial window preparations (table 1). An important improvement may be achieved by optical techniques which do not require a craniectomy. Infrared spectroscopy, through the closed skull, has been shown to be an applicable approach for measurement of oxygen saturation and metabolic events in the brain. Further applications of this technique will follow. As to the reactivity of small cerebral resistance vessels, however, cranial window techniques have yielded basic physiological knowledge and will not be replaced easily by newer techniques.

REFERENCES

Armstead, W.M., Mirro, R., Busija, D.W., and Leffler, C.W. (1989). Vascular responses to vasopressin are tone-dependent in the cerebral circulation of the newborn pig. Circ. Res. 64, 136-144.

Beart, P.M., Sheehan, K-A.M., and Manallack, D.T. (1988). Absence of N-Methyl-D-Aspartate receptors on ovine cerebral microvessels. J. Cereb. Blood Flow Metab. 8, 879-882.

Busija, D.W. and Leffler, C.W. (1989). Dilator effects of amino acid neurotransmitters on piglet arterioles. Am. J. Physiol. 257, H1200-H1203.

Dora, E. and Kovach, G.B. (1983). Effect of topically administered epinephrine, norepinephrine, and acetylcholine on cerebrocortical circulation and the NAD/NADH redox state. J. Cereb. Blood Flow Metab. 3, 161-169.

Edvinsson, L. (1991). Innervation and effects of dilatory neuropeptides on cerebral vessels. New aspects. Blood Vessels 28, 35-45.

Faraci, F.M. and Breese, K.R. (1992). Does nitric oxide mediate metabolic vasodilation of the cerebral microcirculation? Faseb J. 6, A1462.(Abstract)

Faraci, F.M. and Heistad, D.D. (1990). Regulation of large cerebral arteries and cerebral microvascular pressure. Circ. Res. 66, 8-17.

Furchgott, R.F. and Zawadzki, J.V. (1980). The obligatory role of endothelial cells in the relaxation of arterial smooth muscle by acetylcholine. Nature 288, 373-376.

Gross, P.M., Heistad, D.D., Strait, M.R., Marcus, M.L., and Brody, M.J. (1979). Cerebral vascular responses to physiological stimulation of sympathetic pathways in cats. Circ. Res. 44, 288-294.

Haberl, R.L., Anneser, F., Villringer, A., and Einhäupl, K.M. (1990). Angiotensin II induces endothelium-dependent vasodilation of rat cerebral arterioles. Am. J. Physiol. 258, H1840-H1846.

Kuschinsky, W., Wahl, M., Bosse, O., and Thurau, K. (1972). Perivascular potassium and pH as determinants of local pial arterial diameter in cats. Circ. Res. 31, 240-247.

Kuschinsky, W. and Wahl, M. (1977). Interactions between perivascular norepinephrine and potassium or osmolarity on pial arteries of cats. Microvasc. Res. 14, 173-180.

Kuschinsky, W. and Wahl, M. (1980). Letter to the editor. Microvasc. Res. 19, 385-386.

Levasseur, J.E., Wei, E.P., Raper, A.J., Kontos, A.A., and Patterson, J.L. (1975). Detailed description of a cranial window technique for acute and chronic experiments. Stroke 6, 308-317.

MacFarlane, R., Tasdemiroglu, E., Moskowitz, M.A., Uemura, Y., Wei, E.P., and Kontos, H.A. (1991). Chronic trigeminal ganglionectomy or topical capsaicin application to pial vessels attenuates postocclusive cortical hyperemia but does not influence postischemic hypoperfusion. J. Cereb. Blood Flow Metab. 11, 261-271.

Marshall, J.J. and Kontos, H.A. (1990). Endothelium-derived relaxing factors. A perspective from in vivo data. Hypertension 16, 371-386.

Mayberg, M., Langer, R.S., Zervas, N.T., and Moskovitz, M.A. (1981). Perivascular meningeal projections from cat trigeminal ganglia: Possible pathway for vascular headaches in man. Science 213, 228-230.

McCulloch, J. and Edvinsson, L. (1984). Cerebrovascular smooth muscle reactivity: a critical appraisal of in vitro and in situ techniques. J. Cereb. Blood Flow Metab. 4, 129-139.

Navari, R.M., Wei, E.P., Kontos, H.A., and Patterson, J.L. (1978). Comparison of the open skull and cranial window preparations in the study of the cerebral microcirculation. Microvasc. Res. 16, 304-315.

Rosenblum, W.I. (1986). Endothelial dependent relaxation demonstrated in vivo in cerebral arterioles. Stroke 17, 494-497.

Rosenblum, W.I. and Kontos, H.A. (1974). The importance and relevance of studies of the pial microcirculation. Stroke 5, 425-428.

Rosenblum, W.I. and Nelson, G.H. (1990). Tone regulates opposing endothelium-dependent and -independent forces: resistance brain vessels in vivo. Am. J. Physiol. 259, H243-H247.

Wahl, M., Kuschinsky, W., Bosse, O., and Thurau, K. (1972a). Dependency of pial arterial and arteriolar diameter on perivascular osmolarity in the cat. Circ. Res. 32, 162-169.

MULTIPARAMETRIC IMAGING OF MICROREGIONAL CIRCULATION OVER THE BRAIN CORTEX BY VIDEOREFLECTOMETRY

A. Eke

Experimental Research Department and 2nd Institute of Physiology, Semmelweis University of Medicine, Budapest, Hungary and Departments of Neurology and Pathology, University of Alabama, Birmingham, USA

INTRODUCTION

One of the primary tasks of parenchymal circulation is to distribute blood flow as it flows through an organ among increasingly larger number and smaller volumes of tissue elements until the level of nutritive microcirculation is reached. It seems, the only adequate approach to characterize this process is to map the distribution of blood flow at the level of microcirculation over tissue microareas. It has to be recognized, however, that the term *blood flow* is rather inadequate to describe the passage of its two components, *erythrocyte and plasma flow* through the complex system of capillary networks of tissue microareas where the diameter of capillaries and the traversing erythrocytes are comparable. The methodology of *multiparametric imaging of tissue microcirculation by computerized videoreflectometry* provides the means of obtaining high resolution maps of local erythrocyte and plasma volumes, mean transit times, volume flows, local tube and discharge hematocrit, blood volume and flow in a repeatable manner. Current state of instrumentation and technology (Eke 1992) has made multiparametric assessment of cerebrocortical circulation for these parameters possible typically at a spatial resolution of 10000 sites/10 mm^2 and a frequency of up to 1 measurement/minute. These features seem adequate to assess the complexity of tissue microcirculation in the brain cortex in the spatial and temporal domain alike.

METHODS

This nondestructive methodology has evolved through a number of intermediate steps (Eke et al. 1979; Eke 1982, 1983a, 1983b, 1984; Eke and Halsey 1984; Eke 1988a, 1992) representing an interplay between the actual level of understanding of how microcirculation can be assessed by intraparenchymal indicator-dilution and the level

Optical Imaging of Brain Function and Metabolism, Edited by
U. Dirnagl *et al.*, Plenum Press, New York, 1993

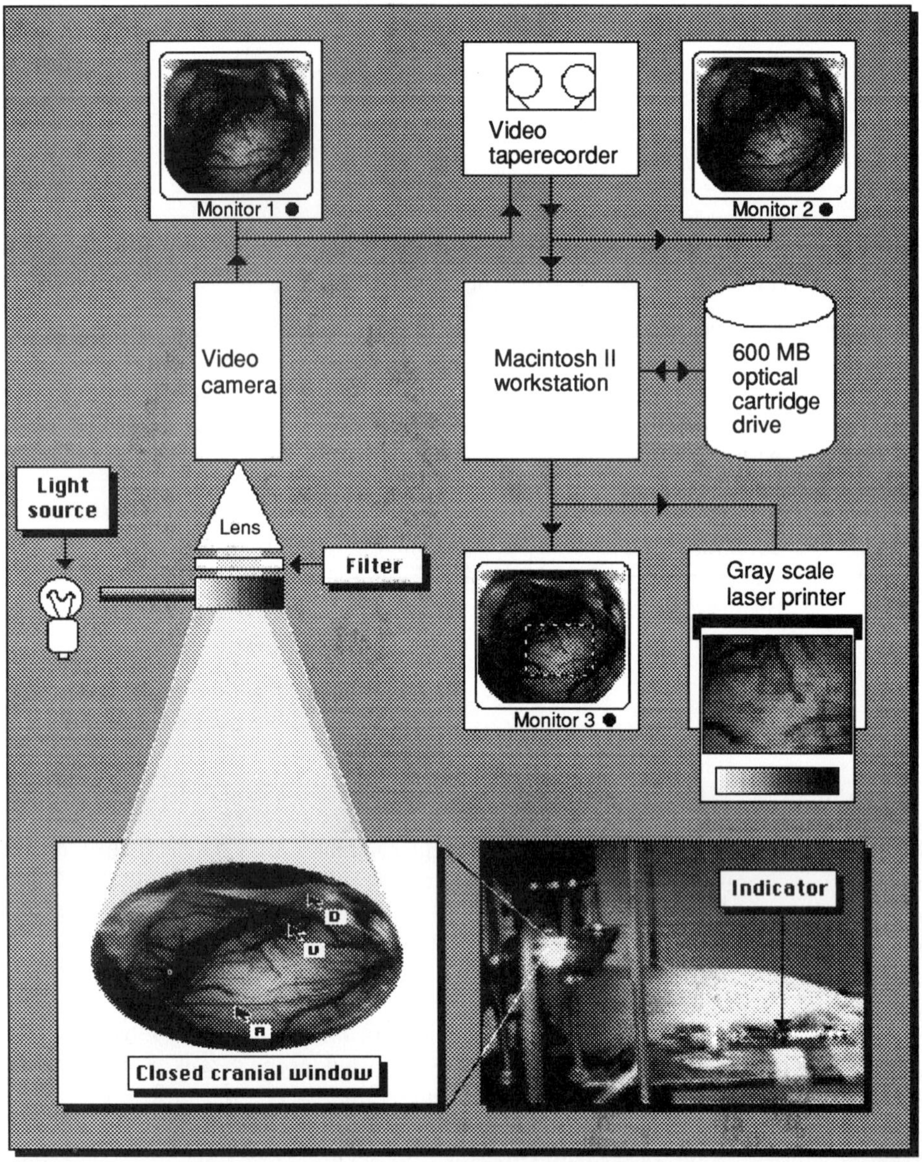

FIGURE 1. Schematic representation of the computerized videoreflectometric setup for multiparametric imaging of the cerebrocortical microcirculation. The area of the brain cortex to be mapped is exposed and epiilluminated through a cranial window of 4 mm in diameter with white, cold light (left insert; A=artery, V=vein, D=dura cut and folded). The insert on the right shows an anesthetized rat under television epiillumination reflectometry prepared for indicator injection into the common carotid artery. Cooled halogen lightsource via annular fiber optics provides an even intensity profile across the field. A television camera scans the epiilluminated area for a square array of backscattered light intensities (reflectance) during the transit of the indicators through tissue microregions at 589 nm, one of the isosbestic points of hemoglobin. Reflectance images during indicator-perfusions are stored by a VHS cassette tape recorder. The Macintosh II workstation performs reflectance image digitizing, densitometric and methodological calculations, image reconstruction and displaying under computer control. Monitor-1 is for on-line, Monitor-2 for off-line viewing of the video image. Monitor-3 (system monitor) shows a region of interest selected for data acquisition.

of instrumentation and technology available to do quantitative spatial imaging as the means of detecting intraparenchymal indicator dilution from capillary beds of the intact brain cortex.

A closed cranial window-technique and in situ quantitative television densitometry were used to record and process the television images of an epiilluminated area of the brain cortex during bolus perfusion by erythrocyte and plasma indicators (isosmotic saline and Evans blue solutions respectively) for maps of the aforementioned microcirculatory parameters according to an updated protocol (Eke 1988). The schematic representation of the quantitative videoreflectometric instrumentation and experimental arrangement is shown in Fig. 1. Intraparenchymal indicator dilutions for erythrocytes and plasma can be nondestructively detected by epiillumination television reflectometry when erythrocyte and plasma indicators (transparent and dense to light; see Fig. 2) are being monitored in their transit through a 100x100 square array of cylindrical volumes of tissue each of which measures 0.0003 - 0.007 cubic mm and are located superficially in the brain cortex in terms of their changing the tissue's reflectance in proportion to their local, intracapillary concentration (Eke et al. 1979; Eke 1982, 1988a). The erythrocyte-indicator, saline, is a diluent solution for erythrocytes. In essence, however, the endogenous red blood cells themselves act as indicators. Dyes bound to plasma proteins - like Evans blue - perform as plasma-indicators.

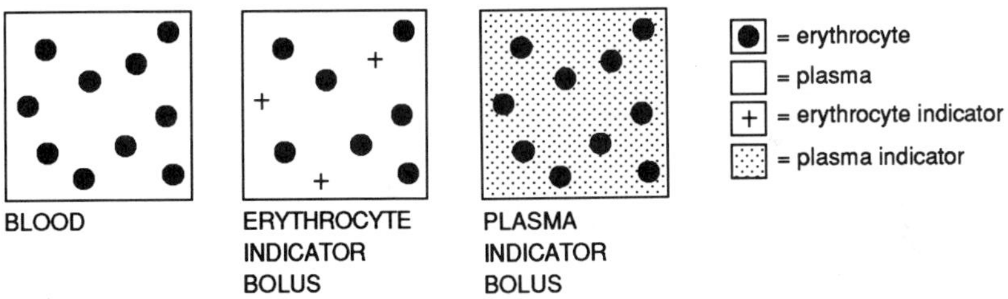

FIGURE 2. Indicators used by the multiparametric imaging method of the cerebrocortical microcirculation.

The magnitude of the video signal intensity at any given pixel location corresponds to the reflectance intensity from the tissue voxel imaged by that pixel (Fig. 3). The Macintosh II workstation shown in Fig. 1 performs reflectance image digitizing, densitometric and methodological calculations, image reconstruction and displaying under sofware control. The Macintosh computer is ideally suited for implementing the complex tasks of the multiparametric imaging method, which requires extreme computational power and flexible imaging capabilities. These tasks begin with videoreflectometric data preparation, quality control and digitization implemented by the *Input Software Module* based on a public domain software called "Image" created by Dr. Wayne Rasband (NIH, Bethesda, Maryland, U.S.A.). Real-time video image digitization of the reflectance record results in a stack of reflectance images, essentially a digital movie of the microregional transit of the indicators from which the microregional indicator-dilution curves can be reconstructed.

The *Data Processing Software Module* developed by the author takes care of the data processing and methodological calculations (Eke 1988) utilizing stacks of reflectance images generated by the Input Software Module from the video records of microregional indicator-transit. First, it reconstructs a microregional indicator-dilution curve for each pixel of the reflectance image stack. Then, it performs a number of methodological calculations on each of these microregional indicator-dilution curves resulting in the aforementioned microcirculatory parameters. This is a substantial task in terms of the amount of data to be processed. It can only be done efficiently under software control, which does not require human assistance for computing maps of local erythrocyte and plasma volumes, mean transit times, flows, tube and discharge hematocrits, blood volume and flow from the reflectance data set. The result can be hundreds of microcirculatory parameter images - like the ones shown in Fig. 4 - automatically databased for further numerical or statistical analyses.

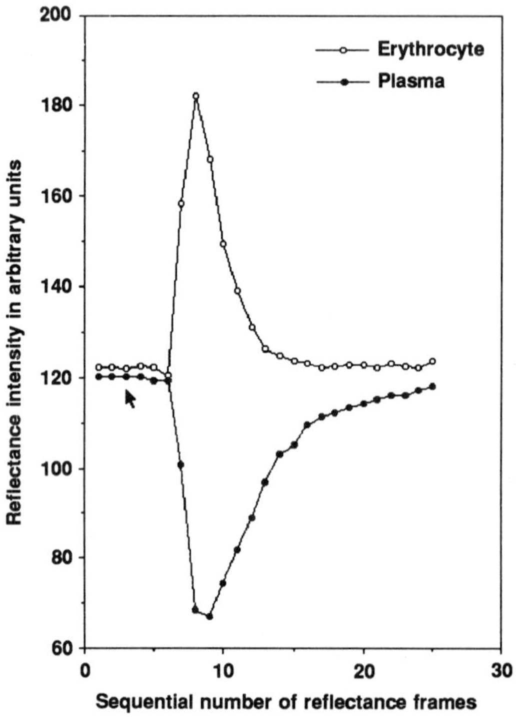

FIGURE 3. Indicator dilution curves obtained by the multiparametric imaging method during the transit of an erythrocyte and plasma indicator bolus through a microarea of the cerebrocortical circulation. Indicator injection in form of a slug injection of these indicators were done at the time indicated by the arrow (t_o). Indicator dilution curves were superimposed at their respective t_o.

Microcirculatory parameter image data can be handled with ease and in a very complex manner using the graphic tools of the Macintosh II computer with 8 megabyte of core memory and a 600 megabyte storage capacity of a rewritable optical cartridge drive (Fig. 1). Program "Image" can be very effectively used to implement the tasks of this final segment of the methodological protocol, the *Output Software Module*: plotting profiles across regions of interest, generating histograms, differential maps, animations, etc. (See Figs. 5, 6 and 8) (Eke 1992).

Animation of Microcirculatory Parameter Maps

Recent implementation of the reflectometric methodology on the Macintosh II workstation has provided the platform for creating animations of the aforementioned local parameter maps by program "Image" that show them react to an ischemic stress in the brain cortex (Fig. 8). These intensity coded digital movies show spatial and temporal events in the parameter maps as they evolve over the time span of ischemia (90 minutes) and reperfusion (30 minutes) compressed into a 10 second time window of the animation. The multiparametric nature of the reflectometric methodology in itself can provide the means for a very complex analysis of the cerebrocortical microcirculation. Earlier, reconstruction of time-function and thereby a dynamic analysis has been possible only for single elements of the parameter maps or a lumped group of these elements representing changes in a single tissue microarea (Eke 1988b). The parameter maps representing the complex spatial relations of these local variables could only be studied in a static manner. Animation, however, by opening up the dimension of time for a set of parameter maps organized in a "stack" for the computer to play as a digital movie allows a dynamic perception and processing of their spatial and temporal composition to take place in our brain thereby enhancing our understanding of the most complex aspects of the observed microcirculatory phenomenon.

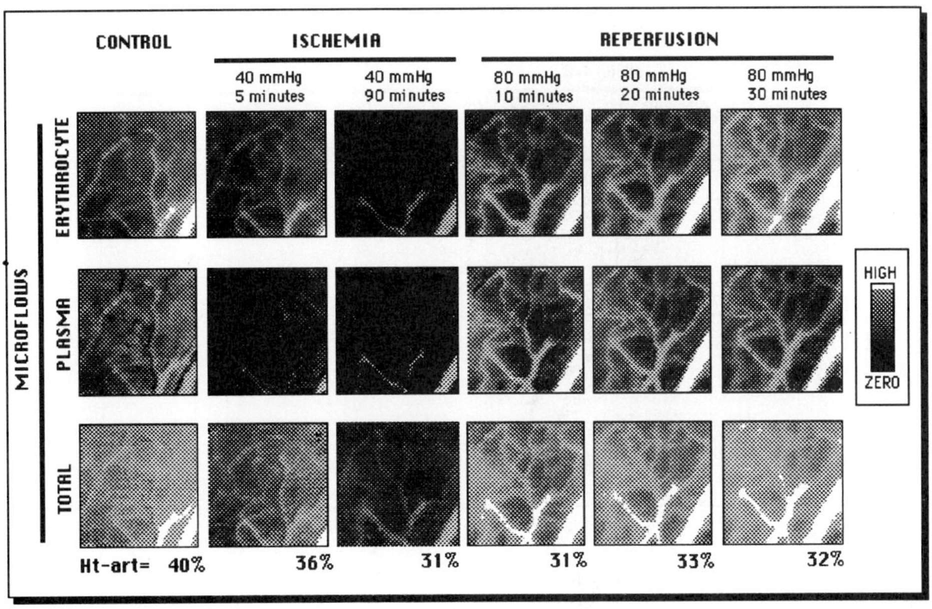

FIGURE 4. Erythrocyte, plasma and total microflow images obtained in an anesthetized cat within a cerebrocortical area of 3x3 mm at a spatial resolution of 100x100 image elements (voxels). These maps were acquired in control condition and at given time intervals during global cerebral ischemia induced by central arterial hemorrhage followed by a period of reperfusion. Image data are presented in intensity code by the scale shown on the right. Central arterial hematocrit values are shown as "Ht-art". In most cases, flow within pial vascular structures stands out from the mean of the map characteristic to flow within the intraparenchymal compartments (capillary beds). The large vessel in the lower right hand corner of the images is a larger pial artery that remains patent at all time during the ischemic stress.

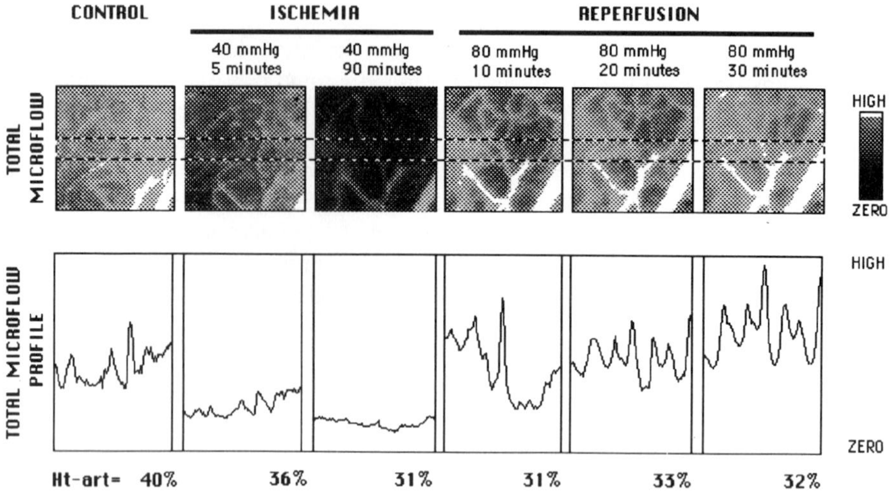

FIGURE 5. Total microflow maps and profiles for the maps shown in Fig. 4. Profiles were generated by averaging the image data within the area between the two dashed lines. Central arterial hematocrit values are shown as "Ht-art".

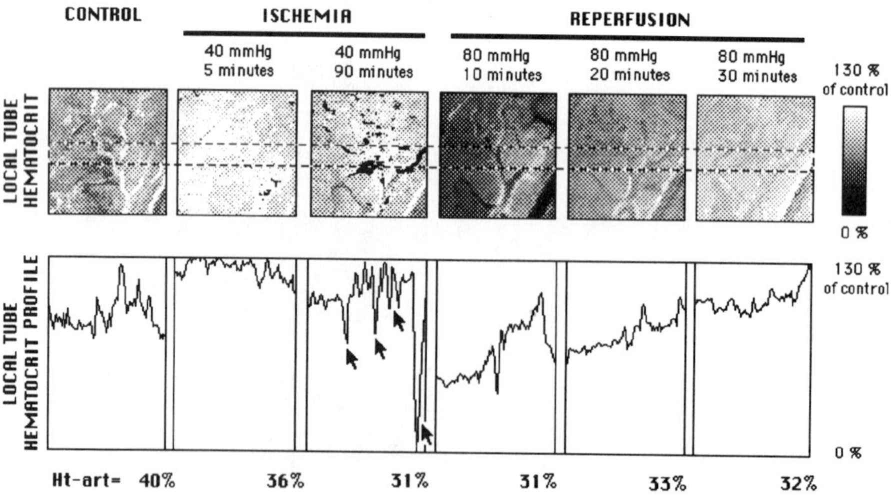

FIGURE 6. Local tube hematocrit maps and profiles obtained simultaneously with the microflow maps shown in Fig. 4. Profiles were generated by averaging the image data within the area between the two dashed lines. Central arterial hematocrit values are shown as "Ht-art". The maps are being composed of 100 x 100, that is 10000 individual, simultaneous measurements of local tube hematocrit. Lighter areas indicate a rise in hematocrit. Solid black areas on the second and third maps from the left indicate microareas where valid values of local hematocrit could not be obtained due to severely stagnant flow conditions (stasis). These spots are also noted by the arrows in the corresponding profiles.

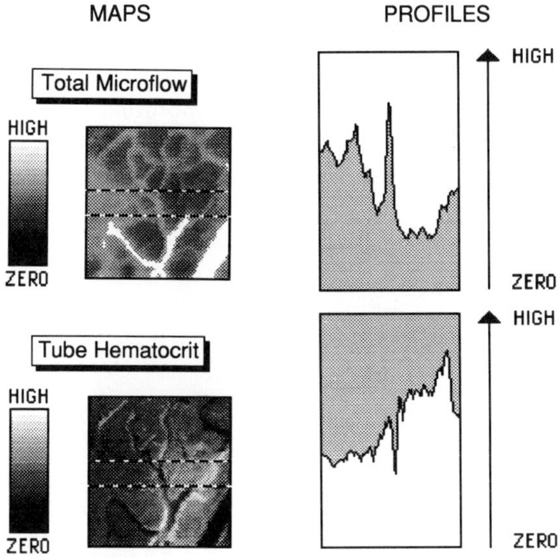

FIGURE 7. Total microflow and local tube hematocrit maps and profiles taken from Figs. 5 and 6 showing the reciprocal relation of these two parameters across a 3 mm segment of the cerebrocortical parenchyma and pial vasculature.

RESULTS

High resolution maps and profiles of microregional (local) erythrocyte, plasma and total flows along with that of local tube hematocrit obtained within a cerebrocortical area of 3 x 3 mm in an anesthetized cat are shown in Figures 4-8. These maps were acquired in control condition and at given time intervals during global cerebral ischemia induced by central arterial hemorrhage followed by a period of reperfusion. Arterial hemorrhage was carried out by letting the animal's blood via a catheter introduced into the abdominal aorta into a pressurized reservoir under computer control so that the mean arterial blood pressure could be lowered to and maintained at 40 mmHg for a period of 90 minutes sufficient to produce global cerebral ischemia. An analysis of hematocrit changes in the extra- and intraparenchymal circulation under these experimental conditions has been published earlier (Eke 1988b) based on regionally intergrated values. Current level of sophistication allows us to look at more closely to local variations in the microcirculatory parameters for the maps are being composed of 100 x 100, that is 10000 individual, simultaneous measurements.

In the selected experiment, erythrocyte, plasma as well as total microflow abruptly decreased upon lowering the mean arterial blood pressure (MABP) from control to 40 mmHg (Figs. 4 and 5). By the 90th minute at this level of MABP, highly focal areas of stagnant flow or stasis have developed (Figs. 4 and 6). These areas are surrounded by microareas of high local tube hematocrit but of total flow still at the level of the map's mean. Some of these areas got later promptly reperfused, but some showed a delayed recovery in its hematocrit. It is worthwhile to note that erythrocyte and plasma flow neither in regional nor in local areas did recover in a parallel manner following global ischemia. This can result in an asymmetry in the distribution of local tube hematocrit within the imaged tissue area as can be seen in the hematocrit map and profile taken at the 10th minute of the reperfusion period (Fig. 6). Total microflow in the same area showed an inverse pattern and profile (Fig. 7). Despite of the preliminary nature of these data, they indicate the importance of local tube hematocrit in affecting local flow in the capillary networks of tissue microareas in the brain cortex.

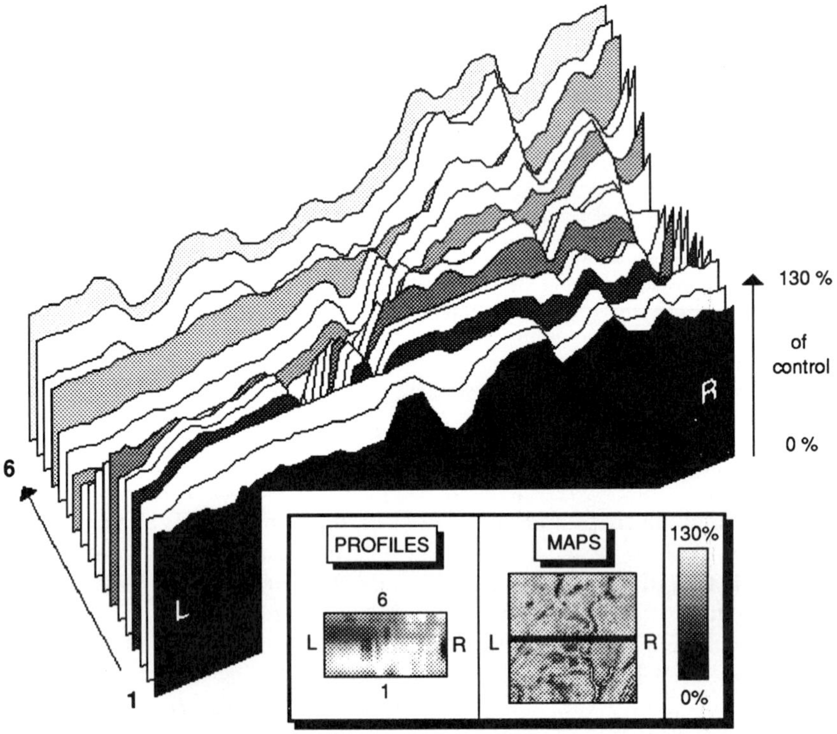

FIGURE 8. Three dimensional (3-D) surface plot of time dependent changes of local tube hematocrit maps shown in Fig. 6 along a profile of 3 mm between points L(eft) and R(ight) during global cerebral ischemia followed by reperfusion. The purpose of this figure is to emphasize the integrative analysis in the space and time domain that has become possible with the multiparametric reflectometric imaging method. The sequence of the analysis as done by the "Image" program was as follows. First, the local tube hematocrit maps shown in Fig. 6 were organized into a stack shown with the control map on top in the insert. Then a profile between points L and R was defined across the image and image data were extracted along this profile for all of the 6 maps in the stack. Data of these profiles were intensity coded and in the L to R (space) and 1 to 6 (time) order placed into an image shown on the left in the insert. This data set was then plotted as 3-D surface plot. Note that for improving the perception of the changes of the data from one profile to the other, additional profiles have also been generated by extrapolating in between two original profiles in the stack identified by varying shades of gray. Note the asymmetry in the local tube hematocrit profile as it develops during reperfusion. This presentation is the closest to animation of the microcirculatory parameter maps that the reflectometric method can support but that cannot be reproduced in printed form, for it is based on computer controlled updating of the intensity or color coded computer display of microcirculatory parameter maps from a stack of these maps acquired in time sequence.

ACKNOWLEDGMENTS

This work has been jointly supported from personal funds of the author, by grants-in-aid of the Departments of Neurology and Pathology of the University of Alabama at Birmingham, Birmingham, U.S.A., the Experimental Research Department and 2nd Institute of Physiology of the Semmelweis University of Medicine, Budapest, Hungary and in part by OTKA Grant 2040. The cooperation of Dr. Wayne Rasband at the NIH, Bethesda, Maryland, U.S.A. is greatly appreciated in incorporating features of the Input and Output Software Modules in his "Image" program. The author can be reached for further technical details and with inquiries on installation of the method's hardware and software on the following telephone/fax number: (36-1) 134-3162.

REFERENCES

Eke A. (1982) Reflectometric mapping of microregional blood flow and blood volume in the brain cortex. J Cereb Blood Flow Metabol 2:41-53.

Eke A. (1983a) Integrated microvessel diameter and microregional blood content as determined by cerebrocortical video reflectometry, in: "The Cerebral Veins," L. M. Auer, and F. Loew, eds., Springer Verlag, Vienna, New York .

Eke A. (1983b) Heterogeneity of cerebrocortical microflow in epileptic seizure, in: "Cerebral Blood Flow, Metabolism and Epilepsy," M. Baldy-Moulinier, D. H. Ingvar, and B. S. Meldrum, eds., John Libbey Eurotext, London, Paris.

Eke A. (1984) Repetitive mapping of tissue hematocrit over microareas of the brain cortex, Intl J Microcirc Clin Exp 3(3/4):548.

Eke A. (1988a) Reflectometric imaging of local tissue hematocrit in the cat brain cortex, in: "Cerebral Hyperemia and Ischemia: From the Standpoint of Cerebral Blood Volume," M. Tomita, T. Sawada, N. Naritomi and W. D. Heiss, eds., Elsevier, Amsterdam.

Eke A. (1988b) Hematocrit changes in the extra- and intraparenchymal circulation of the feline brain cortex in the course of global cerebral ischemia, Adv Exp Med Biol 248:439-449.

Eke A. (1992) Instrumentation and technology for multiparametric mapping of intraparenchymal circulation in the brain cortex. Adv Exp Med Biol 317:671-678.

Eke A., Halsey J. H. (1984) Distribution of cerebrocortical microflow in normo- and hypertensive rats. Adv Exp Med Biol 180:203-210.

Eke A., Hutiray Gy., Kovach A. G. B. (1979) Induced hemodilution detected by reflectometry for measuring microregional blood flow and blood volume in cat brain cortex. Am J Physiol 236(5):H759-H768.

TOWARDS IMAGING OF CEREBRAL BLOOD FLOW AND METABOLISM ON A MICROSCOPICAL SCALE IN VIVO

A. Villringer and U. Dirnagl

Department of Neurology
University of Munich
Munich, Germany

INTRODUCTION

Brain cell function and cerebral blood flow are tightly coupled (Schmidt and Hendrix 1938, Olesen 1971, Leniger-Follert and Hossmann 1979, Sokoloff 1981, Frostig et al. 1990). However, the mechanisms how brain cells and cerebral blood vessels do interact in order to adapt cerebral blood flow to different functional and metabolic needs are unknown, although a variety of potential ways of interactions have been proposed (Lassen 1959, Kuschinsky et al. 1972, Rubio et al. 1975, Lou et al. 1987, Heistad and Kontos, 1983, Iadecola et al. 1991, Dirnagl et al. 1993a). One important reason for this situation is the lack of a suitable technique to study this interaction. An ideal method should assess both aspects, i.e. the functional and/or metabolic state of brain cells on the one hand, and cerebral blood flow on the other hand simultaneously. The spatial resolution of such a method should be at least on the order of the objects of investigation, i.e. brain cells and microvessels, and in order to study the dymamics of the interaction and to avoid artifacts induced by an ex vivo approach, the method preferably should be performed in vivo.

In this communication, we propose an optical method, which employs confocal laser scanning microscopy (Minsky 1957, Fine et al. 1988) in a closed cranial window model of the rat in vivo (Villringer et al. 1989, Dirnagl et al. 1991a). Using this method, we demonstrate the feasibility to assess intracellular ion concentrations as a marker for brain cell function and/or metabolism and cerebral blood flow, simultaneously.

MATERIALS AND METHODS

Animal preparation

Male Wistar rats (250 - 350 g) were anesthetized (100 mg/kg body weight thiobutabarbiturate, Inactin^R, Byk Gulden), tracheotomized, and artificially ventilated,

Optical Imaging of Brain Function and Metabolism, Edited by
U. Dirnagl *et al.*, Plenum Press, New York, 1993

and the femoral artery and vein were cannulated. A craniectomy (5 mm x 5 mm) was performed, and the dura mater incised and removed. In order to reestablish the integrity of the skull and ensure maintenance of normal gas tensions and extravascular pressure, a closed cranial window was implanted. Arterial blood pressure, intracranial pressure (ICP), endexspiratory pCO_2, and body temperature were monitored continuously, while blood gases and hematocrit were determined intermittently. Body temperature was maintained at 37.5 - 38.5 °C with a heating pad; arterial pCO_2, pO_2, and pH were controlled at 30 - 35 mmHg, 80 - 110 mmHg, and 7.35-7.45, respectively, by adjusting the ventilator and the F_iO_2. By adjusting the level of the outflow catheter of the cranial window, ICP was regulated between 2 and 5 mmHg. Animals with blood pressures lower than 80 mmHg and higher than 110 mmHg were excluded. The animals were placed in a stereotactic frame beneath the confocal laser scanning unit (Bio-Rad MRC 600).

Labelling of intravascular space

Depending on the experimental paradigm, plasma was labelled by intravenous injection of sodium fluorescein (MW 376, Sigma Chemicals) or Texas Red conjugate (MW 40 000, Molecular Probes) 1-2 mg/100 g bodyweight in 0.5 ml 0.9% saline each. These dyes do not penetrate the intact blood brain barrier. Blood cells appear in negative contrast.

Labelling of brain cells

The labeling procedure is described by Them et al. in this volume in more detail. In brief, the brain surface was exposed to AM-esters of ion sensitive fluorescent dyes (BCECF, Ca-Crimson, Fura Red, Fluo-3, all Molecular Probes, Oregon, USA) solved in artificial CSF-solution. These lipophilic agents penetrate cell membranes, and after intracellular cleavage of the ester, they are trapped intracellularly (Tsien 1981, Tsien 1989).

Confocal Laser Scanning Microscopy

The principles of confocal microscopy have been described in great detail elsewhere (Wilson and Carlini 1987, Wilson 1989, Brakenhoff 1989, Carlsson 1990, Fine et al. 1988, Inoue 1990). We used a BioRad MRC 600 CLSM unit (BioRad Microscience Division, UK) attached to a Nikon Optiphot microscope with a Zeiss water immersion lens (x 40, numerical aperture 0.75, working distance 1.6 mm, corrected for a cover slip). The light source was an Argon/Krypton Laser (Ion Laser Technologies) with principal lines at 488 nm (blue), 568 nm (yellow), and 647 nm (red). Confocal microscopy was performed through the cranial window which was made of a microscope cover glass. Appropriate filter sets were used according to the fluorescent dye(s) employed. Dual channel detection permitted the optical separation of dyes with differing optical properties.

Image acquisition time of a two dimensional image was between 1 s and a minimum of 1/16 s depending on the number of pixels/frame. To quantify blood cell flow dynamics, a single line approach (Hernandez-Cruz et al. 1990) was employed (Figure 4), trading spatial for temporal resolution. The data acquisition time for a single line was 2 milliseconds. If such a line is scanned repetitively and subsequent acquisitions are displayed beneath each other in a two-dimensional image, a space (horizontal)-time (vertical) plot is generated. If the line passes a capillary and a brain cell labelled with a fluorescent indicator of intracellular ion concentration blood cell perfusion within the capillary and the time course of the intracellular ion concentration are assessed simultaneously.

Laser Doppler Flowmetry

A Laser Doppler Flow probe (Dirnagl et al. 1989, Haberl et al. 1989) was placed on the cranial window irradiating an area on the brain surface 2 mm from the focus of the confocal microscope. This setup permitted continuous monitoring of local CBF within a brain volume of approximately 1 mm^3.

RESULTS

In noninvasive optical sections, individual *brain cells* that were labelled with ion sensitive dyes were resolved up to a depth of 100 μm beneath the brain surface. Depending on the dye employed, ion concentration dependent fluorescent signal was obtained from 100 to 200 μm (in the case of BCECF) beneath the surface. We used the pH sensitive dye BCECF (Figure 1), and the Calcium-concentration sensitive dyes Fluo 3 (Figure 2), Calcium Crimson (Figure 3) and FuraRed (Figure 4).

The *microcirculation* was assessed either by observing the width of pial vessels which are seen unlabeled on the brain surface (Figure 1), by laser Doppler flowmetry (Figure 2, Haberl et al. 1989, Dirnagl et al. 1989) or by monitoring blood cell and plasma perfusion in subsurface intracortical capillaries with the CLSM (Figure 4).

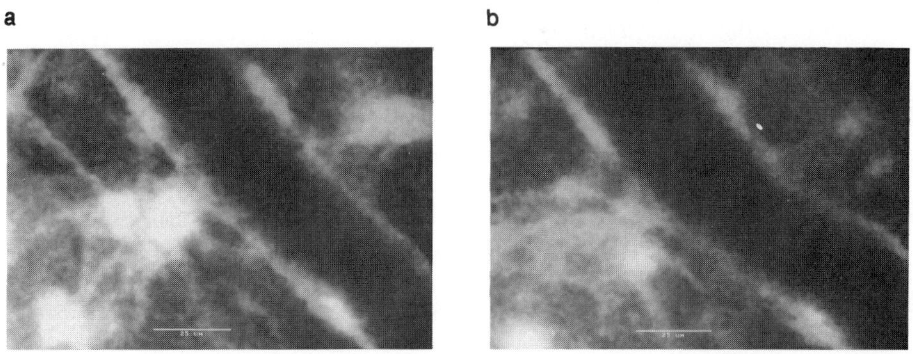

FIGURE 1. Optical section 20 μm beneath the brain surface. Labelling of brain cells was achieved by 30 min superfusion of the brain surface with the 10μM solution (in artificial CSF) of the pH indicator BCECF. Figure 1 a is 10 min before and Figure 1 b 10 min after transient global ischemia (occlusion of both carotid arteries and hypotension by exposing the body to subathmospheric pressure for 5 min, Dirnagl et al. 1993b). Scale bar is 25 μm.

Various ways to *combine* these approaches for the simultaneous assessment of *intracellular ion concentrations and the cerebral microcirculation* are given in Table 1, and the feasibility of these combinations is illustrated in Figures 1-4.

Figure 1 shows an optical section 20 μm beneath the brain surface. Brain cells are labeled with the pH indicator BCECF and pial microvessels are visible unstained in negative contrast. An image at the same position was acquired before (Fig 1 a) and

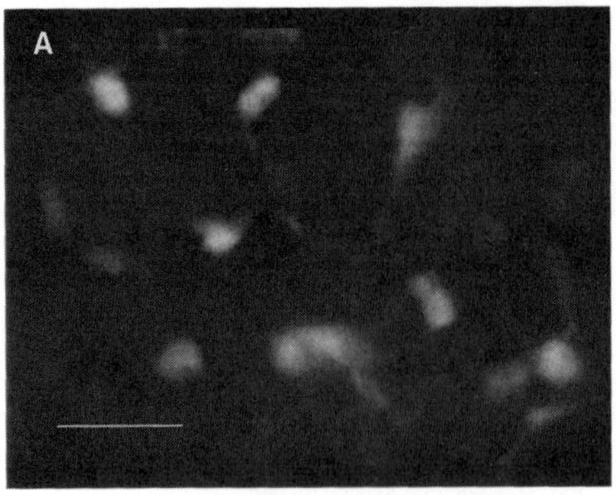

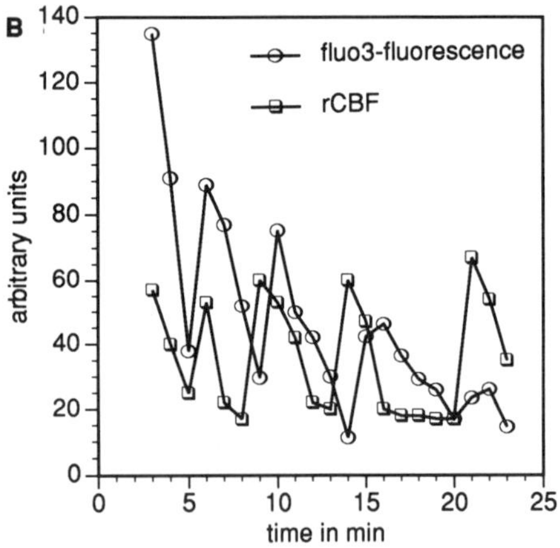

FIGURE 2. A: Optical section 30 μm beneath the brain surface. In order to label brain cells the brain surface was superfused with 10μM Fluo 3 in artificial CSF for 30 min. **B:** Time course of intracellular signal intensity of brain cell labelled with Fluo 3 and laser Doppler measured cerebral blood flow in a distance of 2 mm during waves of spreading depression. Spreading depression was induced by exposing the frontal brain surface to 50 mM KCl.

10 minutes after transient global ischemia (Figure 1b). The drop in cellular fluorescence correponds to a lower intracellular pH, the pial arteriole is dilated after the ischemic event ("postischemic hyperperfusion").

Figure 2 a gives an optical section 30 μm beneath the brain surface. The brain cells were labelled with the Calcium indicator Fluo 3. In this experiment cerebral blood flow was assessed by Laser Doppler flowmetry 2 mm adjacent to the focus of the confocal microscope. The time course of intracellular Calcium concentration related

TABLE 1. Approaches for simultaneous assessment of intracellular ion concentrations and cerebral microcirculation

Cell Label	Vascular label	Visualization of	Measurement parameter
intracellular fluoresc. dyes (BCECF, Fluo-3, Ca-Crimson, Fura Red)	none	brain cells, pial vessels in negative contrast	intracellular ion concentrations width of pial arterioles
intracellular fluoresc. dyes (BCECF, Fluo-3, Ca-Crimson, Fura Red)	laser doppler flowmetry	brain cells, pial vessels in negative contrast	intracellular ion concentrations width of pial arterioles, local CBF
intracellular fluoresc. dyes (BCECF, Fluo-3, Ca-Crimson, Fura Red)	intravascular fluoresc. dye (fluorescein, Texas-Red)	brain cells, pial vessels in positive contrast, subsurface capillaries: plasma in positive, blood cells in negative contrast	intracellular ion concentrations, width of pial arterioles, capillary blood cell flow, capillary plasma flow

fluorescent signal and of laser Doppler blood flow during waves of spreading depression (induced by potassium chloride solution applied onto the brain surface through a second cranial window located over the frontal cortex) is given in Figure 2b.

Double labeling of the intracellular and intravascular space is shown in Figure 3. The optical section given in Figure 3 is 30 μm (A/B) and 100 μm (c/D) beneath the brain surface. Brain cells are labeled with Calcium Crimson (A/C), the intravascular space with fluorescein sodium (B/D). The separation of the emission and excitation wavelengths was provided by appropriate filter sets.

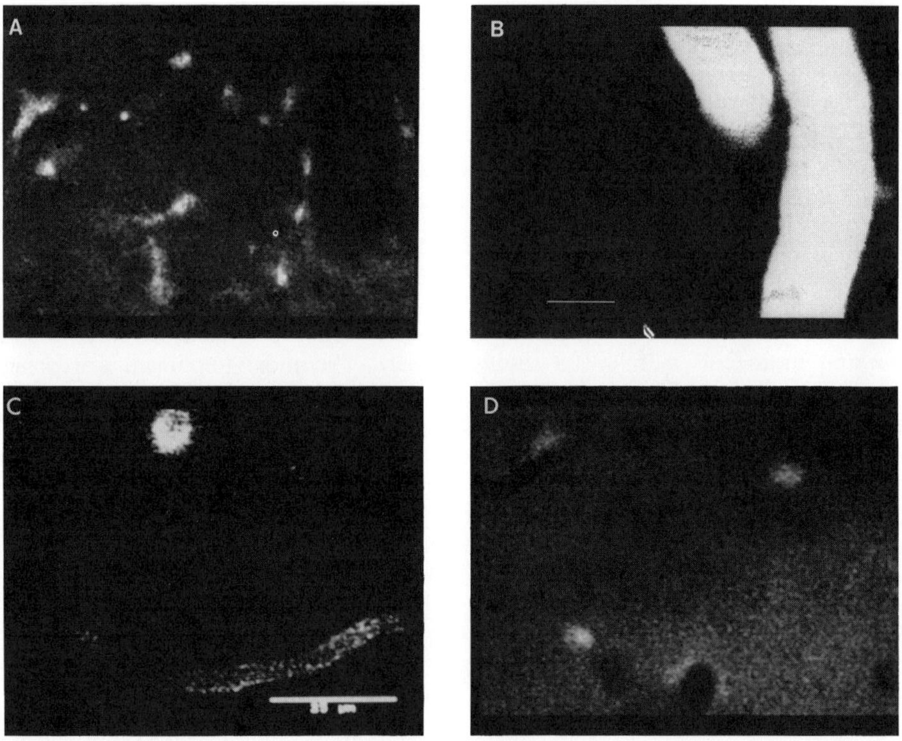

FIGURE 3. Double labeling of brain cells and the intravascular compartment. Brain cells labelled with Calcium Crimson (A/C, 30 min superfusion of the brain surface with 10μM solution in artificial CSF). The intravascular space was labeled by bolus intravenous injection of 1 mg/100 g body weight sodium fluorescein (B/D). **A/B:** Optical section 30 μm beneath the brain surface. Scale bar is 50μm **C/D:** Optical section 100 μm beneath the brain surface. Scale bar is 25μm.

In order to assess blood cell flow in a single capillary, a one-dimensional imaging approach was employed by scanning a single line (temporal resolution 2 ms) that is placed across a capillary and a Fura Red (intracellular calcium indicator) labeled cell. Figure 4 show a space time plot (subsequent lines displayed on the TV-screen) during a wave of spreading depression. The high flow condition in the capillary corresponded to low signal within the brain cell corresponding to high intracellular Calcium concentration (Fig. 4 b) and vice versa (Fig. 4 a).

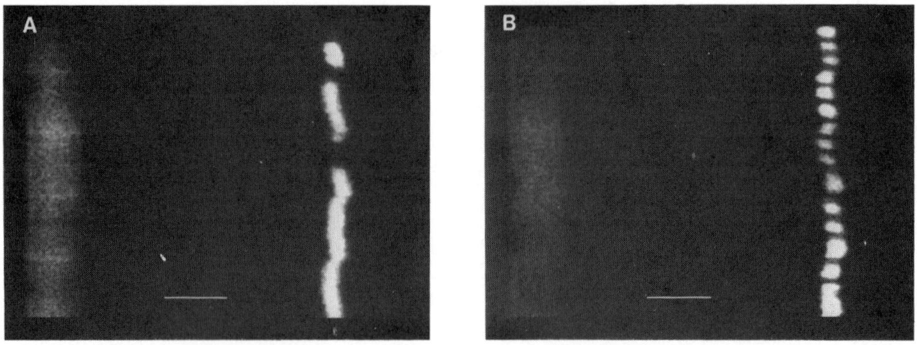

FIGURE 4. Simultaneous assessment of blood cell flow in capillary and intracellular ion Calcium concentration. A horizontal scan line was placed across a FuraRed-labelled (30 min superfusion of brain surface with 10μM solution in artificial CSF) brain cell (left colummn) and a fluorescein-labelled capillary (column on right side). Subsequent single line scans are placed beneath each other to form a space (horizontal axis) - time (vertical axis) plot. Figure 4 a and Figure 4 b were acquired during different phases of a spreading depression wave. Figure 4 a: low flow phase (as assessed by comcomittant laser Doppler flowmetry). Figure 4 b: high flow phase (as assessed by concomittant laser Doppler flowmetry).

DISCUSSION

A number of different methods have been proposed to measure cerebral blood flow and brain cell metabolism simultaneously. The most important are positron emission tomography (Phelps and Mazziotta 1985, Knudsen 1993), magnetic resonance based approaches (Corbett et al. 1991, Frahm et al. 1992) and ex vivo autoradiographic methods (Mies et al. 1986, Mies et al. 1992). All these have significant limitations: PET has a spatial resolution on the order of millimeters and it is difficult to perform sequential examinations, magnetic resonance spectroscopy has a poor spatial resolution on the order of 1 cm^3 (Frahm et al. 1991, 1992), and ex vivo approaches (Mies et al. 1986, Mies 1993), are not able to monitor dynamic events and post mortem artifacts may hamper the validity of the findings.

Optical techniques (Chance et al. 1992) theoretically offer high spatial resolution imaging and high temporal resolution. They have been used to monitor (1) brain cell activity and metabolism as well as (2) the cerebral microcirculation:

(1) The development of new loading techniques for fluorescent dyes (Tsien 1981, Tsien 1987) has enabled the non-invasive optical measurement of an array of biologically relevant ions in various cells. Those dyes have been used in microscopy of brain cells in culture (Cohan et al. 1987, Connor et al. 1986, Connor et al. 1988, Cornell Bell et al. 1990, Glaum et al. 1990, McDermot et al. 1986, Hernandez-Cruz et al. 1990, Holliday and Spitzer 1990, Jensen and Chiu 1991, McNaughton et al. 1990, McNaughton et al. 1990b, Michaels and Rothman 1990, Nedergaard et al. 1990, Przywara et al. 1991, Revest et al. 1991) brain slices (Kudo et al. 1990, Dani et al. 1991) and brain tissue in situ (Uematsu et al. 1988, Dirnagl et al. 1991b). Membrane potential and hence electrical acitivity changes of the brain tissue have been measured in real time (millisecond-range) with potential sensitiv dyes using macro lenses (Kauer 1988) or diode arrays (Grinvald et al. 1984). Recently, we have been able to demonstrate that using our model of CLSM in a closed cranial window preparation,

individual brain cells labelled with intracellular fluorescent dyes can be imaged and assessed during functional alterations (Dirnagl et al. 1991b, Them 1993).

(2) Using conventional wide-field microscopes the microcirculation on the brain surface was assessed in a number of studies (e.g. Morii et al. 1986, Haberl et al 1988), however, the poor depth penetration and discrimination of conventional microscopes prevented systematic studies of deeper vascular structures. Using laser Doppler flowmetry (Dirnagl et al. 1989) the microcirculation in an volume of appr. 1 mm^3 can be assessed. CLSM in a closed cranial window preparation (Villringer et al. 1989) permits the assessment of the three-dimensional morphology of the microvascular network up to 250 μm beneath the brain surface (Villringer 1990a, Dirnagl et al. 1991a), the assessment of erythrocyte and leukocyte flow within capillaries (Dirnagl et al 1992, Villringer et al. 1991), as well as plasma flow within capillaries (Villringer et al. 1990b).

In the present feasibility study, we combined above mentioned approaches for the assessment of intracellular ion concentration and cerebral blood flow. We demonstrate that the morphological as well as functional simultaneous assessment of these two compartments is possible. Various approaches to achieve this are demonstrated, differing in the intracellular ion measured (Hydrogen, Calcium) and in the method for the assessment of regional CBF (indirect by width of pial arteriole, laser Doppler flowmetry, capillary blood cell flow) (see also Table 1). We are currently investigating the usefulness of these methods in the assessment of blood flow and intracellular pH changes in transient global ischemia (Dirnagl et al. 1991b) and intracellular Calcium alterations and capillary blood cell flow during somatosensory stimulation of the parietal cortex.

ACKNOWLEDGEMENTS

This study was supported by the Sander Stiftung, Munich, the Deutsche Forschungsgemeinschaft (Di 454/4-1), Bonn, and the Friedrich-Baur Stifung, Munich.

LITERATURE

Brakenhoff GJ, van der Voort HTM, van Spronsen EA, Nanninga N (1989) Three-dimensional imaging in fluorescence by confocal scanning microscopy. J Microscopy 153:151-159(1989)

Carlsson K (1990) Scanning and detection techniques used in a confocal scanning laser microscopy. J Microscopy 157:21-27

Chance B, Villringer A, Dirnagl U, Einhäupl KM (1992) Optical Imaging of Brain Function and Metabolism. J Neurol 239:359-360

Cohan CS, Connor JA, Kater SB (1987) Electrically and chemically mediated increases in intracellular calcium in neuronal groth cones. J Neurosci 7:3588-3599

Connor JA (1986) Digital imaging of free calcium changes and of spatial gradients in growing processes in single mammalian centra nervous system cell. Proc Natl Acad Sci 83:6179-6183

Connor JA, Wadman WJ, Hockberger PE, Wong RKS (1988) Sustained dendritic gradients of Ca^{2+} induced by excitatory amino acids in CA1 hippocampal neurons. Science 248:649-653(1988)

Corbett RJT, Laptook AR, Olivares E (1991) Simultaneous measurement of cerebral blood flow and energy metabolites in piglets using deuterium and phosphorus nucrleas magnetic resonance. J Cereb Blood Flow Metabol 11:55-65

Cornell-Bell AH, Finkbeiner SM, Cooper MS, Smith SJ (1990) Glutamate induces calcium waves in cultured astrocytes: long-range glial signaling. Science 247:470-473

Dani JW, Chernjavsky A, Buchanan J, Smith SJ (1991) Neuronal activity elicits astrocyte Ca^{2+} waves and oscillations within hippocampal slices. Soc Neur Abstr 17: 57

Dirnagl U, Villringer A, Gebhard R, Haberl RL, Schmiedek, P, Einhäupl KM (1991a) Three-dimensional reconstruction of the rat brain cortical microcirculation in vivo.J Cereb Blood Flow Metab 11:353-360

Dirnagl U, Villringer A, Einhäupl KM (1991b) Imaging of intracellular pH in normal and ischemic rat brain neocortex using confocal laser scanning microscopy. J Cereb Blood Flow Metabol 11,Suppl2:S206

Dirnagl U, Villringer A, Einhäupl KM (1992) In-vivo confocal scanning laser microscopy of the cerebral microcirculation. J Microscopy 165:147-157

Dirnagl U, Villringer A, Haberl Rl, Einhäupl KM (1990) In vivo confocal Laser scanning microscopy (CLSM): laser light induced alterations of erythrocyte flow in rat brain capillaries sensitized with intravascular fluorescein. Transactions of the Royal Microscopical Society, New Series Volume 1:337-340

Dirnagl U, Kaplan B, Jacewicz M, Pulsinelli W (1989) Continuous measurement of cerebral cortical blood flow by laser-Doppler flowmetry in a rat stroke model. J Cereb Blood Flow Metab 9:589-596

Dirnagl U, Lindauer U, Villringer A (1993a) Role of Nitric oxide in the coupling of cerebral blood flow to neuronal activation. Neurosci Let 149:43-46

Dirnagl, U., Thoren, P., Villringer, A., Sixt, G., Them, A., and Einhäupl, K.M. (1993b). Global forebrain ischemia in the rat: controlled reduction of cerebral blood flow by hypobaric hypotension and two vessel occlusion. Neurol Res (in press)

Fine A, Amos WB, Durbin RM, McNaughton PA (1988) Confocal microscopy: applications in neurobiology. Trends Neurosci 8:346-351

Frahm J, Bruhn H, Hänicke W, Merboldt KD, Mursch K, Markakis E (1991) Localized proton NMR spectroscopy of brain tumors, using short-echo time STEAM sequences. J Comput Assist Tomogr 15:991-922

Frahm J (1992) Nuclear magnetic resonance studies of human brain in vivo: anatomy, function, and metabolism. Adv Exp Med Biol (this issue)

Frostig RD, Lieke EE, Ts'o DY, Grinvald A (1990) Cortical functional architecture and local coupling between neuronal activity and the microcirculation revealed by in vivo high-resolution optical imagin of intrinsic signals. Proc Natl Acad Sci USA 87:6082-6086

Glaum SR, Holzwarth JA, Miller RJ (1990) Glutamate receptors activate Ca^{2+} mobilization and Ca^{2+} influx intor astrocytes. Proc Natl Acad Sci 87:3454-3458

Haberl Rl, Heizer ML, and Ellis EF (1987) Effect of Thromboxane A_2 mimetic U 46619 on pial arterioles of rabbits and rats. Stroke 18:796-800

Haberl RL, Heizer M, Ellis EF (1989) Laser-Doppler assessment of brain microcirculation: effect of local alterations. Am J Physiol 25:H1255-H1260

Heistad DD, Kontos HA (1983) in: Handbook of Physiology, J.T. Shepherd and F.M. Abboud, Eds. (The William & Wilkins Company, Baltimore). vol. III, pp. 137-182.

Hernandez-Cruz A, Sala F, Adams PR (1990) Subcellular calcium transients visualized by confocal microscopy in a voltage clamped vertebrate neuron. Science 247:858-862(1990)

Holliday J, Spitzer NC (1990) Spontaneous calcium influx and its roles in differentiation of spinal neurones in culture. Dev Biol 141:13-23

Iadecola C (1991) Intrinsic and extrinsic neural regulation of the cerebral circulation. in: Schmiedek P, Einhäupl K, Kirsch CM (editors), Stimulated cerebral blood flow, experimental findings and clinical significance, Springer Verlag Berlin, Heidelberg, New-York. pp 19-36

Inoue S (1990) Foundations of confocal scanned imaging in light microscopy. in: Handbook of Biological confocal microscopy. Ed: Pawley JB. Plenum Publishing Corp. New York, pp 1-14

Jensen AM, Chiu SY (1991) Differential intracellular calcium responses to glutamat in type 1 and type 2 cultured brain astrocytes. J Neurosci 11:1674-1684

Kuschinsky W, Wahl M, Bosse O, Thurau K (1972) Pervascular potassium and pH as determinants of local pial arterial diameter in cats. A microapplication study. Circ Res 31:240-247

Knudsen GM (1993) Blood brain barrier transport measurements using PET-scanning and intavenous double indicator technique. Adv Exp Med Biol (this volume)

Kudo Y, Takeda K, Yamazaki K (1990) Quin 2 protects against neuronal cell death due to Ca-overload. Brain Res 528:48-54

Lassen NA (1959) Cerebral blood flow and oxygen consumption in man. Physiol Rev 39:183-238

Leniger-Follert E, Hossmann KA (1979) Simultaneous measurement of microflow and evoked potentials in the somato-motor-cortex of the cat during specific sneory activation. Pflügers Arch 380:85-89

Lou HC, Edvinsson L, MacKenzie (1987) The concept of coupling blood flow to brain function: revision required ? Ann Neurol 22:289-297

MacDermott AB, Mayer ML, Westbrook GL, Smith SJ, Barker JL (1986) NMDA-receptor activation increases cytoplasmic calcium concentration in cultured spinal cord neurones. Nature 321:519-522

McNaughton LA, Lagnado L, Hunter SP, McNaughton PA (1990a) Use of the confocal microscope to measure changes in free intracellular Calcium in type 1 astrocytes cultured from rat cerebral cortex. J Physiol 424, 5P

McNaughton LA, Lagnado L, Socolovsky M, Hunt SP, McNaughton PA (1990b) Glutamate elevates free intracellular Calcium in type 1 astrocytes from rat cerebral cortex. J Physiol 424:48P

Michaels RL, Rothman SM (1990) Glutamate neurotoxicity in vitro: Antagonsit pharmacology and intracellular ion concentrations. J Neurosci 10:283-292

Minsky M. (1957) US Patent No. 3013467

Mies G, Bodsch W, Paschen W, Hossmann K-A (1986) Triple-tracer autoradiography of cerebral blood flow, glucose utilization, and protein synthesis in rat brain. J Cereb Blood Flow Metab 6:59-60

Mies G (1993) Autoradiographic and biochemical imaging in cerebral ischemia. Adv Exp Med Biol, this issue

Morii S, Ngai AC, Winn R (1986) Reactivity of rat pial arterioles and venules to adenosine and carbon dioxide: With detailed description of the closed cranial window technique in rat. J Cereb Blood Flow Metab 6:34-41

Nedergaard M, Desai S, Pulsinelli W (1990) Dicarboxy-dichlorofluorscein: a new fluorescent probe for measuring acidic intracellular pH. Anal Biochemistry 187:109-114

Olesen J (1971) Contralateral focal increase of cerebral blood flow in man during arm work. Brain 94:635-646

Phelps ME, Mazziotta JC (1985) Positron Emission Tomography: Human Brain Function and Biochemistry. Science 228:799-899

Przywara DA, Bhave SV, Bhave A, Wakade TD, Wakade AR (1991) Stimulated rise in neuronal calcium is faster and greater in the nucleus than in the cytosol. FASEB J 5:217-222

Revest PA, Abbott NJ, Gillespie JI (1991) Receptor-mediated changes in intracellular Calcium-concentration in cultured rat brain capillary endothelial cells. Brain Res 549:159-161

Rubio R, Berne RM, Bockmann EL, Curnisk RR (1975) Relationship between adenosine concentration and oxygen supply in rat brain. Am J Physiol 228:1896-1902

Schmidt C, Hendrix J (1938) Action of chemical substances on cerebral blood vessels. Res Publ Assoc Res Nerv Ment Dis 18:229-276

Sokoloff L (1981) Relationship among local functional activity, energy metabolism, and blood flow in the central nervous system. Federation Proc 40

Them A (1993) Intracellular ion concentrations in the brain: approaches towards in situ confocal imaging. Adv Exp Med Biol (this issue)

Tsien RY (1981) A non-disruptive technique for loading calicum buffers and indicators into cells. Nature 290:527-528

Tsien RY (1989) Fluorescent indicators of ion concentrations. Methods in Cell Biology 30:127-156

Uematsu D, Greenberg JH, Reivich M, Karp A (1988) In vivo measurement of cytosolic free calcium during cerebral ischemia and reperfusion. Ann Neurol 24:420-428

Villringer A, Haberl RL, Dirnagl U, Anneser F, Verst M, Einhäupl KM (1989) Confocal Laser Microscopy to study microcirculation on the rat brain surface in vivo. Brain Research 504:159-60

Villringer A, Dirnagl U, Gebhardt R, Haberl RL, Einhäupl KM (1990a) Non-invasive optical sectioning of structures in the living rat brain. Transactions of the Royal Microscopical Society, New Series Volume 1:353-356

Villringer, A., Dirnagl, U., Haberl, R.L., Schürer, L., Büttner, U., and Einhäupl, K.M. (1990b). Visualization of erythrocytes; leukocytes; and blood plasma in the rat brain microcirculation in vivo using confocal laser scanning microscopy. Soc Neurosci Abstr 16, Part 1, 24.

Villringer A, Dirnagl U, Them A, Schürer L, Krombach F, Einhäupl KM (1991) Imaging of Leukocytes in the rat brain cortex in vivo. Microvasc Res 42:305-315

Villringer A, Dirnagl U, Piepgras A, Schmiedek P, Einhäupl KM (1992a) Erythrocyte flow in cerebral capillaries under resting and stimulated conditions. in: Schmiedek P, Piepgras A, Einhäupl KM (eds): Cerebral Blood Flow under Stimulated Conditions, Springer Verlag, Berlin, Heidelberg, New-York, pp. 61-65

Villringer A, Dirnagl U, Einhäupl KM (1992b) Microscopical visualisation of the brain in vivo. in: Kriete (ed): Visualisation in biomedical microscopies, VCH, Weinheim, 161-181

Wilson T (1989) Trends in confocal microscopy. Trends in Neurosci 12:486-493

Wilson T, Carlini AR (1987) Three-dimensional imaging in confocal imaging systems with finite sized detectors. J Microscopy 149:51-66

DETECTION OF BRAIN FREE OXYGEN RADICAL GENERATED PHOTONS IN VIVO: PRELIMINARY RESULTS

U. Dirnagl, U. Ködel, H.W. Pfister, A. Villringer, L. Schleinkofer‡, and K.M. Einhäupl

Department of Neurology, University of Munich, Germany, and
‡ Hamamatsu Photonics Deutschland, Herrsching, Germany

INTRODUCTION

Oxygen free radicals have been implicated as a major factor of organ tissue damage in an array of diseases (heart: McCord 1988, intestine: Kaminski and Proctor 1990, endothelium: Matsubara and Ziff 1986, muscle: Freischlag and Hanna 1991, lung: Till et al. 1991, liver: Reinke et al. 1991). In the nervous system, oxygen free radicals have been incriminated as mediators of neurodegeneration (Adams and Odunze 1991), brain trauma (Ikeda et al. 1989), ischemia-reperfusion (Traystman et al. 1991) and inflammation (Fantone and Ward 1982, McCord 1987).

Thus far, direct detection of oxygen free radicals in complex biological systems has not been possible, and most of the evidence for the association of brain damage and oxygen free radicals was derived from studies using indirect methods (i.e. amelioration of damage by treatment with free radical scavengers in animal models, for review see Ikeda and Long 1990, Traystman et al. 1991). To study the pathophysiology of free radical damage, an experimental system is required to continuously and non-invasively measure the generation of oxygen free radical species in the brain in vivo.

In biological tissues, oxydative metabolism, including oxygen free radical formation, is closely related to the production of photons of specific wavelengths, particularly in the near-infrared and infrared (ultraweak chemiluminescence, Slawinski 1988, Kanovsky 1989). Chemiluminesence thus potentially provides us with a noninvasive, continuous means of oxygen free radical detection.

The aim of this study was to test the feasibility of in-vivo chemiluminescence detection to study the pathophysiology of brain ischemia and inflammation.

METHODS

Anesthetized and ventilated male Wistar rats (250-350g) were equipped with a closed cranial window over the parietal cortex (0.25 cm^2 exposed) (Dirnagl et al. 1991). Intracranial pressure (ICP), mean arterial blood pressure (MABP), body temperature,

Optical Imaging of Brain Function and Metabolism, Edited by
U. Dirnagl *et al.*, Plenum Press, New York, 1993

and endexpiratory pCO_2 were monitored and continuously recorded on a PC-computer (Asyst, MacMillan Software, New York, USA). Arterial blood gases were determined intermittently.

The animals were shielded with aluminium foil and housed in a dark box (fig.1). A reflector was fitted to the cranial window, and the opening of the detector (cooled photon counting camera Hamamatsu C2400-25 connected to Argus-100 processor, Hamamatsu Photonics, Hamamatsu City, Japan) was focussed to the reflector. Chemiluminescence was enhanced with 10^{-2} M lucigenin (1 ml/h i.v., Sigma Chemicals, Deisenhofen FRG). Two animals were studied in each of the following experimental groups: controls, hyperoxia (100% O_2), hypoxia (5% O_2, 95% N_2), global forebrain ischemia (see below), and pneumococcal meningitis (see below).

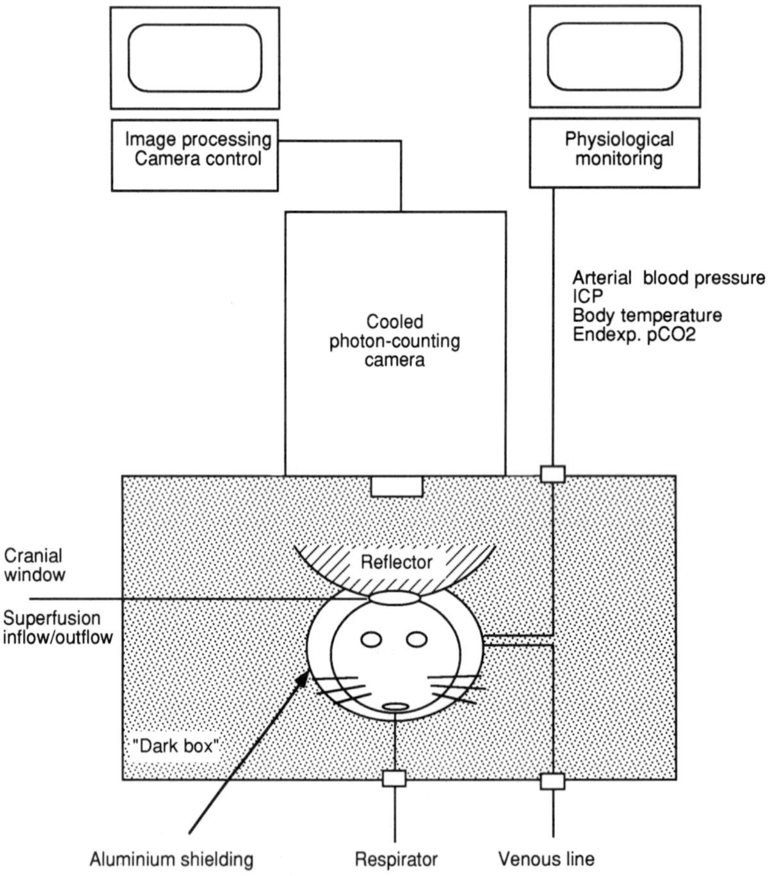

FIGURE 1. Experimental setup

Temporary global forebrain ischemia was induced by reversible occlusion of both common carotid arteries, combined with hypobaric systemic hypotension (50 mm Hg) for ten minutes (Dirnagl et al. 1993). For the induction of ischemia, the animal had to be removed from the dark box.

Pneumococcal meningitis was induced by injection of 75 μl of pneumococcal suspension (10^7 colony forming units/ml, kindly provided by the Department of Microbiology, University of Munich) into the cisterna magna via a microsurgically placed catheter (Pfister et al. 1990).

All values are expressed as means ± SD. Photon counts are presented as counts per minute (CPM) and were not corrected for dark count.

RESULTS

The dark count of the detector was 234 ± 14 CPM (n=10). Imaging of biophotons (unenhanced/enhanced) with macro lenses failed.

Using the reflector focussed to the opening of the detector, 672 ± 196 CPM (n=10) from the brain were counted . I.v. infusion of lucigenin led to an enhancement to 969 ± 346 CPM (n=10). Lucigenin infusion had no measurable effect on MABP, ICP, cortical cerebral blood flow as measured with Laser-Doppler flowmetry (Dirnagl et al. 1989), endexpiratory pCO_2, and blood gases. In 2 animals the count was followed over 3h (control) and did not change with time (fig.2).

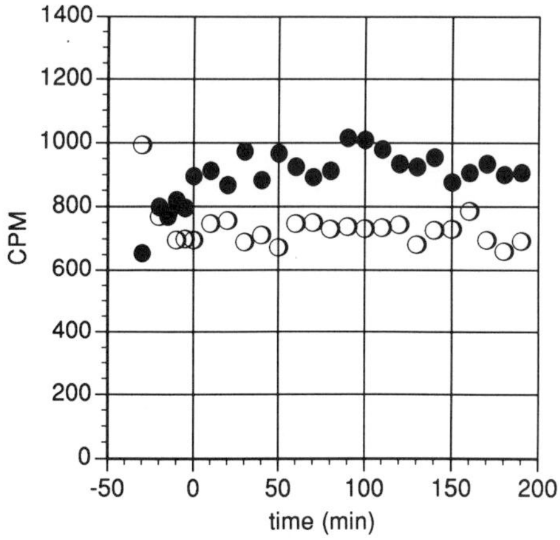

FIGURE 2. Brain chemiluminescence: control (n=2), CPM: counts per minute

Hyperoxia (n=2) increased the count from 1153 ± 652 to 1647 ± 716 CPM (this and the following data are lucigenin-enhanced counts) within 25 minutes. Treatment with the free radical scavengers superoxide-dismutase (Cu-Zn-SOD) or catalase (both Sigma Chemicals, Deisenhofen, FRG) decreased the hyperoxic count (fig.3).

Hypoxia (n=2) for 5 minutes led to an immediate drop in the count rate from 1374 ± 56 CPM to 1014 ± 260 CPM, followed by a delayed increase to 1771 ± 130 CPM within 60 minutes (fig.4).

Global forebrain ischemia (n=2) for 10 minutes followed by reperfusion lead to an increase in the count rate from 720 ± 12 CPM to 1238 ± 382 CPM (fig.5). For technical reasons, photons could not be counted during the ischemic period.

Induction of pneumococcal meningitis (n=2) raised the count from 817 ± 160 to 1417 ± 143 CPM within 220 minutes. Treatment with Cu-Zn-SOD decreased the count to 1238 ± 382 CPM (fig.6).

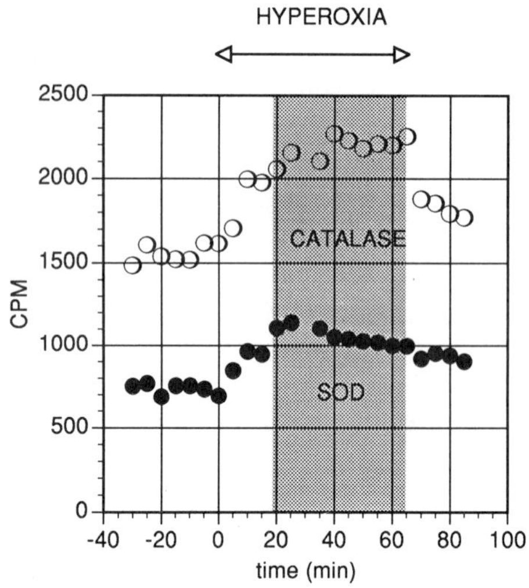

FIGURE 3. Brain chemiluminescence: hyperoxia (n=2) CPM: counts per minute, SOD: superoxide dismutase

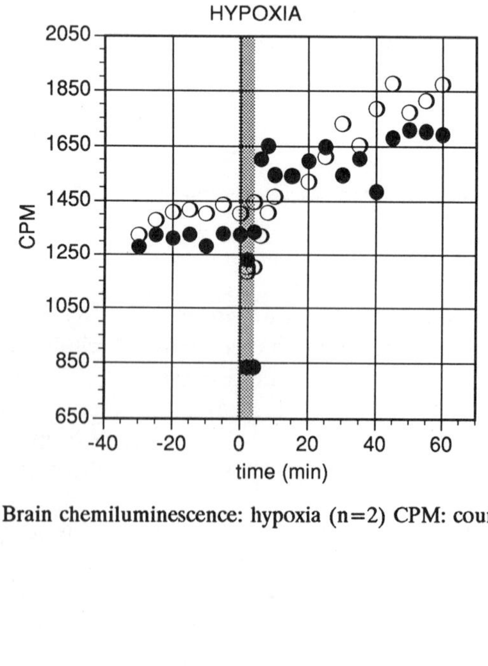

FIGURE 4. Brain chemiluminescence: hypoxia (n=2) CPM: counts per minute

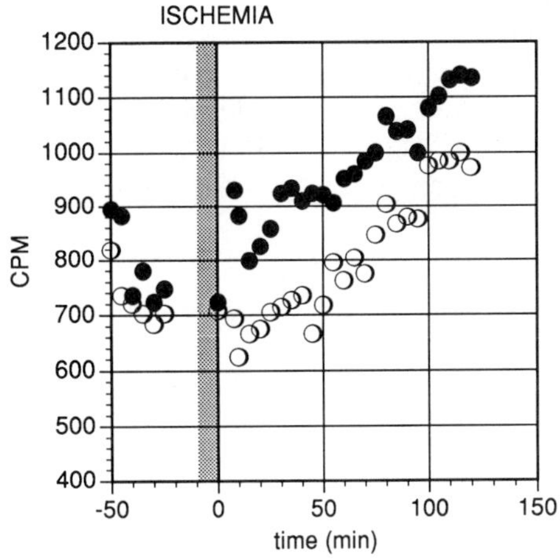

FIGURE 5. Brain chemiluminescence: global forebrain ischemia (n=2) CPM: counts per minute

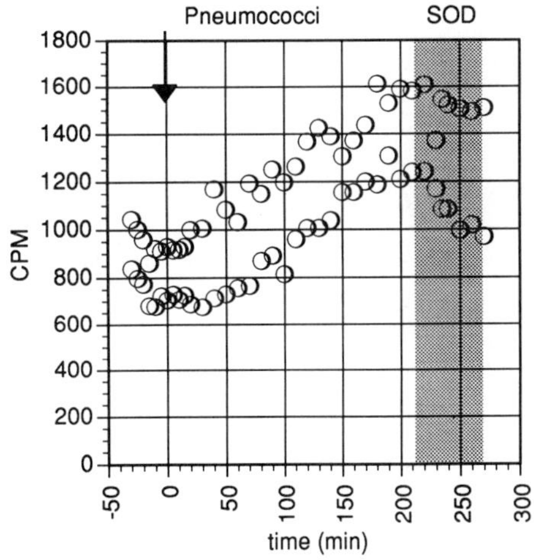

FIGURE 6: Brain chemiluminescence: meningitis (n=2) CPM: counts per minute, SOD: superoxide dismutase

DISCUSSION

From Oxygen to Photon

In biological tissues, photons of various wavelengths are generated by oxydative metabolism (Cadenas and Sies 1984, fig.7).

In brief, the sequence of events from oxygen to photon is as follows: Under physiological conditions, oxygen free radicals (superoxide anion, hydroxyl radical, hydrogen peroxyide etc.) are generated in biological tissues by a variety of enzymatic and nonenzymatic systems (the "electron-leak" in the respiratory chain, the homolytic scission of hydroperoxides, cyclooxygenase activity etc.). With brain pathology, especially

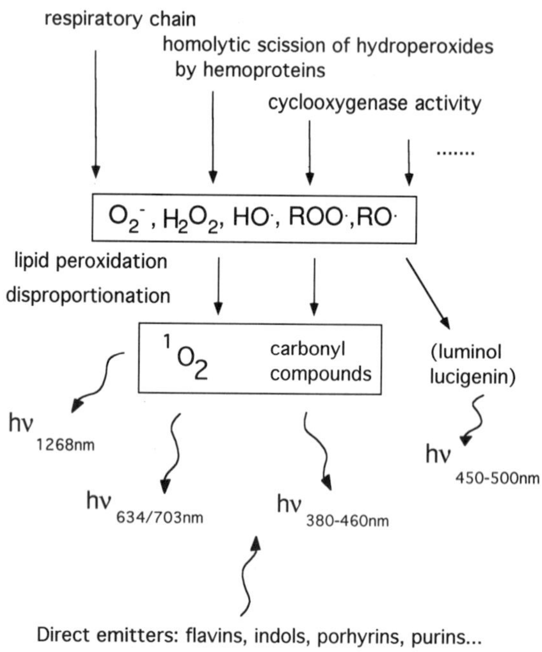

FIGURE 7. Photon generating mechanisms of oxygen free radicals

ischemia and reperfusion, additional oxygen free radicals may be generated by leukocytes (Weiss 1989), endothelium (Betz 1985), purine metabolites and xanthinoxydase (Braughler and Hall 1989), norepinephrine- and dopamine-β-hydroxylase (Cohen 1984), nitric oxide (Beckman et al. 1990), and free fatty acids and the arachidonic acid cascade (Abe et al. 1989).

Even under normal conditions, where effective mechanisms of defense (Brawn and Fridovich 1980, Demopoulos 1973) prevent tissue damage from oxygen free radicals, oxygen species are converted to the highly reactive, short-lived singlet oxygen and carbonyl compounds via lipid peroxidation and disproportionation. Relaxation of the electronically activated singlet oxygen to the ground state produces photon emission at 1268 nm (monomol emission). The collision of two singlet oxygen molecules with simultaneous relaxation leads to emission at 634 and 703 nm (dimol emission). Triplet-singlet transitions of the excited carbonyl compounds emit at 380-460 nm.

Enhanced Chemiluminescence

Synthetic chemiluminescent compounds (luminol, lucigenin, etc.), emit light when oxidized by oxygen free radicals. Such compounds can be used to enhance chemiluminescence in biological systems. Lucigenin (bis-N-methylacridinium nitrate) is particularly sensitive to the superoxide anion, and emits at 450 - 500 nm (Müller-Peddinghaus 1984).

With the use of a photon counting camera and macro lenses, we were unable to image the production of photons at the surface of the brain. This was probably due to the extremely low rate of photon production, the reduction in sample volume by the increase in spatial resolution, insufficient coupling of brain and detector, and the low numerical aperture of macro lenses (NA<0.1). However, by using the camera as a counting device and focussing the surface of the brain to the aperture of the detector we were able to count photons from the brain without enhancing compound. Due to the extremely small number of photons, the reaction to hypoxia/hyperoxia was not reproducible in this mode, and the enhancing compound lucigenin was used. Enhanced chemiluminescence presents the advantage of lowering the level of detection for biophotons. However, certain disadvantages are associated with the use of synthetical enhancing compounds: 1) Potential intrinsic actions of the compound on cerebral metabolism, 2) low specifity for particular oxygen radical species, and 3) the possibility that the enhancing compound accumulates in the brain after breakdown of the blood brain barrier. Future studies will optimize the detector and coupling to the brain to avoid the use of enhancing compounds.

Detectors for Ultralow Light Levels

In ultralow light level applications the choice of detector is crucial. Only very few studies have reported values for spontaneous chemiluminescence of the brain. Adamo et al. (1989) counted 130 cps/cm^2 from the brain of young rats, Boveris et al. (1980) 72 cps/cm^2 from the brain of aging rats, and Chance (personal communication) 30 cps/cm^2 from dog brain . We have counted 45 cps/cm^2 without lucigenin, and 65 cps/cm^2 in the enhanced mode. With such photon fluxes and an imaging system with 128x128 resolution, each pixel will display a signal due to a photon approximately once every 5 minutes. In addition, since each type of detector also exhibits noise the likelihood of the observed signal being real must be weighed against the possibility of it being solely due to random noise. The solution to this problem at ultralow photon flux rates can only be to sacrifice temporal and spatial resolution by integrating over time without imaging. The most suitable and economical detectors for this purpose are cooled photomultiplier tubes with direct optical coupling of brain and detector. Cooled photomultipliers for photon counting have low dark currents (<20 cps), good quantum yield (0.2 - 0.3), and are relatively inexpensive.

We have used a cooled photon counting camera to investigate the prospects of in-vivo chemiluminescence imaging with or without enhancement. Photon counting cameras have imaging as well as photon counting capabilities, a favourable spectral sensitivity (300-950 nm), and a very low dark current (4 cps in our study), but they are extremely expensive. Liquid nitrogen cooled charge coupled devices (CCDs) would be attractive detectors because of their high quantum efficiency (0.4-0.8), high spectral sensitivity from the blue to the near infrared, and the fact that the spatial resolution of those detectors can be adapted to the signal to noise ratio by combining charges from adjacent pixels into superpixels (binning). Binning could be used to combine all the pixels on the chip, which then is used as a photon counting device with long integration times before readout. However, even for the new MPP-mode CCDs with very low dark currents, the application of CCDs in ultralow light level applications is hindered by the dark current (f.e. 0.03 e^-/pixel/s x 512 pixel x 512 pixel = 7.9×10^3 e^- per second for the whole chip).

Chemiluminescence in vivo

Since the pioneering work of Tarusov et al. (1961) and Chance and coworkers (Boveris et al. 1980), ultraweak chemiluminescence has been recorded from a variety of isolated organs and homogenated tissues (lung: Turrens et al. 1989, heart: Barsacchi et al.1989; liver: Cadenas et al. 1981; brain: Imaizumi et al. 1984). Enhanced chemiluminescence was used to study free radical formation in the isolated perfused heart (lucigenin, Henry et al. 1990), lung (lucigenin and luminol, Archer et al. 1989), and liver (luminol, Okuda et al. 1991). Adamo et al. (1989) recorded spontaneous brain chemiluminescence in control and hyperthyreoid rats in situ.

CONCLUSION

In this pilot study we have shown for the first time that it is feasible to use enhanced chemiluminesence to study in-vivo the role of oxygen free radicals in two experimental models of important disease processes: ischemia-reperfusion and meningitis.

The count rate was increased by inhalation of pure oxygen and decreased by hypoxia and i.v. administration of free radical scavengers. We therefore conclude that the photons were derived from oxygen free radical producing systems. There was a steady increase in the photon count after 5 minutes of hypoxia, after 10 minutes of global forebrain ischemia, and after induction of pneumococcal meningitis, suggesting the production of oxygen free radicals in the acute phase of hypoxia-ischemia, ischemia-reperfusion, and inflammation. Using an improved detection system, future studies will correlate the time course of the photon production with tissue damage and the potential of free radical scavengers to block the photon production and/or the tissue damage.

ACKNOWLEDGEMENT-This study was supported by the Wilhelm Sander-Stiftung and the DFG (Pf 246/3-1).

LITERATURE

Adams JD, Odunze IN (1991) Oxygen free radicals and Parkinsons disease. Free Rad Biol Med 10:161-169

Abe K, Yuki S, Kogure K (1988) Strong attenuation of ischemic and postischemic brain edema in rats by a novel free radical scavenger. Stroke 19:480-485.

Adamo AM, Llesuy SF, Pasquini JM, Boveris A (1989) Brain chemiluminescence and oxydative stress in hyperthyroid rats. Biochem J 263:273-277.

Archer SL, Nelson DP, Weir EK (1989) Simulateous measurement of O2 radicals and pulmonary vascular reactivity in rat lung. J Appl Physiol 67:1903-1911.

Barsacchi R, Pelosi G, Camici P, Bonaldo L, Maioino M, Ursini F (1984) Glutathione depletion increases chemiluminescence emission and lipid peroxidation in the heart. Biochim Biophys Acta 804:356-360.

Beckman JS, Beckman TW, Chen J, Marshall PA, Freeman BA (1990) Apparent hydroxyl radical production by peroxynitrite: implications for endothelial injury from nitric oxide and superoxide. Proc Natl Acad Sci 87:1620-1624.

Betz AL (1985) Identification of hypoxanthine transport and xanthine oxidase activity in brain capillaries. J Neurochem 44:574-579.

Boveris A, Cadenas E, Reiter R, Filipkowski M, Nakase Y, Chance B (1980) Organ chemiluminescence: noninvasive assay for oxydative radical reactions. PNAS 77:347-351.

Brawn K, Fridovich I (1980) Superoxide radical and superoxide dismutases: threat and defense. Acat Physiol Scand Suppl 492:9-18.

Braughler JM, Hall ED (1989]) Central nervous system trauma and stroke I.Biochemical considerations for oxygen radical formation and lipid peroxidation. Free Rad Biol Med 6:289-301.

Cadenas E, Sies H (1984) Low-level chemiluminescence as an indicator of singlet molecular oxygen in biological systems. Meth Enzymology 105:221-232.

Cadenas E, Varsavsky AI, Boveris A, Chance B (1981) Oxygen or organic peroxide-induced chemiluminescence of brain and liver homogenates. Biochem J 198:645-654.

Cohen G (1984) Oxy-radical toxicity in catecholamine neurons. Neurotoxicology 5:77-82.

Demopoulos HB (1973) Control of free radicals in biological systems. Fed Proc 32:1903-1908.

Dirnagl U, Kaplan B, Jacewicz M, Pulsinelli W (1989) Continuous measurement of cerebral blood by laser-Doppler flowmetry in a rat stroke model. J Cereb Blood Flow Metab 9:589-596.

Dirnagl U, Villringer A, Gebhardt R, Haberl RL, Einhäupl KM (1991) Three-dimensional reconstruction of the rat brain cortical microcirculation in vivo. J Cereb Blood Flow Metab 11:353-360.

Dirnagl U, Thoréen P, Villringer A, Sixt G, Them A, Einhäupl KM (1993) Global forebrain ischemia in the rat: controlled reduction of cerebral blood flow by hypobaric hypotension and two vessel occlusion. Neurol Res (in press).

Fantone JC, Ward PA (1982) Role of oxygen-derived free radicals and metabolites in leukocyte-dependent inflammatory reactions. Am J Pathol 107:397-418.

Freischlag JA, Hanna D (1991) Superoxide anion release after ischemia and reperfusion. J Surg Res 50:565-568.

Henry TD, Archer SL, Nelson D, Weir EK, From AHL (1990) Enhanced chemiluminescence as a measure of oxygen-derived free radical generation during ischemia and reperfusion. Circ Res 67:1453-1461.

Ikeda Y, Brelsford KL, Ikeda K, Long DM (1989) Oxygen-free radicals in traumatic brain oedema. Neurol Res 11:213-216.

Ikeda YI, Long DM (1990) The molecular basis of brain injury and brain edema: the role of oxygen free radicals. Neurosurgery 27:1-11.

Imaizumi S, Kayama T, Suzuki J (1984) Chemiluminescence in hypoxic brain: The first report. Stroke 15:1061-1065.

Kaminski PM, Proctor KG (1990) Actions of adenosine on nitro blue tetrazolium deposition and surface pH during intestinal reperfusion injury. Circ Res 66:1713-1719.

Matsubara T, Ziff M (1986) Increased superoxide anion release from human endothelial cells in response to cytokines. J Immunol 137:3295-3298.

McCord JM (1988) Free radicals and myocardial ischemia: overlook and outlook. Free Radic Biol Med 4:9-14.

Müller-Peddinghaus R (1984) In-vitro determination of phagocyte activity by luminol- and lucigenin-amplified chemiluminescence. Int J Immunopharmac 6:455-466.
Okuda M, Ikai I, Chance B, Kumar C (1991) Oxygen radical production during ischemia-reperfusion in the isolated perfused rat liver as monitored by luminol enhanced chemiluminescence. Biochem Biophys Res Commun 174:217-221.
Pfister HW, Ködel U, Haberl RL, Dirnagl U, Feiden W, Ruckdeschel G, Einhäupl K (1990) Microvascular changes during the early phase of experimental bacterial meningitis. J Cereb Blood Flow Metab 10:914-922.
Reinke LA, Kotake Y, McCay PB, Janzen EG (1991) Spin-trapping studies of hepatic free radicals formed following the acute administration of ethanol to rats: in vivo detection of 1-hydroxyethyl radicals with PBN. Free Radic Biol Med 11:31-39.
Sies H (1987) Intact organ spectrophotometry and single-photon counting. Arch Toxicol 60:138-143.
Slawinski J (1988) Luminescence research and its relation to ultraweak cell radiation. Experientia 44:559-571.
Tarusov BN, Polivoda AI, Zhuravlev AI (1961) (in russian). Radiobiologiya 1:150-151.
Till GO, Friedl HP, Ward PA (1991) Lung injury and complement activation: role of neutrophils and xanthine oxidase. Free Radic Biol Med 10:379-386.
Traystman RJ, Kirsch JR, Koehler RC (1991) Oxygen radical mechanisms of brain injury following ischemia and reperfusion. J Appl Physiol 71(4):1185-1195.
Turrens JT, Giulivi C, Pinus CR, Lavagno C, Boveris A (1988) Spontaneous lung chemiluminescence upon paraquat administration. Free Rad Biol Med 5:319-323.
Weiss SJ (1989) Tissue destruction by neurophils. N Engl J Med 320:365-376.

DIFFUSION PROPERTIES OF BRAIN TISSUE MEASURED WITH ELECTRODE METHODS AND PROSPECTS FOR OPTICAL ANALYSIS

C. Nicholson and L. Tao

Department of Physiology and Biophysics
New York University Medical Center
New York, USA

INTRODUCTION

In the context of this paper we shall discuss methods of measuring the diffusion of small molecules through the extracellular space of the brain. The study of such diffusion properties has relevance to the transport of metabolites, the spread of neuromodulators and the application of drugs to the brain. The extracellular space (ECS) is a component of the brain cell microenvironment (Schmitt and Samson 1969, Nicholson 1980, Cserr 1986). The movement of neuromodulators and neuroactive substances through this environment - so-called volume transmission - has been the topic of a recent symposium (Fuxe and Agnati 1991).

The ECS is the narrow interstitium between cell membranes and has a geometry that resembles the soap phase of an aggregate of soap bubbles. The ECS is a channel for the transfer of metabolites and for the hypothesized transmission of chemical signals between cells. It also provides essential isolation between cellular elements. The chemical composition of the bathing medium of the ECS is typically low in K^+ (3mM) and Ca^{2+} (1.2 mM) and high in Na^+ (160 mM), Cl^- (145 mM) and HCO_3^- (21 mM) (Nicholson and Phillips 1981). The latter salt combines with CO_2 to provide essential buffering of extracellular pH (Chesler 1990).

This paper will briefly discuss the diffusion properties of the brain and the fabrication and use of ion-selective microelectrodes (ISMs) to measure diffusion characteristics. Finally some comments will be made on the application of optical techniques to this problem.

DIFFUSION IN A COMPLEX MEDIUM

The diffusion properties of the ECS can be estimated by releasing a suitable substance from a micropipette and allowing it to diffuse in the ECS. In the method discussed in the first part of this paper, an ISM sensitive to the diffusing substance, located at a distance of 100-200 μm away from the source, records the local substance concentration which is calibrated and analyzed by fitting an appropriate diffusion curve. From the fitting procedure the parameters that define the diffusion process are extracted (Nicholson and Phillips 1981, Nicholson and Rice 1991). This method was first used by Lux and Neher (1973) in an attempt to measure K^+ diffusion in brain tissue. As shown later (Nicholson et al 1979), K^+ permeates brain cell membranes too easily to allow the method to work with this ionic species and a relatively impermeable ion, such as tetramethylammonium (TMA^+), is required. In comparison to earlier methods based on radiotracers (Fenstermacher and Kaye, 1988), the advantages of the ISM method are that changes in the ECS can be recorded in real time and the method can be applied to very small regions, including brain slices and isolated brain preparations (Rice and Nicholson 1991), as well as in vivo preparations (Nicholson and Phillips 1981, Nicholson and Rice 1991).

Fick's Second Law of diffusion for a complex medium

Diffusion in complex media, such as porous aggregates or brain ECS, obeys Fick's law with two important modifications. First, the diffusion coefficient, D, is reduced by the square of a parameter called the tortuosity, λ, to an apparent diffusion coefficient $D^* = D/\lambda^2$. This results from the encounters of a diffusing ion with membranous obstructions as it moves randomly in the intercellular clefts. Second, the source strength i.e. the magnitude of the flux of material leaving the injecting microelectrode, is divided by the volume fraction, α, of the ECS so that a given quantity of released substance becomes more concentrated than it would have been in a free medium (Nicholson and Phillips 1981). The volume fraction is simply the ratio of the volume of the ECS to the total volume of the tissue that contains it.
When material diffuses in brain tissue, a third factor, termed non-specific uptake, k' (sec^{-1}), appears in the diffusion equation, for loss of material from the ECS, primarily across the cell membranes (Nicholson and Phillips 1981, Nicholson 1992).

The diffusion equation incorporating α, λ and k' is (Nicholson and Phillips 1981):

$$D^* \nabla^2 C + Q/\alpha - k'C = \partial C/\partial t \qquad (1)$$

where C is concentration (mM), t is time (sec), ∇ is the gradient operator and Q is the source term (mM x sec^{-1}).

Expressions for iontophoresis and pressure ejection

There are two methods of releasing ions from a source micropipette: iontophoresis and pressure ejection. In the following discussion we shall consider the ejection of a small ion, such as TMA^+ having a molecular weight of 74.1. Typical values of D (cm^2 x sec^{-1} x 10^{-6}) for TMA^+ in a free aqueous medium depend on the temperature and are: 7.6 (10°C), 10.0 (25°C) and 13.0 (37°C).

Assuming spherical symmetry of the diffusion around the source point, concentration as a function of time and space for iontophoresis is given by the solution (Nicholson and Phillips 1981) to Equation 1, with $Q = In/zF$ where I is iontophoresis current (amps) applied to the iontophoresis electrode, n is the transport number of the iontophoresis pipette, z is the number of charges on the ion and F is Faraday's electrochemical equivalent (coulombs x $mole^{-1}$).

When an iontophoresis pulse is applied for duration d (sec), then $C = G(t)$ when $t < d$, for the rising phase of the curve and $C = G(t) - G(t-d)$ when $t > d$, for the falling phase of the curve (Fig. 1A).

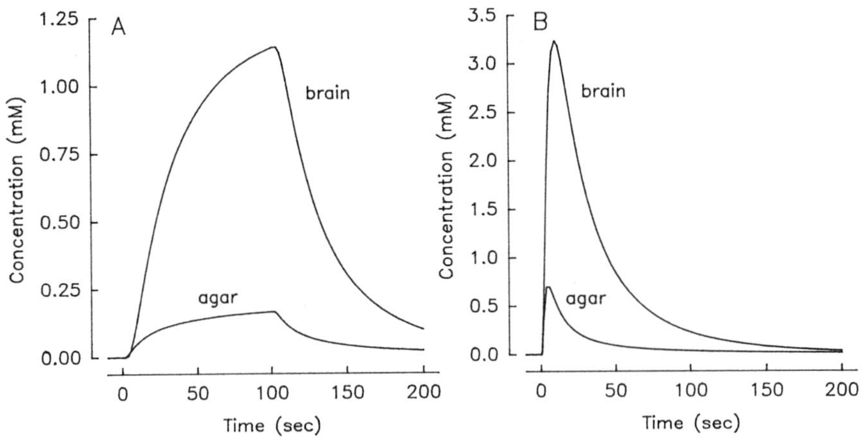

FIGURE 1. Comparison of concentration versus time curves for iontophoresis and pressure injection for a free medium (agar) and brain. The spacing between the source electrode and the recording position was 0.02 cm. The agar was characterized by $\alpha = 1$, $\lambda = 1$ and $k' = 0$. The brain was characterized by $\alpha = 0.2$, $\lambda = 1.6$ and $k' = 0.01\ sec^{-1}$. Both environments were assumed to be at a temperature of 37 °C so a value of $D = 13 \times 10^{-6}\ cm^2.sec^{-1}$ was used. Panel A shows curves using iontophoresis. A current of 100 nA was applied from time zero to 100 seconds and the transport number (n) of the iontophoretic electrode was 0.3. Panel B shows pressure injection. A volume of 7.773×10^{-10} liters of TMA^+ at a concentration of 100 mM was instantaneously injected at the origin. This released the same quantity of material as in the iontophoretic case shown in Panel A. The radius of the injected material in agar was 57.04 μm because $\alpha = 1$ for this medium. In the brain, where $\alpha = 0.2$, the radius of injected material was 97.53 μm.

The function G(u) is evaluated by substituting t or t-d for u in the following expression (Nicholson and Phillips 1981, Expression A14):

$$G(u) = (Q\lambda^2/8\pi D\alpha r)\{\exp[r\lambda(k'/D)^{1/2}]\mathrm{erfc}[r\lambda/2(Du)^{1/2} + (k'u)^{1/2}]$$

$$+ \exp[-r\lambda(k'/D)^{1/2}]\mathrm{erfc}[r\lambda/2(Du)^{1/2} - (k'u)^{1/2}]\}. \qquad (2)$$

The term "erfc" is the complementary error function. The transport number, n, and free diffusion coefficient, D, are estimated by repeating the procedure in a free medium without convection, such as dilute agar or agarose gel. For such a gel $\alpha=1$; $\lambda=1$. The transport number n must be measured for each electrode but D is well established for specific ions. Figure 1A shows typical curves generated from Expression 2 for diffusion in an agar gel and in brain tissue. In this figure the data represent what would be seen by an ISM at a distance of 200 μm from the source electrode (note that, for convenience, the concentration axes in Figure 1 are calibrated in linear increments; the response of an ISM is actually logarithmic, as detailed below). There is a striking disparity between the curve in agar and brain, most obvious here in the greater amplitude of the brain curve. The amplitudes differ by a factor of approximately λ^2/α which amounts to 12.8 given the typical values of λ and α (see legend to Figure 1). The time courses of the two curves also differ since the argument of the complementary error functions differ by a factor of λ, but this is hardly visible in the present figure.

Another way to depict the diffusion from an iontophoretic point source is to show the concentration versus distance curves. Such curves are what would be derived from an optical image (see below) by measuring the density along a diameter of the image. Figure 2A shows a set of such profiles at different times during a 100 second iontophoretic pulse. Since Expression 2 has a singularity at the origin during iontophoresis, the concentration is theoretically undefined there (in practice the source electrode has a finite diameter of 1-2 μm so that a different solution is required as the source is closely approached). Interestingly, after cessation of the pulse it can be shown (Nicholson 1992) that the falling phase is finite at the origin and is described by the Expression:

$$C(0, t > d) = (Q/4\pi(D^*)^{3/2}\alpha)\{(1/\pi^{1/2})[\exp(-k'(t-d))/(t-d)^{1/2})-\exp(-k't)/t^{1/2}] + (k')^{1/2}[\mathrm{erf}((k'(t-d))^{1/2}) - \mathrm{erf}((k't)^{1/2})]\}. \quad (3)$$

This case is graphed in Figure 2B for five separate times after the end of the current.

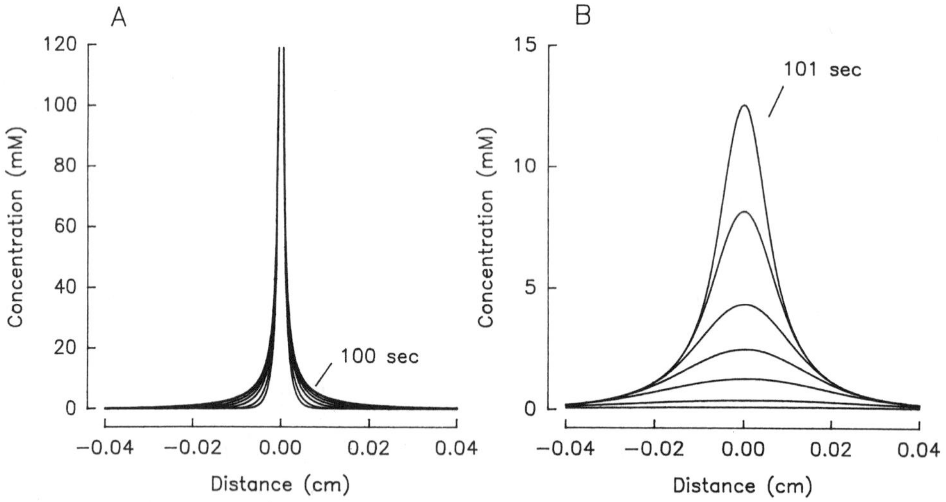

FIGURE 2. Concentration versus distance diffusion curves generated during the iontophoresis current in typical brain tissue. The curves were generated using the same parameters as Figure 1 and represent the concentration values that would be seen along a diameter through the spherically symmetric concentration distribution at different times. Panel A shows curves during the period of the iontophoretic current. The curves represent the times 1, 2, 5, 10, 20, 50 and 100 seconds after the onset of the current. Note that the region $0 < r < 0.0005$ cm was excluded because of the singularity at the origin. Panel B shows concentration versus distance diffusion curves generated after cessation of an iontophoresis current in typical brain tissue. The curves represent the times 101, 102, 105, 110, 120, 150 and 200 seconds after the onset of the 100 second long iontophoresis current. Note that there is no singularity at the origin for finite times after the end of the iontophoresis current (see text for further details).

The TMA^+ also can be ejected from a microelectrode that is subjected to a brief pulse of increased gas pressure. The expression for a brief pressure pulse is given by (Nicholson 1985, Nicholson and Rice 1991, Nicholson 1992):

$$C = \{UC_f\lambda^3/(8(D\pi t)^{3/2}\alpha)\}\exp[(-r^2\lambda^2/4Dt)- k't] \quad (4)$$

where a volume U at concentration C_f is ejected. For this solution to be valid it is assumed that all the material is released in a very tiny volume (or that the observation point is so far away that the release site acts like a point source i.e. a mathematical delta function).

When the ejected material is not so restricted but instead fills a finite region of the ECS and occupies a spherical volume of radius b within the tissue, then from the expression for the volume of a sphere, $b = (3U/4\pi\alpha)^{1/3}$, and the solution to Equation 1 is (Nicholson 1985, 1992):

$$C = (C_f/2)\{\mathrm{erf}[\lambda(r+b)/2(Dt)^{1/2}] - \mathrm{erf}[\lambda(r-b)/2(Dt)^{1/2}] - (2/(\lambda r))(Dt/\pi)^{1/2}[\exp[-\lambda^2(r-b)^2 /(4Dt)] - \exp[-\lambda^2(r+b)^2 /(4Dt)]]\}\exp(-k't). \quad (5)$$

In Expression 5 "erf" represents the error function. Typical concentration versus time curves are shown in Figure 1B. The total amount of released material in the pressure case was set equal to that used in the iontophoretic case (Figure 1A) to permit direct comparison of the data. Again the disparity in amplitudes is evident. The curves generated with the pressure pulse paradigm are briefer in duration because it is assumed that the ejected material is instantaneously deposited in the agar or brain; this also accounts for the higher amplitude in the pressure case compared to the iontophoretic case, at the chosen distance.

Figure 3 shows concentration versus distance curves for a diameter through the concentration distribution generated by the pressure pulse paradigm. The dotted lines show the initial distribution and the figure clearly depicts the "collapse" of the initial distribution of material under the dissipative diffusion process.

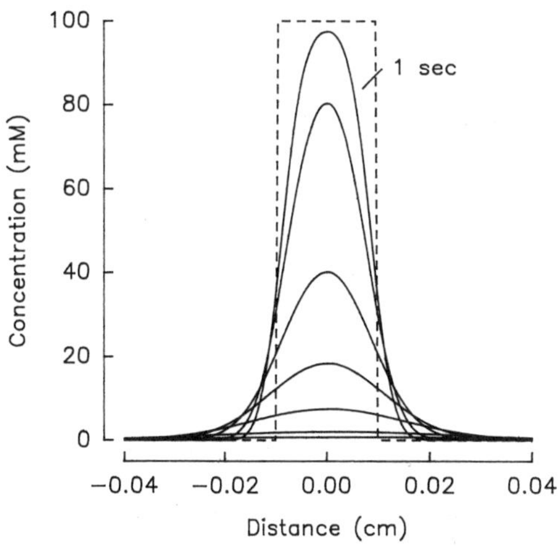

FIGURE 3. Concentration versus distance diffusion curves generated by the pressure injection paradigm described in Figure 1. The dotted lines indicate distribution immediately after injection. The curves represent the times 1, 2, 5, 10, 20, 50 and 100 seconds after the pressure injection which took place at time zero.

If the infiltrating solution forms a solution-filled cavity in the brain ($\lambda = 1$; $\alpha = 1$ within the cavity), then the solution to Equation 1 will be different again. A numerical solution has been derived (Nicholson 1985) but experiments suggest that this situation is uncommon.

In comparing iontophoresis and pressure ejection, the latter method has several advantages. First, the concentration of the released substance just outside the tip of the ejecting electrode is defined by the concentration in the electrode (the solution can be made isotonic by the addition of NaCl). Second, it is feasible to release a solution containing two or more compounds. Another advantage of pressure ejection is that uncharged molecules can be released, this is not very relevant for ISMs since they generally do not sense neutral compounds. With the advent of optical recording, to be outlined below, the movement of fluorescently tagged neutral molecules can be analyzed. One disadvantage of pressure ejection is that it is difficult to calibrate the released quantity reliably. The amount released by a given pressure-duration combination in agar may differ from the amount released under the same conditions in brain tissue.

Construction and use of ion-selective microelectrodes

A modern ISM consists of a glass microelectrode with a liquid ion exchanger in the tip. The ion exchanger contains carriers selective for an ion, such as K^+ or Ca^{2+}, or, as in the situation to be discussed here, an exogenous ion that can be used to probe the diffusion properties of the ECS. The electrode is back-filled with an electrolyte containing the ion to be sensed. When the tip of the ISM is exposed to a different concentration of the ion, a Nernst potential is generated across the exchanger which can be recorded with a high impedance buffer amplifier. Further information on the fabrication and calibration of ISMs can be found in Ammann (1986), Nicholson and Rice (1988).

For diffusion measurements we typically use the cation TMA^+. A suitable exchanger to sense this ion was originally sold by Corning as a K^+ exchanger (Corning 477317) and consists of 3% by weight potassium tetrakis (*p*-chlorophenyl)borate dissolved in 2,3-dimethyl-nitrobenzene, see (Oehme and Simon, 1976; the chemical components are available from Fluka, Buchs, Switzerland / Ronkonkoma, New York). This type of charged carrier lacks selectivity for K^+ and has generally been superseded by neutral carriers but remains an excellent sensor for tetralkylammonium compounds.

Since ISMs are not completely selective for one ion, their response is described by an empirical modification of the Nernst equation, termed the Nicolsky equation:

$$E = (RT/zF)\ln(C + A) \tag{6}$$

where E is the potential difference across the exchanger, R is the universal gas constant, T absolute temperature, z the valence, C the concentration of the ion and A the interference due to other ions. Where one primary interfering compound exists at a concentration C_A then $A = K_A C_A$ where K_A is the interference (Ammann 1986).

A single-barreled ISM will also sense any endogenous potential in the vicinity of the tip. This would add algebraically to E, so a second barrel is added to the ISM that consists of a conventional potential sensing micropipette. The signal from this so-called reference barrel can be electronically subtracted from the ion signal.

In our experiments, the ISM signals are acquired by an analog-to-digital converter and appropriate calibration and diffusion curve fitting procedures carried out using the program VOLTORO developed in this laboratory. The program runs on an

IBM PC or equivalent computer. The program fits Expressions 2, 4 or 5 to the recorded signal using the downhill simplex method (Press et al 1986) to obtain the three parameters α, λ and k'. After sufficient data have been acquired from several experiments, α, λ, k' are analyzed statistically and displayed as quantile plots (Rice and Nicholson 1991).

Some representative values of the diffusion parameters

Over the last several years studies from this laboratory and others have established values for the diffusion parameters in a variety of tissues and preparations using TMA^+ and the ISM method. Some of these data are recorded in Table 1.

Table 1. Measurements of extracellular space geometry using TMA^+ and ISMs.

Preparation	α	λ	k'(sec^{-1})	Reference
Rat cerebellum in vivo*	0.21	1.55	n/a	a
Rat cerebral cortex in vivo	0.18	1.57	0.006	b
Skate cerebellum in vivo*	0.24	1.62	0.007	c
Turtle isolated cerebellum*	0.21	1.65	n/a	d
Hippocampal slice CA1	0.12	1.67	0.013	e
Hippocampal slice CA3	0.18	1.83	n/a	e
Rat neostriatum slice	0.21	1.54	0.01	f

a: Nicholson and Phillips 1981; b: Cserr et al 1991; c: Nicholson and Rice 1986b; d: Nicholson and Rice 1986a; e: McBain et al 1990; f: Rice and Nicholson 1991. n/a - not available.
*Recent studies suggest that the molecular layer of the cerebellum is anisotropic and so it will have three values of λ (Rice, Okada and Nicholson, unpublished).

PROSPECTS FOR OPTICAL RECORDING

While ISMs have proved a reliable method of accurately measuring the diffusion properties of brain tissue they suffer from some limitations. The most evident is that they only record at a single location in the brain and the data can only be interpreted if it is assumed that the diffusing molecules are moving in a homogeneous medium so that, generally, the diffusion pattern is symmetric around the source point. Fortunately it appears that this assumption is usually valid. A second problem is that, while individual curves fit the solution to the diffusion equation, the actual values of the parameters obtained from each curve-fit show variation. In practice as many as 50 measurements are often required to get an accurate mean value so that the data acquisition is time consuming. Finally, the ISM method is restricted to small, charged molecules or ionized species, whereas many practical problems of interest center on the movement of large, sometimes neutral, molecules.

Quantitative Optical Imaging (QOI) has the potential to overcome many of the limitations inherent in the microelectrode method. The goal is to digitize an image of a substance migrating through living tissue and use the quantitative information to determine diffusion parameters. This has only been attempted in a few instances to date (for capillary beds, Nugent & Jain, 1984a; tumors, Nugent & Jain, 1984b and for intracellular cytoplasmic diffusion, Safranyos & Caveney, 1985). Optical signals have long been recorded from living tissue in other contexts; for example studies with voltage sensitive dyes (reviewed by Cohen et al. 1989, Lieke et al. 1989, Blasdel 1989) and ion sensitive dyes (Tsien 1989). The detectors have evolved from single photomultipliers through photodiode arrays to television and charge-coupled device camera systems (Spring & Lowy 1989, Aikens et al. 1989). Other investigators have applied imaging techniques to study the behavior of intracellular fluorescent probes in a variety of systems (Taylor & Wang 1989, Wang & Taylor 1989). Many of these studies have concentrated on solving the technical problems associated with accurate quantification of fluorescent images and the production of fluorescent probes for a variety of tasks. There are limitations to QOI. Recording is restricted to a region not far from the tissue surfaces for example; however this is often a region of considerable interest. A second constraint, common to all methods, is that there are only a limited number of substances that can be detected, in this case fluorescently tagged molecules, and transmitters are not among them yet. In contrast, microelectrode techniques based on voltammetry instead of ISMs can be used to measure the diffusion of such neurotransmitters as dopamine (Rice and Nicholson 1989, 1991). Factors such as these imply that microelectrode methods will remain complementary to optical methods.

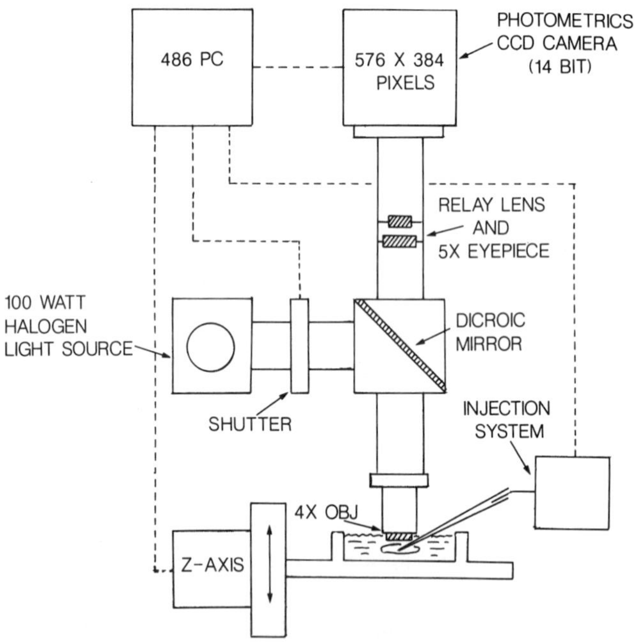

FIGURE 4. Quantitative Optical Imaging System. A Zeiss Standard upright compound microscope is shown schematically with a low power 4x objective. Illumination is provided by a 100 watt tungsten-halogen epi-illuminator fitted with a shutter on the light source to minimize photobleaching and a standard Zeiss dichroic mirror system. The camera is attached to the trinocular housing of the microscope by a standard camera mount with a 5x eyepiece and relay lens. The camera itself is a Photometrics cooled CCD with 14 bit resolution and 576 x 384 pixels. The output of the camera is transferred directly to the memory of a 486 PC. The computer also controls the source electrode for dye injection, the shutter on the light source and the z-axis of the microscope which is moved by a stepping motor. The preparation consists of either an agar or agarose gel or brain slice, mounted on the stage of the microscope which is also equipped with a remotely controlled micromanipulator.

Preliminary Optical Measurements

Our QOI system consists of a Zeiss Standard microscope equipped with a low power objective of 2.5x, 4x or 6.3x. Illumination is provided by a 100 watt tungsten--halogen epi-illuminator and a standard Zeiss dichroic mirror system. The camera is attached to the trinocular housing of the microscope by a standard camera mount with a 5x eyepiece and relay lens. The camera itself is a Photometrics cooled CCD with 14 bit resolution and 576 x 384 pixels. This CCD has 30% to 40% efficiency in the red wavelengths and much lower performance toward the blue, so it is advantageous to use red fluorophores. We have chose Sulforhodamine B or Texas Red labelled molecules (obtained from Molecular Probes, Eugene, OR). The output of the camera is transferred directly to the memory of a 486 PC. The computer also controls the source electrode for dye injection, a shutter on the light source to minimize photobleaching, and the z-axis of the microscope which is moved by a stepping motor (Fig. 4, see previous page).

Our preliminary experiments have been carried out on agarose gel using either an iontophoretic or pressure source. Figure 5 shows data from one test using iontophoresis and fitted using Expression 3 written in normalized form.

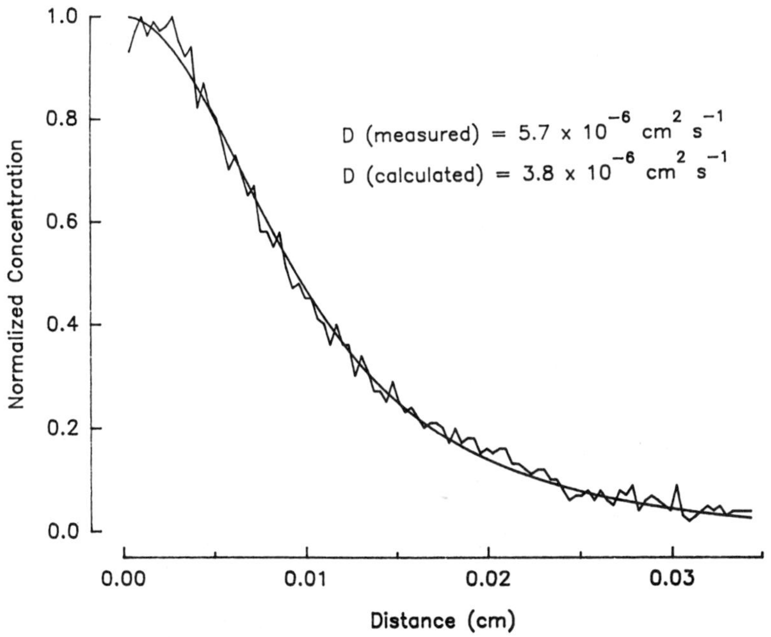

FIGURE 5. Diffusion data for Sulforhodamine B obtained with QOI. The noisy curve is the density trace obtained 2 seconds after the cessation of a 40 second iontophoretic current pulse from a microelectrode filled with 150 mM Sulforhodamine B dye dissolved in water. The injection was made into 0.3% agarose dissolved in 150 mM NaCl. For details of the curve fitting see text.

Fig. 5 shows the data obtained using Sulforhodamine B along one radius from the iontophoretic source 2 seconds after cessation of the iontophoretic current. The smooth line is the best-fit of the normalized expression and this occurs for a D value of $5.7 \times 10^{-6} \times cm^2 \times sec^{-1}$. A rough calculation based on the formula: D x (Molecular

$Wt)^{1/2} = 9 \times 10^{-5}$ (Nicholson & Rice 1991) yields a value of $3.8 \times 10^{-6} \times cm^2 \times sec^{-1}$, since the molecular weight of this dye is 558. There are several types of error involved in fitting this expression to the data and it only represents a first approximation, given here to illustrate the general concepts of our method. A forthcoming paper will present the theory and results in much greater detail.

ACKNOWLEDGEMENT- This work was supported by Grant NS 28642 from NINCDS.

REFERENCES

Aikens, R. S., Agard, D. A. & Sedat, J. W. (1989). Solid-state imagers for microscopy. In: Methods in Cell Biology. Volume 29. "Fluorescent Microscopy of Living Cells in Culture. Part A. Fluorescent Analogs, Labeling Cells, and Basic Microscopy". eds. Wang, Y.-L. and Taylor, D. L. San Diego: Academic. pp 291-313.

Ammann, D. (1986). "Ion-Selective Microelectrodes". Berlin: Springer-Verlag.

Blasdel, G. G. (1989). Visualization of neuronal activity in monkey striate cortex. Ann. Rev. Physiol. 51, 561-581.

Cserr, H. F. (1986) The neuronal microenvironment. Ann. N. Y. Acad. Sci., 481: 1-391.

Cserr, H. F., DePasquale, M., Nicholson, C., Patlak, C. S., Pettigrew, K. D. and Rice, M. E. (1991) Extracellular volume decreases while cell volume is maintained by uptake of ions in rat cerebral cortex during acute hypernatremia. J. Physiol. 442: 277-295.

Chesler, M. (1990) The regulation and modulation of pH in the nervous system. Prog Neurobiology 34: 401-427.

Cohen, L., Hopp, H-P., Wu, J-Y. and Xiao, C. (1989). Optical measurements of action potential activity in invertebrate ganglia. Annual. Rev. Physiol. 51: 527-541.

Fenstermacher, J. D. and Kaye, T. (1988). Drug "diffusion" within the brain. Ann. N. Y. Acad. Sci. 531: 29-39.

Fuxe, K. and Aganati, L. F. (1991) "Volume Transmission in the Brain: Novel Mechanisms for Neuronal Transmission". Raven, New York.

Lieke, E. E., Frostig, R. D., Arieli, A., Ts'o, D. Y., Hildesheim, R. and Grinvald, A. (1989). Optical imaging of cortical activity. Ann. Rev. Physiol. 51: 543-559.

Lux, H. D. and Neher, E. (1973) The equilibration time course of $[K^+]_o$ in cat cortex. Exp Brain Res 17: 190-205.

McBain, C. J., Traynelis, S. F. and Dingledine, R. (1990) Regional variation of extracellular space in the hippocampus. Science: 249, 674-677.

Nicholson, C. (1980). Dynamics of the brain cell microenvironment. Neurosci. Res. Prog. Bull. 18: 177-322.

Nicholson, C. (1985). Diffusion from an arbitrary volume of a substance in brain tissue with arbitrary volume fraction and tortuosity. Brain Res. 333: 325-329.

Nicholson, C. (1992). Quantitative analysis of extracellular space using the method of TMA^+ iontophoresis and the issue of TMA^+ uptake. Can.J.Physiol. Pharmacol. (in press)

Nicholson, C. and Phillips, J. M. (1981). Ion diffusion modified by tortuosity and volume fraction in the extracellular microenvironment of the rat cerebellum. J. Physiol. 321: 225-257.

Nicholson, C., Phillips, J. M. and Gardner-Medwin A. R. (1979) Diffusion from an iontophoretic point source in the brain: role of tortuosity and volume fraction. Brain Res 169: 580-584.

Nicholson, C. and Rice, M. E. (1986a). The migration of substances in the neuronal microenvironment. Ann. N. Y. Acad. Sci. 481: 55-71.

Nicholson, C. and Rice, M. E. (1986b) Diffusion characteristics of skate (Rajaerinacea) cerebellum measured with tetramethylammonium and ion- selective microelectrodes. Mt. Desert Island Biol. Bull. 25: 54 -55.

Nicholson, C. and Rice, M. E. (1988). Use of ion-selective microelectrodes and voltammetric microsensors to study brain cell microenvironment. In: "Neuromethods; The Neuronal Microenvironment". Eds. Boulton, A. A., Baker, G. B. & Walz, W. Clifton, NJ: Humana. pp 247-361.

Nicholson, C. and Rice, M. E. (1991) Diffusion of ions and transmitters in the brain cell microenvironment. In: "Volume Transmission in the Brain". Eds. K. Fuxe and L. F. Agnati. New York: Raven. pages 279-294.

Nugent. L. J. and Jain, R. K. (1984a). Plasma pharmacokinetics and interstitial diffusion of macromolecules in a cappillary bed. Am. J. Physiol. 246: H129-H137.

Nugent. L. J. and Jain, R. K. (1984b). Extravascular diffusion in normal and neoplastic tissues. Cancer Res. 44: 238-244.

Oehme, M. and Simon, W. (1976) Microelectrode for K^+ based on a neutral carrier and comparison of its characteristics with a cation exchanger sensor. Analytica.chim. Acta. 86, 21-25.

Press, W. H. Flannery, B. P.,Teukolsky S. A., and Vetterling, W. T. (1986) "Numerical Recipes". Cambridge University Press, London. Chapter 10.

Rice, M. E. and Nicholson, C. (1989) Measurement of nanomolar dopamine diffusion using low-noise perflorinated ionomer coated carbon fiber microelectrodes and high-speed cyclic voltammetry. Analyt. Chem. 61: 1805-1810.

Rice, M. E. and Nicholson, C. (1991) Diffusion characteristics and extracellular volume fraction during normoxia and hypoxia in slices of rat neostriatum. J. Neurophysiol. 65: 264-272.

Safranyos, R. G. A. and Caveney, S. (1985). Rates of diffusion of fluorescent molecules via cell-to-cell membrane channels in a developing tissue. J. Cell Biol. 100: 736-747.

Schmitt, F. O. (1984). Molecular regulators of brain function: A new view. Neuroscience 13: 991-1001.

Schmitt, F. O. and Samson, F. E. Jr. (1969) Brain cell microenvironment. Neurosci. Res. Progr. Bull. 7, 277-417.

Spring, K. R. and Lowy, R. J. (1989) Characteristics of low light level television cameras. In: Methods in Cell Biology. Volume 29. "Fluorescent Microscopy of Living Cells in Culture. Part A. Fluorescent Analogs, Labeling Cells, and Basic Microscopy". eds. Wang, Y.-L. and Taylor, D. L. San Diego: Academic. pp 269-289.

Taylor, D. L. and Wang, Y.-L. eds.(1989). Methods in Cell Biology. Volume 30. "Fluorescent Microscopy of Living Cells in Culture. Part B. Quantitative Fluorescence Microscopy - Imaging and Spectroscopy". San Diego: Academic.

Tsien, R. Y. (1989). Fluorescent indicators of ion concentration. In: Methods in Cell Biology. Volume 29. "Fluorescent Microscopy of Living Cells in Culture. Part A. Fluorescent Analogs, Labeling Cells, and Basic Microscopy". eds. Wang, Y.-L. and Taylor, D. L. San Diego: Academic. pp 127-156.

Wang, Y.-L. and Taylor, D. L. eds. (1989). Methods in Cell Biology. Volume 29. "Fluorescent Microscopy of Living Cells in Culture. Part A. Fluorescent Analogs, Labeling Cells, and Basic Microscopy". San Diego: Academic.

MEASURING OXYGEN USING OXYGEN DEPENDENT QUENCHING OF PHOSPHORESCENCE: A STATUS REPORT

D.F. Wilson

Department of Biochemistry and Biophysics
Medical School, University of Pennsylvania
Philadelphia, USA

Oxygen dependent quenching of phosphorescence has recently been recognized as providing an optical method for measuring oxygen pressure in biological systems (see for examples Vanderkooi et al, 1986; 1987; Wilson et al, 1987; 1988; 1991; Robiolio et al, 1989; Rumsey et al, 1989). When light is absorbed by a phosphorescent molecule it is raised to an excited singlet state, with the electron remaining in the same spin state it had in the ground state. It can return from the excited singlet state to the ground state with emission of light (fluorescence), a process consistent with excited state lifetimes of a few nanoseconds. In some molecules there is, however, rapid "intersystem crossing" in which the spin state of the electron in the excited state is "flipped" into the opposite spin, yielding a triplet state. The excited triplet is a lower energy state containing unpaired electrons and for which return to the ground state can occur only with a simultaneous change in electron spin state, i.e. return to the ground state is spin forbidden. Triplet states are therefore generally much longer lived than singlet states. When the excited electron of the triplet state returns to the ground state with emission of light, the emitted light is called phosphorescence. Triplet states can have very long lifetimes, but for measuring oxygen it is convenient to use ones with lifetimes of hundreds of microseconds to milliseconds.

Both fluorescence and phosphorescence can be quenched by molecular oxygen. This quenching generally occurs through energy transfer from the excited state molecule to molecular oxygen during collision of the two molecules. The lifetime of the excited singlet state is too short for a significant fraction of the population to collide with oxygen, at least at physiological oxygen pressures. The oxygen dependent quenching of fluorescence is therefore small. Excited triplet states, with lifetimes in the absence of oxygen of nearly a millisecond, have approximately 10^5 as long to collide with oxygen than do singlet states. Thus, the oxygen dependent quenching of phosphorescence is large enough to provide an effective measure of oxygen pressure. The following discussion will be restricted to phosphorescence but is equally relevant to fluorescence.

Optical Imaging of Brain Function and Metabolism, Edited by
U. Dirnagl *et al.*, Plenum Press, New York, 1993

Quenching of phosphorescence by oxygen is limited by the probability of collision between the triplet state molecule and oxygen. It is therefore a diffusion limited reaction which follows the Stern-Volmer relationship:

$$I^o/I = T^o/T = 1 + k_Q T^o PO_2 \quad (1)$$

where I^o and T^o are the phosphorescence intensity and lifetime at zero oxygen pressure respectively and I and T are the phosphorescence intensity and lifetime at an oxygen pressure PO_2. The quenching constant, k_Q, is dependent on the frequency of collision between oxygen and the excited state triplet molecule, the probability of a collision resulting in energy transfer and the physical environment of the phosphorescent probe.

Increasing oxygen pressure causes an increase in the rate of decay of phosphorescence (shorter lifetimes) and a decrease in total phosphorescence intensity. The use of phosphorescence lifetime as the measure of oxygen pressure results in a method insensitive to absorbance changes of other chromophores in the system. This is particularly useful when observations are being made in tissue *in vivo* where there are other chromophores which may change absorbance such as myoglobin, hemoglobin and cytochrome. The lifetime measurements are quantitative, rapid (they can require less than 1 msec) and sensitive to oxygen pressures throughout the physiological range.

A survey of phosphorescent probes suitable for use in measuring oxygen in biological samples (see for example Vanderkooi et al, 1987) identified some Pd-porphyrin compounds as having properties being particularly useful for measuring oxygen in biological systems. The criteria for selecting probes was several fold:

1. There was a high quantum efficiency for phosphorescence.
2. They had phosphorescence lifetimes in the absence of oxygen in the range of 100 μsec to 10 msec.
3. Their phosphorescence decay occurred with good fit to a single exponential.
4. The probes bound tightly to serum albumin.

Experimentally, the phosphorescence of the Pd-porphyrins has been found to precisely follow a single exponential over a wide range of oxygen pressures. The oxygen dependence of the phosphorescence intensity and lifetime also accurately follows equation 1 (see for example Vanderkooi et al, 1987; Wilson et al, 1991; Pawlowski and Wilson, 1992). Intersystem crossing is sufficiently rapid that the fluorescence of these compounds is negligible and the phosphorescence quantum efficiency is high. The Pd-porphyrins also bind well to serum albumin, a property which was considered important for several reasons.

A. The albumin provides a well defined physical environment for the probes. As long as excess albumin is present in the biological system being measured, the probe environment remains constant.

B. When the porphyrins are bound to albumin, access to oxygen is decreased. This lowers k_Q from near 10^9 $M^{-1}sec^{-1}$ to near 10^8 $M^{-1}sec^{-1}$. For a probe with a T^o of approximately 1 msec, this results in the value of T at air saturation being about 30 μsec instead of about 3 μsec. Although the sensitivity to oxygen is decreased, the dynamic range is still excellent and the measurements are technically more convenient. Flash lamps with light pulses of less than 5 μsec width at half-maximal height are readily purchased and for reasonable prices. These are of sufficiently short pulse width to measure lifetimes longer than about 20 μsec but cannot be used for lifetimes less than about this value. Currently available light sources with pulse widths of less than a microsecond and of sufficient brightness (such as pulsed lasers) are generally much more expensive, are less durable, and provide less reproducible flash intensities than flashlamps. The latter two limitations are particularly true for the high repetitive flash rates needed for many applications of the oxygen dependent quenching of phosphorescence. The photomultiplier signal needs to be digitized for analysis and A/D conversion rates of 1 MHz with 12 bit resolution are available as relatively inexpensive commercial boards for microcomputers.

C. Probes which bind to albumin are inhibited from binding to membranes and other relatively hydrophobic agents in the biological materials. Binding to agents other than albumin would result in uncertainty concerning the values for T^o and k_Q, since these can be dependent on the physical environment of the probes.

D. When the probes are bound to albumin interaction between probe molecules is decreased. Ground state probe molecules can quench phosphorescence if collision occurs between the triplet state molecule and a ground state molecule. This "self quenching" is greatly suppressed when the probes are bound to albumin. This is due to decreased rate of diffusion (albumin is a large molecule) and to limited possibility of close approach of two probe molecules when both are bound to albumin. As a result there is no measurable dependence of the phosphorescence lifetimes on probe concentration, at least for the concentration range used for *in vivo* or *in vitro* oxygen measurements.

Phosphorescence lifetime can be readily measured using conventional photomultipliers or intensified video cameras. Photomultiplier based instruments (see for example Green et al, 1989; Pawlowski and Wilson, 1992) have excellent sensitivity and can readily make repetitive measurements of oxygen at a very high rate. Thus using a simple microcomputer based instrument we have made phosphorescence lifetime (oxygen) measurements at more than 20 times per second and rates of more than 100/second are readily attainable. Photomultiplier based systems provide measurements which average the phosphorescence decay for the observed volume. In tissue and other samples having distributions of oxygen pressures, the phosphorescence decay is a sum of decay for probes in different oxygen environments (different lifetimes). Deconvolution of the decay curve into its component single exponentials can be used to partially resolve oxygen pressure distributions within the sample (see for example Pawlowski and Wilson, 1992). In general, the ability to resolve the component exponentials is limited by signal to noise to 3-4 different lifetimes within the 30 μsec to 1 millisecond range of probe lifetimes at physiological oxygen pressures (larger ranges of lifetimes increases the number of components which can be resolved). Deconvolution is most effective when there are two to three volumes of similar initial phosphorescence intensities but distinctly different oxygen pressures in the sampled volume.

Better resolution of regions of different oxygen pressure in a sample can achieved by optical methods. The volume of sample which is illuminated and from which phosphorescence is collected can be optically restricted. Observation of sample volumes less than the size of the regions of different oxygen pressure can be combined with a mechanism for stepwise scanning of the sample. This would obviously allow two dimensional resolution of the oxygen pressure differences. In the limiting case, such an optical system would be operationally equivalent to the confocal microscopes currently used for imaging fluorescent probes. Resolution of the two and three dimensional distribution of phosphorescence lifetimes could be obtained at levels similar to those obtained for the distribution of fluorescent probes in samples using confocal microscopes.

Intensified CCD array cameras make it possible to collect two dimensional images of phosphorescence and thus rapidly acquire data on the spatial distribution of oxygen in samples (see Rumsey et al, 1989; Wilson et al, 1991). These cameras, however, require higher phosphorescence intensities than do photomultipliers. Photomultipliers are single detectors and all the light collected from the phosphorescent sample is concentrated on one detector. Intensified video cameras function as arrays of thousands of small detectors. The camera used in the author's laboratory uses a 512 x 480 CCD array, and each CCD must receive sufficient light energy to make a reliable measurement of light intensity. Assuming equal light sensitivity, the video camera thus requires 245,000 times (512 x 480) as much light as would a single detector with a light sensitive area similar to that of the total diode array.

In our laboratory, a Wild-Leitz Macrozoom microscope with an epifluorescence attachment is used for imaging because we are generally interested in resolving structures in tissue areas less than about 1 cm square. A Xybion intensified CCD camera (Xybion Electronics Systems Corp., San Diego, CA) has been chosen. It has excellent light sensitivity and its 2nd generation intensifier can be gated in less than 250 nseconds. When this camera is equipped with a red sensitive photocathode it is well suited for measuring the phosphorescence of the Pd-porphyrins which is at 650-750 nm. The phosphorescence can be excited using light of wavelength either near 400 nm (the Soret absorption band of the Pd-porphyrins) or in the visible region near 530 nm (the alpha and beta absorption bands of the Pd-porphyrins). The depth of penetration of the excitation light into tissue is dependent on the tissue chromophores (particularly hemoglobin, myoglobin and cytochrome). Light of near 400 nm wavelengths (near ultra-violet) penetrates only 0.1-0.2 mm into tissue whereas the 530 nm (green) light penetrates about 1 mm. Phosphorescence, which occurs at wavelengths from 650 nm to 850 nm, passes through much thicker layers of tissue because tissue absorption is weak at these wavelengths. There is no interference from natural phosphores since tissues examined to date have shown no detectable phosphorescence at wavelengths greater than 650 nm before the probe was injected into the animal.

It should be emphasized that although the natural chromophores of tissue determine the depth of tissue penetrated by the excitation light, these chromophores do not interfere with phosphorescence lifetime measurements. The chromophores in tissue do not change during the 1 msec needed to measure the phosphorescence lifetime using the photomultiplier system. With the video imaging system it is necessary for the tissue to be in a steady state during the time required to collect the phosphorescence images i.e. the absorption of the chromophores does not change during the measurement. Under these conditions the measurements are independent of the degree of oxygenation of hemoglobin and myoglobin as well as of the state of reduction of cytochromes.

Oxygen probes such as Pd-coproporphyrin and Pd-mesoporphyrin, have very small dependencies of T^o and k_Q on pH or temperature (see Wilson et al, 1989; Pawlowski and Wilson, 1992). The probe we currently use most extensively is Pd-meso-tetra (4-carboxyphenyl) porphine (Porphyrin Products, Logan, Utah). This probe also has little pH or temperature dependence in the physiological range but becomes pH dependent on the acidic side of pH 7.0. It is substantially less expensive than the coproporphyrin and mesoporphyrin compounds, however, and this has led to its being our preferred oxygen probe. The Pd complex of tetra-(4-carboxyphenyl) porphine, when it is bound to bovine serum albumin and at physiological pH and 38°

C, has a k_Q value of about 325 $Torr^{-1}$ sec^{-1} and a T^o near 600 μsec (see Wilson et al, 1991; Pawlowski and Wilson, 1992).

Measurements of oxygen pressure in tissue *in vivo* have been extensively carried out using the Pd complex of meso tetra-(4-carboxyphenyl) porphine (approx. 20 mg/kg body weight). The probe (approx. 10 mg/ml) is infused through an artery or vein as a complex with bovine serum albumin (Fraction V, ICN ImmunoBiologicals, Costa Mesa, CA; 60-70 mg/ml) dissolved in physiological saline at pH 7.4. Injection of the Pd-porphyrins does not result in measurable alterations in physiological parameters (blood pressure, blood gases, EEG, blood glucose, blood pH) either during injection or during experiments lasting several hours. After a bolus injection, the phosphorescence is stable for several hours, indicating the probe is not being removed from the blood at a significant rate. Similarly, there is no evidence that significant amounts of probe enter the interstitial space during experiments lasting several hours (longer times have not been tested). In our experience there have been no adverse reactions to the probes. The properties of the Pd-porphyrin probes suggest they should not have significant toxicity for animals.

An EG&G 45 watt xenon flashlamp (EG&G, Salem, MA) with a flash duration of less than 5 μsec is mounted in a Leitz lamp housing to provide illumination through the epifluorescence attachment. The flash lamp was controlled by a 16 MHz 80386 microcomputer (Spear Technology, Northbrook, IL) which determined the timing of the flashes and gating of the intensifier on the video camera. One of the protocols used for obtaining the phosphorescence images at different delay times was as follows: Number of flashes averaged for each delay time, 8; Delay times after the flash, 20 μsec, 40 μsec, 80 μsec, 160 μsec, 300 μsec, 600 μsec, and 2,500 μsec; gate width in all cases, 2,500 μsec. The image processor currently in use can average 8 frames for a delay time, display this averaged image and record it on the the hard disk in about 7 seconds. Using the current instrument, a set of 7 images can be acquired in total of about 40 seconds. This includes the time required to save the images on the hard disk. It is important that the tissue not move during the measurements and this is readily achieved for tissues such as brain. Rapid advances are being made in the handling the images on a microcomputer. It is reasonable to expect that within a year all 7 phosphorescence images will be collected in less than 5 seconds, including saving them on the hard disk.

Two different methods have been used to analyze the data:

1. Regions of interest (small areas within the phosphorescence images), are identified. The average phosphorescence intensity in each region is determined for each image. These intensity values are plotted as the logarithm of phosphorescence intensity against delay time. Fit to a straight line is determined. If the decay is biphasic (indicated by a fit to a straight line with a correlation coefficient of less than 0.97) a fit to two or more straight lines can be used. In most cases the fit to a straight line is satisfactory (decay follows a single exponential). The measured decay constants (T) are substituted into equation 1 and the oxygen pressure in each region of interest is calculated, making use of values of k_Q and T^o obtained by calibration. This approach is useful for following oxygen pressure in discrete anatomical features, particularly when there is reason to suspect the oxygen pressures may not be homogeneous (the phosphorescence decay does not follow a single exponential).

2. The images of phosphorescence are digitized as 512 x 480 pixel arrays of data. These data arrays are filtered and used to calculate a phosphorescence lifetime for each pixel of the image set. A 3x3 filter is passed over each image and then the background (the image taken with a 2,500 μsec delay) subtracted from each of the other images of the set. The decrease in intensity at each pixel of the array with increasing delay time is fitted to a single exponential decay curve. The result is a two

dimensional map of the phosphorescence lifetimes in the observed area of the tissue. An oxygen pressure map is then calculated using equation 1 and the known values of k_Q and T^o. Calculation of an oxygen pressure map from a phosphorescence image set such as that described above presently requires about 3 minutes. Alterations currently being made in the procedure will decrease the calculation time to less than 30 seconds.

The images of phosphorescence intensity of the brain cortex (Wilson et al, 1991; 1992A,B) show that under control conditions the veins as relatively bright, well defined vessels on a background of lower phosphorescence intensity. The oxygen probe is dissolved in the blood and phosphorescence intensity is dependent primarily on:

1. The amount of probe in the tissue volume irradiated by the excitation light (this is proportional to the concentration of blood in thickness of tissue penetrated by the excitation light).

2. The oxygen pressure in the blood in the region of observation (phosphorescence intensity increases with decreasing oxygen pressure as described by the Stern-Volmer equation).

3. The intensity of the illuminating light (phosphorescence intensity increases in proportion to the illuminating light intensity). Uneven illumination by the light flash results in phosphorescence intensity which is similarly uneven.

The phosphorescence images therefore show veins as bright vessels because of their high blood content and low oxygen pressure. The capillary beds have only 10% or less of their volume as blood, the rest being cells and interstitial space. Therefore the capillary beds have a lower phosphorescence than veins although the oxygen pressure is similar to that in the veins. Arterioles are not seen without using experimental conditions emphasizing short phosphorescence lifetimes. Although the concentration of blood, and therefore of probe, in the arterioles is high, the vessels are small in diameter and the oxygen pressure is high. The latter quenches the phosphorescence emission from the arterioles to below that of the capillary bed. The excitation light passes through the arteriole and into the capillary bed below. Phosphorescence from the capillaries dominates the images and, as a result, the calculated phosphorescence lifetimes.

Phosphorescence lifetimes are independent of the concentration of phosphorescent probe in the blood and of the amount of blood in the observed volume. In tissue the greatest part (about 75%) of the blood in tissue is contained in the venous ends of the capillaries and in the veins. As a result calculated phosphorescence lifetimes for the veins and capillary beds are normally not very different. Maps of phosphorescence lifetime and oxygen pressure therefore do not show the pattern of vessels in the surface of the brain, although the vessels are readily seen in the images of phosphorescence intensity.

Oxygen dependent quenching of phosphorescence provides a major new tool for studying oxygen delivery to tissue. It has also been shown to effectively report abnormal structures within tissue, particularly when they are associated with abnormal vascularization. The list of different tissues which have been examined using phosphorescence quenchinf techniques and the type of questions addressed is rapidly expanding. In *vitro* measurements have been made in isolated perfused rat liver (Rumsey et al, 1989; Wilson et al, 1989) and isolated perfused carotid body (Rumsey et al, 1991). *In vivo* measurements include the cortex of the brain of newborn piglets (Wilson et al, 1991; 1992B) and adult cats (Wilson et al, 1992A), the carotid body of the cat (Rumsey et al, 1992), the retina and choroid of the cat eye (Shonat et al, 1992), rat heart (Ince et al, 1992), subcutaneous and intramuscular tumors in the rat (Wilson and Cerniglia, 1992). In each case, phosphorescence measurements have provided new and valuable data on the delivery and utilization of oxygen. Phosphorescence quenching also appears likely to become an important new clinical tool for diagnosis and treatment tumors and for evaluation of the degree of pathology and efficacy of treatment of many forms of vascular disease. One can speculate that it may also provide a new tool in treatment of wound healing, which appears to be critically dependent of oxygen pressure in the wound area. Lack of methods for measuring oxygen pressure in the eye has inhibited our understanding of the role of abnormal oxygen delivery in pathology of the eye. Phosphorescence imaging removes this limitation and the possible role of abnormal tissue oxygenation in development and recognition of pathology of the eye will rapidly be determined.

ACKNOWLEDGEMENTS-This work was supported in part by a grant NS-10939 from the National Institutes of Health and N00014-89-J-1243 from the Office of Naval Research.

REFERENCES

Green, T.J., Wilson, D.F., Vanderkooi, J.M., and DeFeo, S.P. (1989) Phosphorometers for analysis of decay profiles and real time monitoring of exponential decay and oxygen concentrations. Analy. Biochem. 174: 73-79.

Ince, C., Ashruf, J., Sanderse, E.A. and Bruining, H.A. (1992) NADH and Pd-porphyrin video fluorometry of the rat heart *in vivo*. Adv. Exp. Med. Biol. in press.

Pawlowski, M. and Wilson, D.F. (1992) Monitoring of the oxygen pressure in the blood of live animals using the oxygen dependent quenching of phosphorescence. Adv. Exp. Med. Biol. in press.

Robiolio, M., Rumsey, W.L., and Wilson, D.F. (1989) Oxygen diffusion and mitochondrial respiration in neuroblastoma cells. Amer. J. Physiol. 256: C1207-C1213.

Rumsey, W.L., Iturriaga, R., Spergel, D., Lahiri, S., and Wilson, D.F. (1991) Optical measurements of the dependence of chemoreception on oxygen pressure in the cat carotid body. Amer. J. Physiol. 261: C614-C622.

Rumsey, W.L., Lahiri, S., Iturriaga, R., Mokashi, A., Spergel, D., and Wilson, D.F. (1992) Optical measurements of oxygen and electrical measurements of oxygen chemoreception in the cat carotid body. Adv. Exp. Med. Biol. in press.

Rumsey, W.L., Robiolio, M., and Wilson, D.F. (1989) Contribution of diffusion to the oxygen dependence of energy metabolism in human neuroblastoma cells. Adv. Exp. Med. Biol. 248: 829-833.

Rumsey, W.L., Vanderkooi, J.M., and Wilson, D.F. (1989) Imaging of phosphorescence: a novel method for measuring oxygen distribution in perfused tissue. Science, Wash. DC 241: 1649-1651.

Shonat, R.D., Wilson, D.F., Riva, C.E., and Pawlowski, M. (1992) Oxygen distribution in the retinal and choroidal vessels of the cat as measured by a new phosphorescence imaging method. Applied Optics in press.

Vanderkooi, J.M., Maniara, G, Green, T.J., and Wilson, D.F. (1987) An optical method for measurement of dioxygen concentration based on quenching of phosphorescence. J. Biol. Chem. 262: 5476-5482.

Vanderkooi, J.M., and Wilson, D.F. (1986) A new method for measuring oxygen in biological systems. Adv. Exp. Med. Biol. 200: 189-193.

Wilson, D.F. and Cerniglia, G.J. (1992) Localization of tumors and evaluation of their state of oxygenation by phosphorescence imaging. Cancer Res. in press.

Wilson, D.F., Pastuszko, A., DiGiacomo, J.E., Pawlowski, M., Schneiderman, R., Delivoria-Papadopoulos, M. (1991) Effect of hyperventilation on oxygenation of the brain cortex of newborn piglets. J. Appl. Physiol. 70(6): 2691-2696.

Wilson, D.F., Gomi, S., Pastuszko, A., and Greenberg, J.H. (1992A) Oxygenation of the cortex of the brain of cats during occlusion of the middle cerebral artery and reperfusion. Adv. Exp. Med. Biol. in press.

Wilson, D.F., Pastuszko, A., Schneiderman, R., DiGiacomo, J.E., Pawlowski, M. and Delivoria-Papadopoulos, M. (1992B) Effect of hyperventilation on the oxygenation of the brain cortex of neonates. Adv. Exp. Med. Biol. in press.

Wilson, D.F., Rumsey, W.L., Green, T.J., and Vanderkooi, J.M. (1988) The oxygen dependence of mitochondrial oxidative phosphorylation measured by a new optical method for measuring oxygen concentration. J. Biol. Chem. 263: 2712-2718.

Wilson, D.F., Rumsey, W.L., and Vanderkooi, J.M. (1989) Oxygen distribution in isolated perfused liver observed by phosphorescence imaging. Adv. Exp. Med. Biol. 248: 109-115.

Wilson, D.F., Vanderkooi, J.M., Green, T.J., Maniara, G., DeFeo, S.P., and Bloomgarden, D.C. (1987) A versatile and sensitive method for measuring oxygen. Adv. Exp. Med. Biol. 215: 71-77.

LASER FLUORESCENCE SPECTROSCOPIC EXPERIMENTS FOR MONITORING MOLECULES IN BRAIN

F. Fink, R. Paul, and A. Pfarrherr*

Laser Labor Adlershof e.V.,
*Institut für Optik und Spektroskopie
Berlin, Germany

INTRODUCTION

The selective detection of biomolecules present in complex organic structures is important, not only for medical research, but also for the development of new diagnostic procedures in all areas of medical practice. Molecules of interest in connection with brain research and diagnosis are, besides others, biomolecules (e.g. co-enzymes of endogenous metabolism, neurotransmitters) and drugs.

By determining the concentration of released neurotransmitters in certain brain areas, it is possible to draw conclusions about the general function of the brain, the influence of drugs on brain functions and the change of the neuronal function due to neurological illnesses or mental disease. Observance of the concentration of co-enzymes of endogenous metabolism, especially NADH, can be used for example in brain tumor diagnosis to decide between normal and tumor tissue.

Chemical, immunological and optical methods can selectively detect biomolecules. For in vivo detection, the methods of voltammetry and microdialysis were developed (Marsden 1983). The employment of lasers has considerably increased the application range of optical methods (Andersson-Engels et al. 1990). Variations of the absorption, Raman, infrared, and fluorescence laser spectroscopy enable safe and highly sensitive selective detection of principally all interesting biomolecules. Significant photophysical properties of the molecules, e.g. absorption, emission, Raman and infrared bands generally determine which laserspectroscopic method should be applied for a highly selective detection. It is easy to detect those molecules which show fluorescence in a visible spectral range. The fluorescence of these molecules can be excited by introducing laser light with the wavelength of the molecules' absorption band. The excited molecule emits part of the absorbed energy as characteristic laser induced fluorescence (LIF), which is analysed. Fluorescence light is characterized by the quantum yield, spectral intensity distribution, and the decay time. These three properties of fluorescence form the basis for the fluorescence spectroscopic procedure for selective molecular detection. Photodetectors connected to spectrometers and to

Optical Imaging of Brain Function and Metabolism, Edited by
U. Dirnagl *et al.*, Plenum Press, New York, 1993

sensitive data processing systems are used to record the fluorescence. Sensitive detecting systems, which simultaneously enable a high temporal resolution of fluorescence detection, are phase-fluorometry in connection with modulated exciting laser irradiation, boxcar and sampling techniques in connection with pulsed laser sources, synchroscan streak techniques, and the time correlated single photon counting in connection with high repetition rate laser sources. Using single photon counting it is possible to detect fluorescent molecules in solutions of very low concentrations ($<10^{-12}$ mol). Further positive properties of the laser induced fluorescence (LIF) are high spatial resolution (μm range), short registration time, and the ability to make measurements without touching the sample.

This paper describes LIF studies of the neurotransmitters serotonin and 5-HIAA, the antiparkinson drug lisuride, and NADH molecules involved in metabolism. The aim of this investigation is to find criteria for their selective detection. The experiments are conducted employing

- the boxcar technique in connection with a spectrometer and a pulsed N_2-laser and
- the time correlated single photon counting in connection with a spectrometer and a mode-locked laser.

In comparison to generally used LIF procedures for the selective detection of molecules the procedure described in this paper employs the simultaneous spectral and temporal resolution of the fluorescence signal. This procedure measures the fluorescence decay time in small wavelength steps within the fluorescence band.

It is possible with this procedure to determine changes of fluorescence decay times within the emission band of samples, which include different fluorescent molecules. The changes of decay times are a consequence of the competiting fluorescence decay processes of different molecules and of the energy transfer between them. Suitable parts of the spectrum and the corresponding decay times can be used for identification of the different molecules, which are contained in the sample.

In this way the reliability of the detection can be improved. Moreover, this method provides a deeper insight into the energy transfer process and the kinetics of energy migration in molecules. The simultaneous measurement of the spectral and temporal intensity distribution is also a prerequisite for the quantitative determination of the fluorescence yield.

The wavelength and time resolved intensity of fluorescence will be provided for lisuride and NADH. Spectral ranges and fluorescence decay times obtained from these spectral and temporal curves which are suitable for selective detection of molecules will be discussed. The knowledge obtained regarding optimal detection conditions for NADH molecules was used to create an experimental arrangement for monitoring the metabolism of brain tissue.

SELECTIVE DETECTION OF NEUROTRANSMITTERS BY TIME RESOLVED LASER INDUCED FLUORESCENCE USING AS AN EXAMPLE SEROTONIN AND ITS METABOLITE 5-HIAA

The selective detection of serotonin and its metabolite 5-HIAA in an aqueous solution is of interest for pharmacological experiments. Experiments to identify both types of molecules in solution with the aid of the fluorescence spectra only were not successful. The fluorescence spectra were recorded with a double prism spectrometer employing the lock-in technique. Fig.1 shows the spectra of the aqueous solution of serotonin and 5-HIAA singularly, and as a mixture of both molecules at a concentration of 10^{-7} mol. It is quite clear from the spectrum of the mixture that no statement can be made regarding its content.

The evaluation of the spectra alone is not a definitive method for detecting serotonin and 5-HIAA in aqueous solutions. If however, the time behavior of the fluorescence is taken into account, then both molecules can be selectively detected in an aqueous solution.

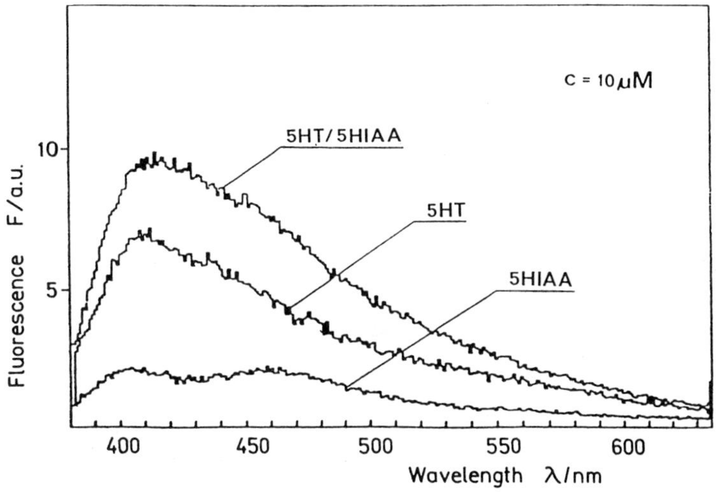

FIGURE 1. Spectra of aqueous solution of serotonin, 5-HIAA, and of a mixture of both molecules at a concentration of 10^{-7} mol.

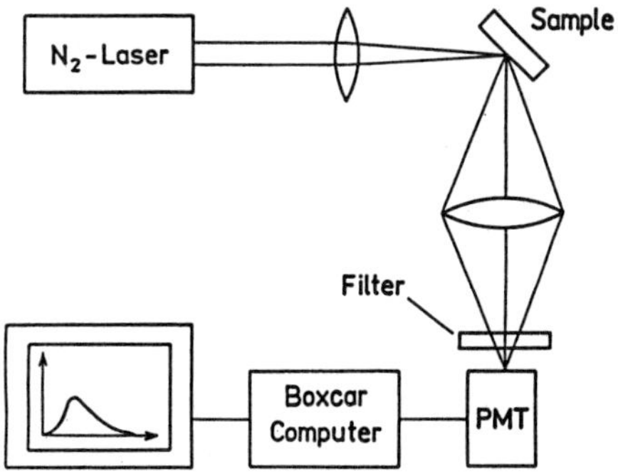

FIGURE 2. Experimental arrangement for time resolved fluorescence measurements. PMT photomultiplier tube.

To measure the time resolved fluorescence the experimental arrangement in Fig.2 was used. A N2-laser generates the excitation pulse, which has a wavelength of 337,1 nm and a duration of 500 ps. The induced fluorescence is separated from the excitation light with the aid of filters and registered with a fast photomultiplier tube. The time resolution of the signal is achieved by the box-car technique. A spectral resolution is not executed by this experimental arrangement. Fluorescence decay time determined using this technique for an aqueous solution of serotonin (10^{-7} mol) is 1.63 ns and for a corresponding 5-HIAA solution 2.07 ns.

Fig.3 shows the time resolved intensity of the fluorescence of the mixture of both molecules (see dotted curve). This course of time resolved fluorescence can not be described by a single exponential curve.

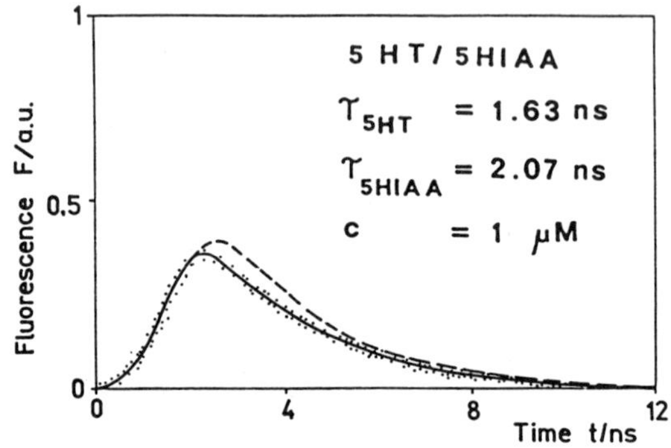

FIGURE 3. Time-resolved fluorescence intensity of an aqueous serotonin/5-HIAA solution.

However, it can be fitted very well by a double exponential curve. This is a significant indicator for the presence of two fluorescent molecules in the solution. Each of the molecules has its own decay time τ_1, τ_2 respectively known from a separate experiment. The course of fluorescence can be described by the following equation:

$$F(t)= A \exp(-t/\tau_1)+B \exp(-t/\tau_2).$$

The terms on the right hand side of the equation represent the single exponential fluorescence decay for each single molecule. The parameter A and B are determined by fitting the calculated curve to the experimental one. Since the quantum yield of the fluorescence is correlated to the concentration of molecules in the solution it is possible to deduce from the parameters A and B the relative concentration of the serotonin and 5-HIAA molecules in the solution. In this way not only a selective detection of serotonin and 5-HIAA in an aqueous solution is possible, but also an estimation of the relative concentration of both molecules in the solution. Serotonin and 5-HIAA molecules can be selectively detected in solutions at concentrations up to 10^{-9} mol using this technique.

SPECTRAL AND TIME RESOLVED LASER-INDUCED FLUORESCENCE OF LISURIDE AND NADH

The method of time resolved fluorescence, without using spectral resolution, as described in the previous section, is used by many groups (e.g. Schneckenburger et al. 1990) and is sufficiently selective for most applications.

However, there are cases of molecular detection in which the time behavior of fluorescence varies within the fluorescence band. This effect can be used for an identification of molecules which are embedded in complex structures.

The experimental arrangement which allows the simultaneous spectral and

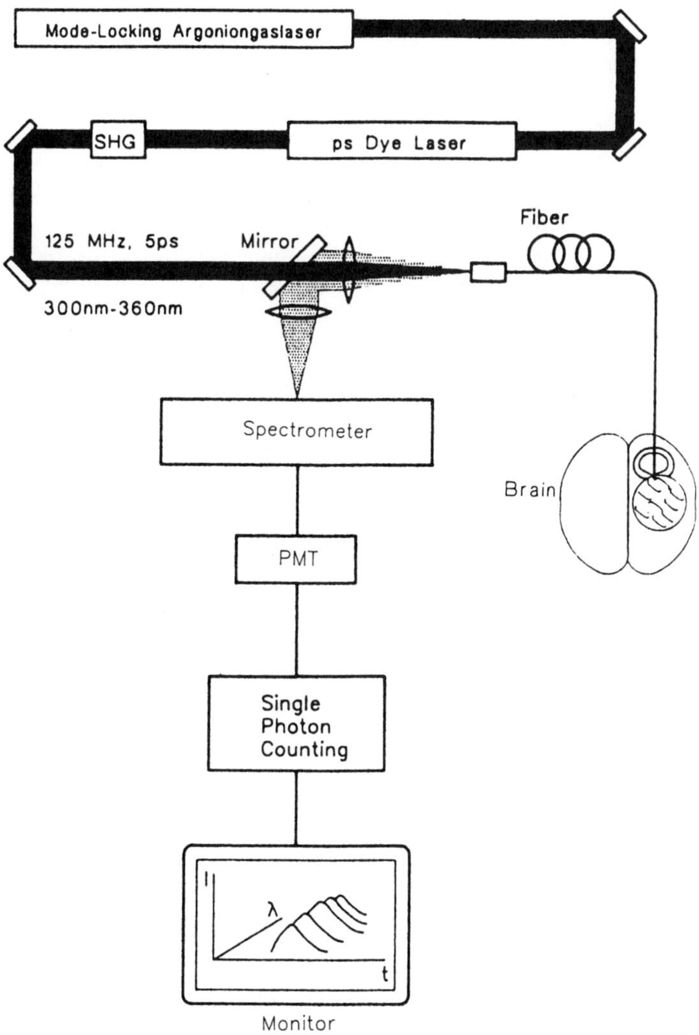

FIGURE 4. Laserfluorometer. SHG second harmonic generation, PMT photomultiplier tube.

temporal resolution of a fluorescence signal is shown in Fig.4. It employs as an excitation source an argon laser which is mode-locked at 476 nm and 514 nm, respectively. By choice a dye laser can be synchronously pumped. By doubling the frequency of the argon laser and the dye laser, pulses can be generated in the spectral range between 257 nm and 350 nm.

Pulses with a duration of 80 ps are generated at a repetition rate of 125 MHz. The time correlated single photon counting in connection with the high repetition rate laser enables a sensitive detection of the fluorescence. The system is so sensitive, that the LIF can even be passed through a spectrograph. Thus it is possible to execute time resolved measurements in spectral steps within the fluorescence band. Combining the fibre technique with this arrangement it is possible to make measurements at selected brain areas, which are very difficult to investigate.

The result of a measurement with this equipment is a temporal and spectral resolved intensity distribution of the fluorescence. An analysis of the fluorescence provides, not only information regarding the conditions for a selective detection, but also information on intra- and intermolecular energy transfer.

Lisuride

This experimental arrangement was used to investigate the fluorescence of an aqueous lisuride solution (ASL) and of lisuride in a brain cell membrane suspension (BCM) after excitation with pulses at 476 nm.

Lisuride was developed to stimulate dopamine receptors. It rapidly achieved an important role in the control of Parkinson's disease and certain pituitary tumors (Calne et al. 1983). The optical detection of lisuride in the brain may be a useful tool for pharmacological research.

Fig.5 shows the spectral and time resolved fluorescence of a) an ASL sample (10^{-5} mol), b) a BCM sample, c) a mixture of ASL (1 ml) and BCM (50 μl) and d) a mixture of BCM (1 ml) and ASL (10 μl).

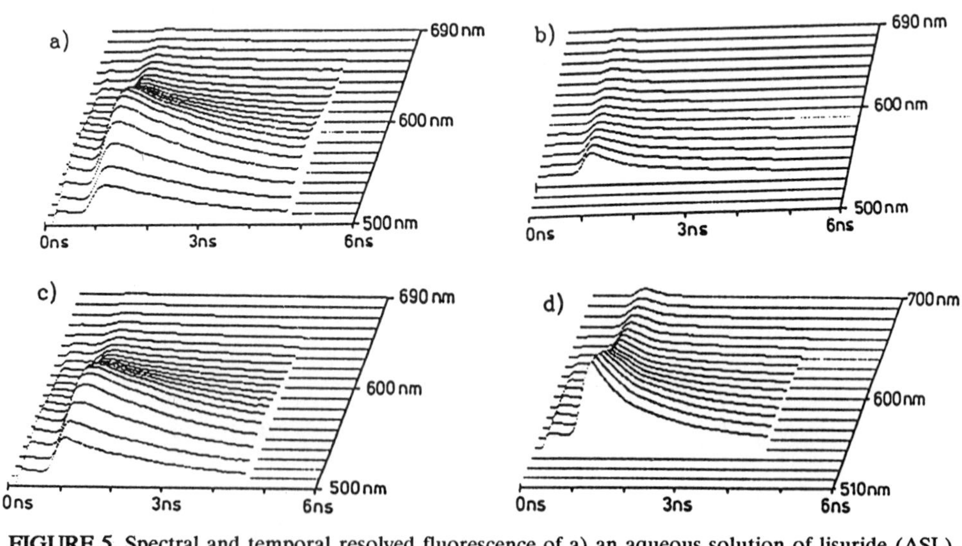

FIGURE 5. Spectral and temporal resolved fluorescence of a) an aqueous solution of lisuride (ASL) (10^{-5} mol), b) a suspension of brain cell membranes (BCM), c) a mixture of ASL (1 ml) and BCM (50 μl)and d) a mixture of BCM (1ml) and ASL (10 μl).

The spectral distribution of the fluorescence seems to be identical for ASL a) and the mixture of ASL (1 ml) and BCM (50 μl) c) and for the BCM samples b) and the mixture of BCM (1 ml) and ASL (10 μl) d). The fluorescence spectra are determined by the substances which are dominantly present in the mixture. The detection of a substance with a lower concentration in the mixture can only be determined to a lower degree of certainty with the aid of the spectrum. If however, we take into consideration the time behavior at different wavelengths as an additional criteria it is possible to obtain a significant identification of the lisuride molecules.

In Fig.6 a)-d) the dependency of the decay times on wavelength is shown for different samples. Marked differences in the course of decay times throughout the wavelength can be found between ASL a), ASL (1 ml)/BCM (50 μl) c) and BCM b) on the one hand and between BCM b) and BCM (1 ml)/ASL (10 μl)(d) on the other hand. Whereas for ASL and ASL (1 ml)/BCM (50 μl) nearly the same course of fluorescence decay throughout the wavelength is determined the corresponding courses for BCM and BCM (1ml)/ASL (10 μl) are completely different.The fluorescence decay of BCM is described by three exponential terms with three characteristic decay times. By adding ASL to the BCM sample only two characteristic decay terms can be found.

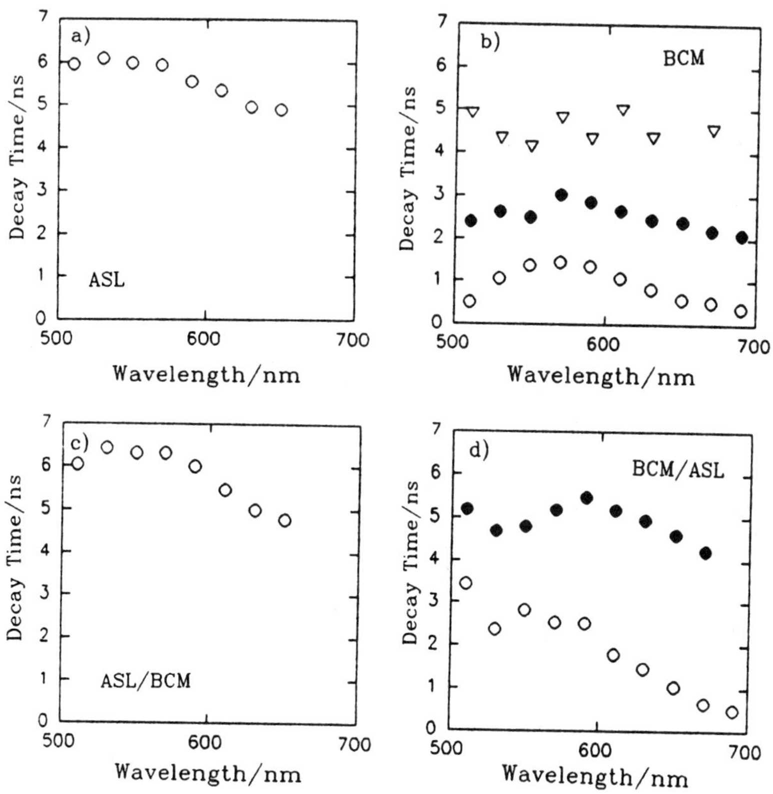

FIGURE 6. Decay times versus wavelength for the samples described in Fig.5.

Obviously, the fast decay component of the pure BCM sample will be quenched by the lisuride molecules in the spectral range between 500 nm and 600 nm. In the spectral range above 650 nm the fast decay component appears again. In the case of high lisuride concentration even the short decay component of Fig.6 d) disappears. This effect can be used to identify the lisuride molecules in BCM samples, and after a calibration of the method, to come to an estimation of its concentration. This experimental arrangement is sensitive enough to detect the fluorescence of lisuride molecules at lower concentrations than 10^{-9} mol.

NADH

A further type of biomolecule investigated by the simultaneous spectral and temporal resolved laser induced fluorescence technique is NADH.

NADH is the reduced form of nicotinamideadenine dinucleotide (NAD). The determination of its concentration in human tissue can be used to characterize the intercellular redox state of this tissue. Such experiments can aid in tumor diagnosis to distinguish healthy and tumor tissue (Lohmann et al. 1990). The observation of changes of NADH concentration is possible by detecting the fluorescence around 460 nm after excitation with N_2-laser pulses (337,1 nm).

Although the NADH fluorescence for metabolism monitoring has been widely used for many years (Chance 1954, Chance and Legallais 1963, Renault et al. 1982, Renault and Py 1989, Schramm and Löwe 1988) there are some unsolved problems. One of these problems is the quantitative determination of the NADH concentration using LIF.

As a contribution to solving this problem we present here for the first time the spectral and temporal distribution of the emitted NADH fluorescence. From these curves it is possible to estimate which part of the fluorescence will be registered by a spectral and temporal gated detector. A detailed discussion of the quantitative determination of the concentration of biomolecules by LIF methods will be given in a further paper.

For the experiments the arrangement shown in Fig.4 was modified. The argon laser and the SPC were replaced by a N_2-laser (MSG 500, Lasertechnik Berlin GmbH)

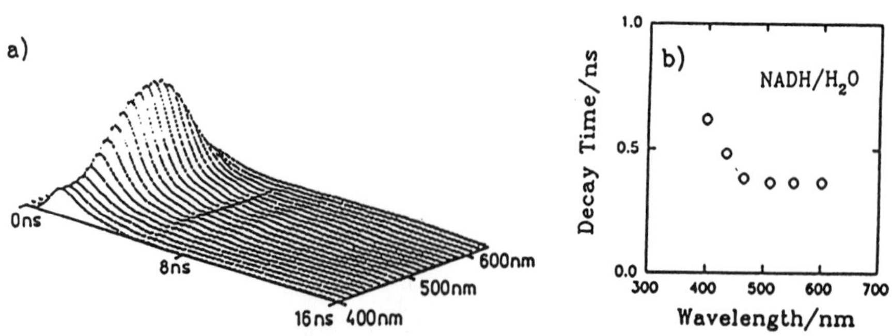

FIGURE 7. Wavelength and time resolved fluorescence distribution of a) an aqueous NADH solution (10^{-7} mol) and b) decay times versus wave-lengths.

and a boxcar computer, respectively. The N_2-laser generates pulses with a duration of 400 ps. Both wavelength and time resolved fluorescence of an aqueous NADH solution (AS-NADH)(10^{-7} mol) are shown in Fig.7 a) (see previous page). The time course of fluorescence can be described by a single exponential decay function. Depending on the wavelength decay times between 600 ps below 500 nm and 450 ps above 500 nm were determined (Fig.7b). For a selective detection of the NADH molecules by using a time window the spectral range of constant decay time between 460 nm and 600 nm should be used.

Fig.8 shows the spectral and temporal intensity distribution of the LIF obtained a) for a sample of pure BCM and b) for a sample of BCM/AS-NADH. It can clearly be seen that in the case of pure BCM a long decay component is dominating, whereas in the case of the BCM/AS-NADH sample the fluorescence decay curve of NADH is dominant. From such representations the spectral and temporal ranges which are optimal for a selective and quantitative detection of the NADH molecules in the brain can be deduced.

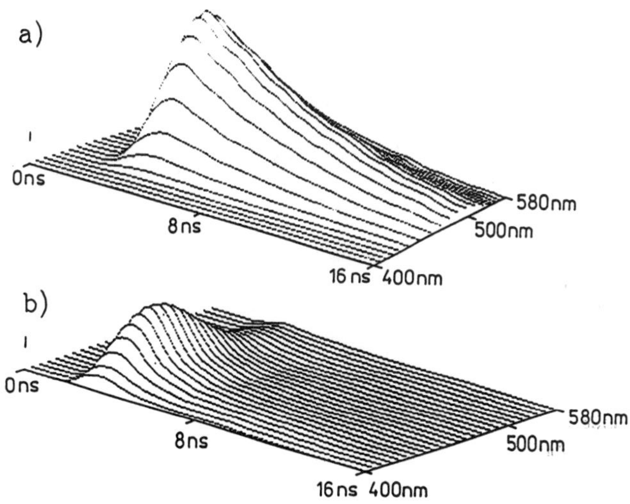

FIGURE 8. Spectral and temporal intensity distribution of LIF a) for a sample of pure BCM and b) for a sample of BCM/AS-NADH.

The first version of a NADH detector for tumor diagnosis in brain areas on the basis of a N_2 laser excited fluorescence analysis was extensively described in a previous paper (Beuthan et al. 1990).

Fig.9 shows the scheme of a device further developed in our laboratory, which can be used in the medical diagnosis and for therapy. The duration of the excitation pulses (350 ps) is optimized for the selective detection of NADH molecules. The excitation pulses and the induced fluorescence are transmitted using an optical fibre (400 μm). This fibre can be used in connection with a catheter. By means of the dielectric mirror only the NADH fluorescence is directed to the detector. An electronic gate of variable time duration (0.2 ns to 1 ns) is used for time selective detection of NADH fluorescence or other excited fluorescence signals.

Fig.10 shows the time course of the radiation which is registered by the detector. The detector is activated precisely at that moment the NADH fluorescence enters it. In this way the NADH fluorescence can be separated from scattered excitation light and from the fluorescence of other molecules. This NADH detector allows to investigate solutions of NADH at concentrations of 10^{-9} mol. For simplification the reference channels used for the consideration of blood flow and properties of tissue are not shown in Fig.9. As a next step this detector system will be optimized for the separate detection of the flavine fluorescence.

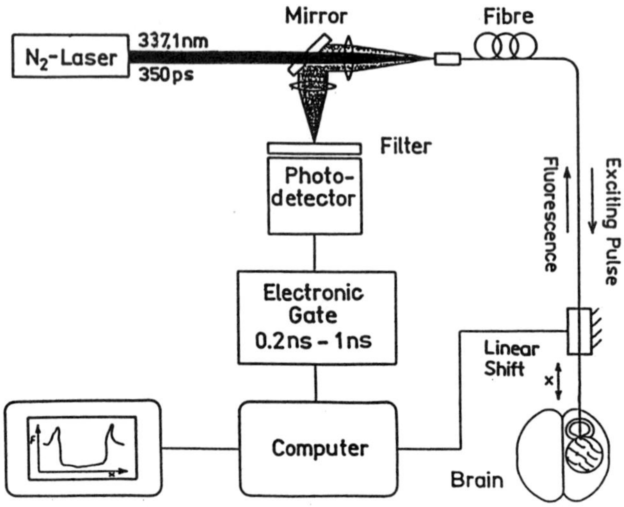

FIGURE 9. NADH detector for medical application.

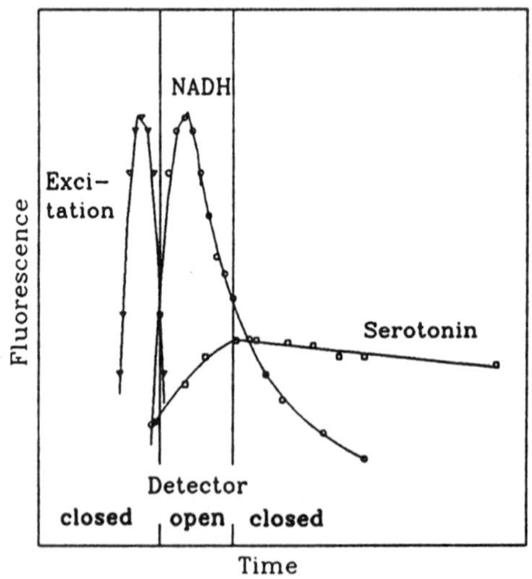

FIGURE 10. Scheme of the time gated detection.

CONCLUSION

In this paper experiments of the selective detection of biomolecules by the laser induced fluorescence technique are described. It is shown that the selective detection of the neurotransmitter serotonin and its metabolite 5-HIAA in an aqueous solution can be achieved by taking into consideration the different decay times of the fluorescence of both molecules after their excitation with pulses of a N_2-laser. An experimental arrangement is described which allows the simultaneous spectral and temporal resolution of the fluorescence induced in molecules by laser pulses. This arrangement was used to investigate the laser-induced fluorescence of the antiparkinson drug lisuride and the NADH molecules. From the analysed fluorescence signals spectral and temporal ranges are deduced suitable for the selective detection of these molecules. The results were used to optimize a NADH detector for medical application.

REFERENCES

Andersson-Engels S., Berg R., Johansson J., Stenram U., Svanberg K., Svanberg S. (1990) Laser spectroscopy in medical diagnostics, in: Photodynamic Therapy: Basic Principles and Clinical Aspects, Marcel Dekker Inc.

Beuthan J., Zur Ch., Fink F., Müller G. (1990) Laser fluorescence spectroscopic experiments for metabolism monitoring. Laser Med. Surg. 6:72-75.

Calne D.B., Horowski R., McDonald R.J., Wuttke W. (1983) Lisuride and other Dopamine Agonists, Raven Press, New York.

Chance B. (1954) Spectrophotometry of intracellular respiratory pigments. Science 120:767-775.

Chance B., Legallais V. (1963) A spectrometer for recording of intracellular oxidation-reduction states. IEEE Trans., BME-10:40-47.

Lohmann W., Hirzinger B., Braun J., Schwemmle K., Muhrer K.-H., Schulz A. (1990) Fluorescence studies on lung tumors, Z. Naturforsch. 45c:1063-1069.

Marsden C.A. (1984) Measurement of neurotransmitter release in vivo. Wiley & Sons, Chichister,New York.

Renault G., Sinet M., Raynal E.,Berthler J.P.,Godard B. (1982) A laser fluorimeter for direct cardiac metabolism investigation. Optics and Laser Techn. 14:143-148.

Renault G., Py A. (1989) In situ monitoring of experimental ischemic biochemical effects in renal tissue by NAD(P)H laser fluorometry. Laser Med. Surg. 5:171-177.

Schneckenburger H., Siedlitz H.K., Haferkamp O. (1990) Energy transfer microscopy for measuring mitochondrial metabolites, Laser Med. Surg. 6:33-37.

Schramm W., Löwe H. (1988) Anordnung zur Erfassung der NADH2- Oberflächenfluoreszenz. Wiss. Z. Ernst-Moritz-Arndt-Univer., Med. Reihe 37:2-3.

INFRARED-INTERFERENCE VIDEOMICROSCOPY OF LIVING BRAIN SLICES

H.-U. Dodt

Clinical Neuropharmacology, Max-Planck-Institute for Psychiatry
Clinical Institute,
Munich, Germany

Infrared radiation penetrates through many kinds of matter more easily than visible light, a fact which has been extensively used in astrophysics (Scheffler and Elsässer 1987). Since brain tissue is also more translucent at longer wavelengths (MacVicar 1984, Svaasand and Ellingsen 1983) we used infrared illumination of the tissue to exploit this property and increase transparency for cellular visualization. Differential interference contrast by Nomarski optics (DIC) (Allen et al. 1968) has been used previously to identify neurons in brain slices (Edwards et al. 1989, Takahashi 1978, Yamamoto 1975). By combining DIC and infrared illumination we reasoned that the natural phase retardation gradients of cellular elements within a slice could be rendered visible (Dodt et al. 1989, Dodt and Zieglgänsberger 1990), and that the contrast of the infrared images could then be further enhanced through image processing with video technology (Allen and Allen 1983, Weiss 1986). The novel combination of infrared and DIC described here permits visualization of neurons in tissue slices of 300-500 μm conventionally employed for in vitro electrophysiological analysis. Contrast and resolution were in addition significantly improved by spatial filtering in the fourier plane of the condensor by inserting a central obstruction and using variable oblique illumination. The combination of these modifications with standard DIC is referred to in the following as interference optics.

The experimental setup consisted of a brain slice chamber with a coverslip bottom mounted on a rigid stage. The slices were kept submerged under a thin film of liquid in standard perfusion medium by holding them gently from above with thin silicon threads. Below the stage an inverted microscope (Zeiss Axiovert) was moved on an x-y table to inspect the slice from below. To permit the use of high numerical aperture oil-immersion objectives for highest resolution, inverted microscopy was selected. For the positioning of electrodes a 0.63 N.A. air condensor was used, while for highest resolution a 1.2 N.A. water immersion condensor front lens was mounted. Near infrared illumination was obtained with a bandpass filter with $\lambda_{max} = 750$ nm, thus excluding visible light and infrared radiation of longer wavelength to prevent heating

Optical Imaging of Brain Function and Metabolism, Edited by
U. Dirnagl *et al.*, Plenum Press, New York, 1993

of the slice. The light source was either a 100 W halogen or xenon lamp, which was coupled with fiber optics to the microscope in order to obtain even illumination of the condensor aperture plane.

The slices were cut with a vibratome at a thickness of 400 μm. Finest neuronal processes were visualized to a depth of 50-80 μm in 400 μm thick slices. In relatively cell free regions such as the stratum radiatum of the hippocampus, neurons were visible up to a depth of 120 μm. Electrophysiological recording was used to control viability of the slices during videomicroscopy.

The image produced by the interference optics was projected on to the target of a Newvicon video camera (Hamamatsu) sensitive to radiation up to 850 nm. After analog contrast enhancement provided by the camera control unit (C 2400, Hamamatsu) the image was displayed on a monitor amd recorded on an S-VHS videorecorder either in real time or time-lapse mode. In addition selected images were stored on the disk of a PC-based digital image processing system. In some cases further digital contrast enhancement was performed by use of electronic sharpen filters and frame averaging. However digital image processing (Allen 1983) is apparently less crucial for infrared microscopy than for standard videomicroscopy. In particular background subtraction was not feasible as a structureless background was difficult to obtain in the thick specimens used.

One of the interesting organizational principles of the neocortex is its modular architecture. It is most easily seen with staining methods in the barrel cortex of the mouse (Woolsey and Van der Loos 1970). In tangentially cut somatosensory cortex of the mouse, barrels were demonstrated with our system without staining at low magnification (5X objective) (Fig. 1A). In the visual cortex the bundling of apical dendrites which has been described in histological sections (Peters and Kara 1987) may represent a similar organisational structure. These dendritic bundles were very prominent in the rat visual cortex with a regular spacing of about 50 μm and were also visualized in other neocortical areas (Fig. 1B).

Comparison of Fig. 1B and Fig. 2A illustrates the importance of the use of high numerical oil-immersion objectives to achieve maximal contrast and resolution. Both figures show neurons in the visual cortex with the same magnification. Fig. 1B was obtained with a 25X 0.8 N.A. objective and additional 4X magnification by a projection lens, whereas Fig. 2A shows the neocortical network as seen with a 100X 1.3 N.A. oil immersion objective. In addition to showing axons and dendrites, the resolution of this objective even allows visualization of structures which resemble spineheads (arrows), in particular if the microscope is focussed through different planes to give a three dimensional impression. Images obtained by infrared videomicroscopy of the hippocampus show the very densely packed pyramidal layer and dendrites in the stratum radiatum. The use of oblique illumination, with enhancement of the shadowcast effect in the direction orthogonal to the shearing direction of the DIC, even allows the dense axonal plexus of the Schaffer collaterals to be visualized (not shown).

As infrared videomicroscopy, in contrast to fluorecence microscopy, has no phototoxic effects, the technique could be used for extended time lapse studies of dynamical morphological changes with maximal intensity of illumination up to 24 hours. By recording the video images of slices of cortex and cerebellum of newborn rats, migration of neurons and microglia could be observed in time lapse mode. Current studies are aimed towards a detailed investigation of neuronal migration and its manipulation by biochemical or electrical factors.

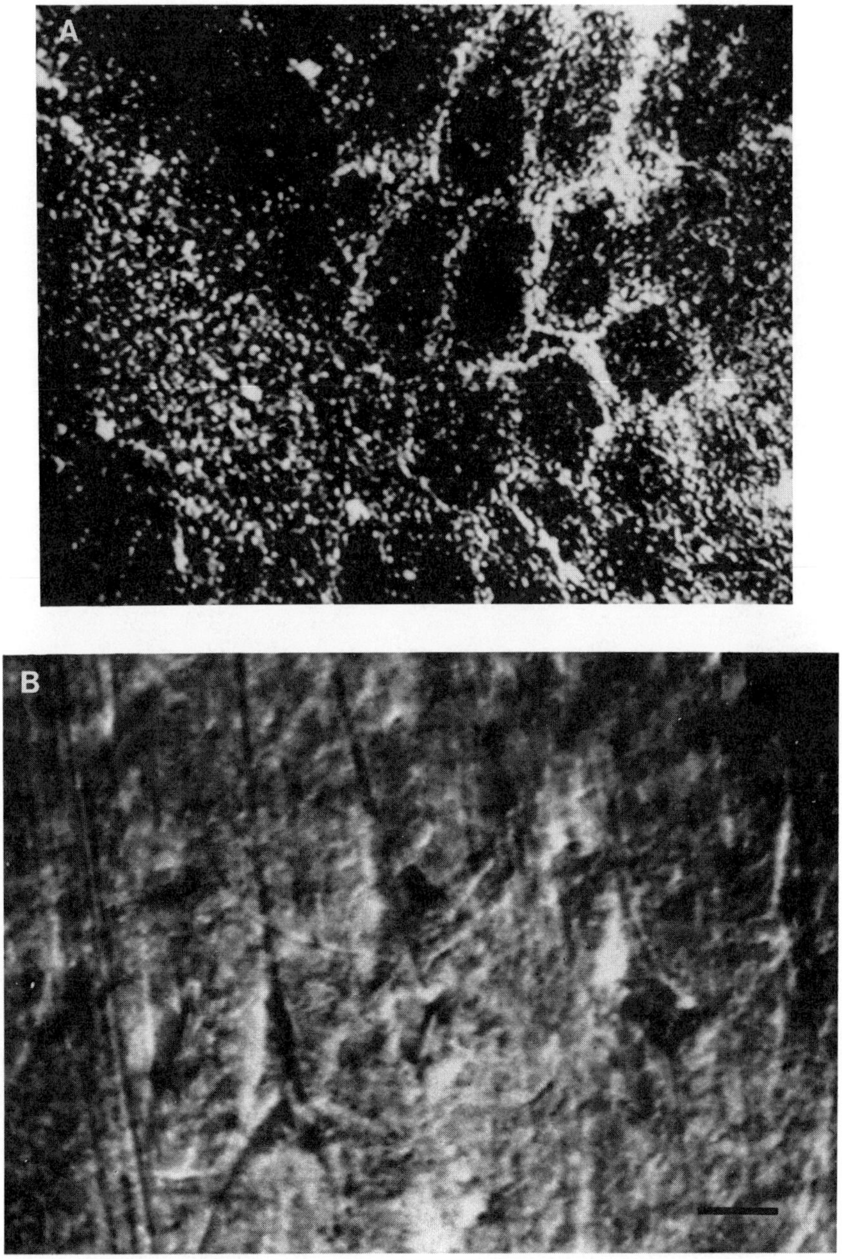

FIGURE 1. A. Neurons in brain slices visualized by infrared-interference videomicroscopy. Barrels in lamina 4 of mouse sensory cortex cut tangentially (5X objective). B: Bundling of apical dendrites in lamina 2/3 of coronally cut rat visual cortex (25X objective, 4X additional magnification). Dimensions of scalebars: 1A = 200 μm, 1B = 10 μm.

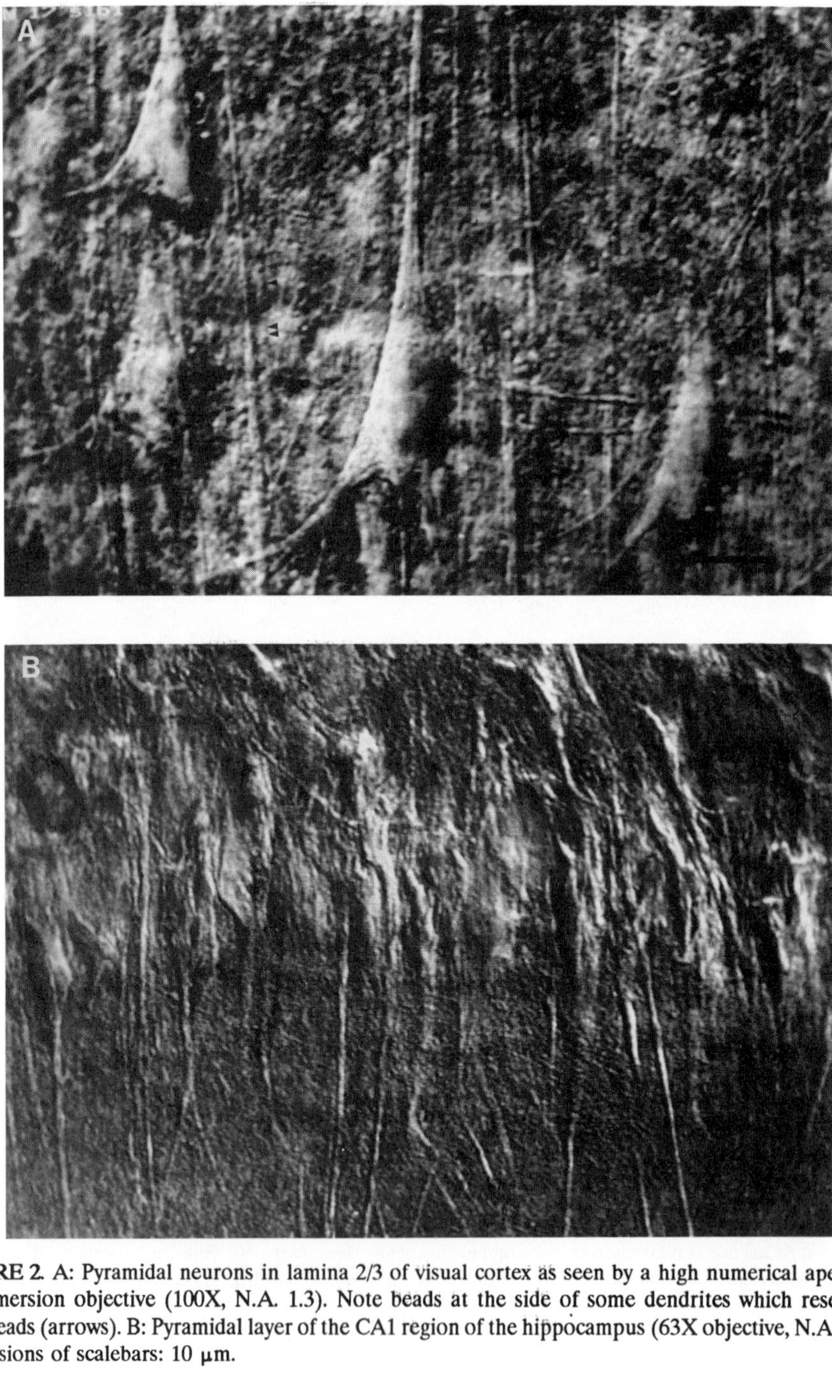

FIGURE 2. A: Pyramidal neurons in lamina 2/3 of visual cortex as seen by a high numerical aperture oil immersion objective (100X, N.A. 1.3). Note beads at the side of some dendrites which resemble spineheads (arrows). B: Pyramidal layer of the CA1 region of the hippocampus (63X objective, N.A. 1.4). Dimensions of scalebars: 10 μm.

The visualisation of neurons in the intact brain would allow many neurobiological problems to be addressed. Thus the application of the infrared technique to the in vivo situation was investigated. In the in vivo situation incident light illuminaton is mandatory, i.e.the neurons have to be illuminated through the objective. By the combination of infrared illumination, DIC and contrast enhancement by video it was possible to visualize neurons with their dendrites in incident light 50 μm deep in cortex slices. In this case background subtraction by the digital image processing system proved to be very helpful. As infrared microscopy in incident light can be combined with confocal microscopy further penetration depth should be achievable. The visualization of neurons in vivo in neocortical lamina 2 of small animals could then become possible.

One of the most exciting prospects of infrared videomicroscopy is the possibility that the clear visualization of neurons should also allow to visualize the spread of excitation in slices with cellular resolution. The image generated by the interference optics will be affected by changes of the refractive index of the specimen, and if these changes are restricted due to local excitation to some neurons, this spread should become visible. Preliminary results obtained with the infrared-interference microscope indicate that the spread of excitation can be visualized by an intrinsic optical signal in the order of seconds after tetanic stimulation. This result suggests that in the future one may be able to use this technique to trace excitatory connections in the brain slice.

REFERENCES

Allen R.D., Allen N.S. (1983) Video-enhanced microscopy with a computer frame memory. J Microscopy 129: 3-17.

Allen R.D., David G.B., Nomarski G.J. (1968) The Zeiss-Nomarski differential interference equipment for transmitted-light microscopy. Z wiss Mikr Mikrotech 69: 193-224.

Dodt H.U., Pawelzik H., Zieglgänsberger, W. (1989) Infrared DIC-videomicroscopy of living brain slices. Soc Neurosci Abstr, 15: 280.

Dodt H.U., Zieglgänsberger W. (1990) Visualizing unstained neurons in living brain slices by infrared DIC-videomicroscopy Brain Res 537: 333-336.

Edwards F.A., Konnerth A., Sakmann B., Takahashi T. (1989) A thin slice preparation for patch clamp recordings from neurones of the mammalian central nervous system. Pflügers Arch 414: 600-612.

MacVicar B.A. (1984) Infrared video microscopy to visualize neurons in the in vitro brain slice preparation. J Neurosci Meth 12: 133-139.

Peters A., Kara D.A. (1987) The neuronal composition of area 17 of rat visual cortex. IV. The organization of pyramidal cells. J Comp Neurol 260: 573-590.

Scheffler H., Elsässer H. (1987) Physics of the galaxy and interstellar matter. Springer, Heidelberg, p. 239.

Svaasand L.O., Ellingsen R. (1983) Optical properties of human brain. Photochem and Photobiol 38: 293-299.

Takahashi T. (1978) Intracellular recording from visually identified motoneurons in rat spinal cord slices. Proc R Soc Lond B 202: 417-421.

Weiss D.G. (1986) Visualization of the living cytoskeleton by video-enhanced microscopy and digital image processing. J Cell Sci Suppl 5: 1-15.

Woolsey T.A., van der Loos, H. (1970) The structural organization of layer IV in the somatosensory region (SI) of mouse cerebral cortex. Brain Research 17: 205-242.

Yamamoto C. (1975) Recording of electrical activity from microscopically identified neurons of the mamalian brain. Experientia 31: 309-311.

BLOOD-BRAIN BARRIER TRANSPORT MEASUREMENTS USING PET-SCANNING AND INTRAVENOUS DOUBLE INDICATOR TECHNIQUE

G.M. Knudsen

Dept. Neurology
Rigshospitalet
Copenhagen, Denmark

INTRODUCTION

The present book deals with different aspects of optical imaging. A more well-established but also more invasive method for monitoring tissue activity, in particular neuronal, is by positron emission tomography (PET). For the sake of completeness, a brief discussion of this method as well as the combination with a more recently developed technique for blood-brain barrier (BBB) measurement will be included in this book. That is, the present paper takes the shape of a review of these methods.

PET scanning allows for three-dimensional imaging of numerous biological in vivo processes and has been under continuous development during the last decade. The positrons, which are positively charged particles, are produced in a cyclotron. Most of them have a very short half time, in the order of minutes. In the biological use, there are several advantages in using positrons. First, the short physical half time makes radiation doses acceptable and gives opportunity for repeated measurements. Secondly, the range of available positrons falls in the category of the most widely distributed atoms in biology: ^{11}C, ^{13}N, and ^{15}O. After production of one of these positrons, it can be further processed and built into virtually any compound of biological importance. Using ^{18}F is convenient because it can be built into ^{18}Fluorodeoxyglucose and because it has a relatively long half time of about 120 minutes. Finally, the inherent nature of the positron decay offers certain advantages for the external detection. After decaying, the particle travels only a very short distance in the tissue. In stopping, it combines with an electron (annihilates), and the energy is carried away by two photons with an energy of 511 keV at an angle of almost exactly 180°. The photons penetrate easily the tissue and are registered outside the body by detectors. Detectors located on opposite sides will, because of the angle of 180°, register the arrival time of the same decay. Next, using an advanced computer system allows for the three-dimensional reconstruction of the decay. During the recent years the resolution of the images has improved with technology. There is, however, a physical limit for how low the resolution can go, but in the more recent PET tomographs this limit is getting closer, the resolution being about 6 mm.

But how is this emission applied practically? The application of the method requires a highly skilled cooperation between the sciences of biology, medicine, chemistry, pharmacology, physics, mathematics, and of computer programming. The development of new tracers and models is time consuming and must be done for each new compound. One of the most well-established and commonly used methods is the deoxyglucose technique, and the practical application has been reviewed more thoroughly in Redies and Gjedde (1989). In the following, it is briefly summarized.

The deoxyglucose technique was originally developed and validated for measurement of regional glucose metabolism in normal brain tissue (Sokoloff et al., 1977). It is based on measurements of local tissue radioactivity by autoradiography or PET with deoxyglucose or fluorodeoxyglucose, respectively. These tracers are chosen instead of labeled glucose since the dephosphorylation rate of these tracers is much smaller than for glucose, so that the tracers essentially are trapped in the cells allowing sufficient sampling time of the radioactivity. The principles of the 2-deoxyglucose technique were later transferred to studies in humans (Phelps et al., 1979; Reivich et al., 1979). The basis of the method is the determination of the steady-state rate of 2-deoxyglucose phosphorylation by the hexokinase reaction. The net rate of glucose phosphorylation is calculated by dividing the net rate of 2-deoxyglucose phosphorylation with an isotope correction factor, the lumped constant, which must be known. The lumped constant incorporates the different affinities of deoxyglucose and glucose for transport across the BBB and phosphorylation by hexokinase. The affinity of deoxyglucose for the hexosetransport carrier located at the BBB is higher than that of glucose, whereas the affinity of deoxyglucose for phosphorylation is lower than that of glucose. The rate constants of glucose and deoxyglucose, in particular k_2 (the transport rate from the brain back to the blood) for deoxyglucose, has until now not been accurately estimated in humans because of uncertainties about the earliest values of tracer kinetics.

METHODS

Measurement of cerebral glucose uptake and metabolism

After injection of about 185 mBq ^{18}Fluorodeoxyglucose, scans with increasing time length are obtained using a PET camera. Arterial samples are drawn from the radial artery. Rate constants are calculated by least square fitting of brain curves derived from different grey and white matter regions based on, e.g., the extended autoradiographic model by Phelps et al. (1979).

Blood-brain barrier permeability measurements using the intravenous double indicator technique

With the intravenous double indicator technique, a 2-4 ml bolus, containing a test compound and several BBB impermeable reference compounds, is injected intravenously through an antecubital catheter. A percutaneous puncture in the neck is made under local anaesthesia and by use of the Seldinger technique, a polyethylene catheter is introduced into one of the internal jugular veins. The tip of the venous catheter is placed in the superior bulb of the internal jugular vein. Starting a few seconds after the injection, a continuous series of 1.0 ml blood samples is collected from the arterial and the venous catheters by means of a sampling machine and deposited into heparinized tubes at fixed time intervals, usually about 1 sample/sec.

Repeated measurements can be done, up to five times in each person, the limit only being set by the amount of blood which can be withdrawn.

Isotopes

The reference substances which usually are used are: $^{24}Na^{+}$, ^{99m}Tc-diethylenetriamine pentaacetic acid (DTPA), and $^{36}Cl^{-}$. As a test substance, a ^{3}H-, ^{99m}Tc-, ^{18}F-, or ^{11}C-labeled substance can, among others, be used.

Measurement of cerebral blood flow

Cerebral blood flow (CBF) is measured in each study using the ^{133}Xe intravenous injection method; 400 MBq of ^{113}Xe dissolved in 1-2 ml of isotonic saline is injected rapidly into an antecubital vein. The clearance of the isotope from the hemisphere is followed and the average hemispheric blood flow is calculated. The ^{133}Xe lung curves and clearance from the brain is recorded by a 10 detector system which integrates detectors and data processing into a mobile unit (Schroeder et al., 1986).

CBF is measured before the BBB measurements. Immediately before the CBF measurement and before each BBB measurement, arteriovenous O_2-differences are measured and CBF is corrected accordingly under the assumption that the total brain metabolism remains constant.

Blood-brain barrier models and calculations

The single membrane (well-mixed) model was used since it has been found suitable for the analysis of glucose (Knudsen et al., 1990a). The applied model is a three-compartment and four-parameter model (Fig. 1).

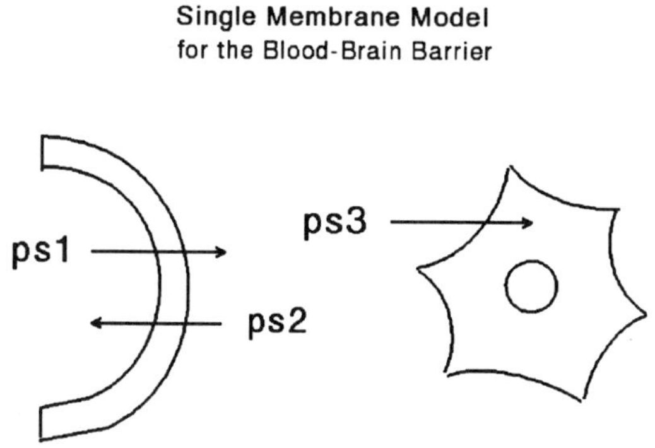

FIGURE 1. Single membrane model for blood-brain barrier transfer parameters. PS_1 is the permeability-surface product (PS) for the transport from the capillary into the brain interstitial fluid; PS_2 designates the transport from the interstitial fluid back to the capillary, and PS_3 designates the cellular uptake.

The difference between this model and the Sokoloff model should be noted here. In the Sokoloff model, the subscripts 3 and 4 designate the glucose phosphorylation and dephosphorylation rates, respectively. With the double indicator technique, which monitors events happening within 30-50 seconds, the phosphorylation rate can not be estimated.

With the use of a transfer function and the single-membrane model, the theoretical output values for the test substance may be calculated if the CBF and reference values are known. The estimates of the model parameters and their standard errors are obtained by minimizing the squares of the differences between the theoretical and the measured output test values.

The operational equation for the single membrane model is

$$C_{test}(t) = \int_0^{\theta_{max}} C_{ref}(\theta)\, C_c[L, t-\tau_l(\theta), \tau_c(\theta)]\, d\theta \qquad (1)$$

where $C_{ref}(t)$ and $C_{test}(t)$ are the amounts of the reference and test substances as measured at the jugular vein at time t. The transit time through the large vessels is τ_l and τ_c is the transit time through the capillary bed. The total capillary transit time is θ while θ_{max} is the value of θ for which $t-\tau_l(\theta) = 0$. If L is the capillary length and x is any distance along the length, then $C_c(L, t, \tau_c)$ is the capillary concentration at $x=L$ and time t for a capillary with the transit time τ_c. With an intravenous injection, the integration in equation (1) is accomplished by replacing $C_{ref}(\theta)$ with the transfer function of the reference material between the artery and the jugular vein. The result is then convoluted with the arterial test concentration to yield theoretical venous values which are then compared to the measured venous test values. Knowledge about the distribution volume, V_e, or the relation between PS_1 and PS_2 is required for calculation of the unknown parameter (Knudsen et al., 1990b).

RESULTS

In our laboratory, ongoing investigations on transfer parameters obtained with PET and the intravenous double indicator technique show that there is a consistent relation between the two sets of parameters. A representative example of an intravenous double indicator curve with a reference substance, with D-glucose and with FDG is shown in figure 2. The brain uptake (PS_1) of FDG is about 70% larger than for D-glucose.

DISCUSSION

Probably the most crucial factor in the correct calculation of the regional glucose metabolism is the correct estimation of the lumped constant. The lumped constant is, however, variable and changes with, e.g., glucose levels. Knowing the kinetics of D-glucose and FDG and the cerebral phosphorylation would be sufficient to predict changes in the lumped constant even under pathological conditions. The ratio between PS_1 for FDG and D-glucose has never before been measured in the same experiment in humans. Earlier studies were done in anaesthetized rats where the relation to the conditions in the awake human is unclear. In our experiments, PS_1 for FDG is about 70% larger than for D-glucose. This is in agreement with animal experiments showing a similar difference (Crane et al., 1983; Fuglsang et al., 1986). In the development of new constrained methods for measurements of glucose metabolism, this ratio has been assumed to be smaller, i.e., 1.1 since only a certain combination of parameters yielded satisfactory values for the net brain influx of glucose (Kuwabara et al., 1990). On the basis of the present findings, this assumption seems erroneous.

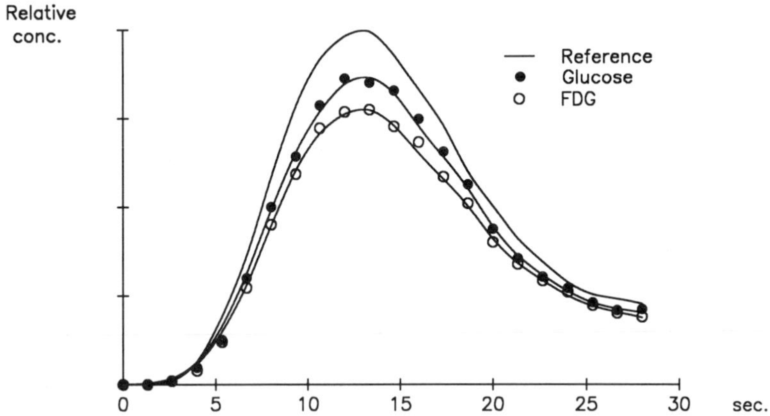

FIGURE 2. A representative example of an intravenous double indicator curve showing the time course of the venous outflow curves of a reference substance, D-glucose and FDG. Note the larger extraction of FDG as compared to D-glucose.

With the increasing complexity of new techniques and models it becomes of major importance to develop supplementary techniques to the existing ones. This is needed in order to confirm models and parameters, which often are assumed values or values obtained from animal experiments. The intravenous double indicator technique offers an example of such a supplementary method for estimation of BBB transport. For substances with a relatively (as compared to the cerebral capillary transit time) slow or extremely fast intracellular uptake, it will not yield data for intracellular uptake (PS_3). Rather, information about the BBB transport after a single capillary passage through the brain is supplied. Due to the short circulation time, metabolism of most tracers is unlikely to happen to any significant degree. The method is rapid and allows for comparison of several test substances evaluated in the same examination as well as repeated examinations within a few minutes.

REFERENCES

Crane P. D., Pardridge W. M., Braun L. D., Oldendorf W. H. (1983) Kinetics of transport and phosphorylation of 2-fluoro-2-deoxy-D-glucose in rat brain. J Neurochem 40:160-167.

Fuglsang A., Lomholdt M., Gjedde A. (1986) Blood-brain transfer of glucose and glucose analogs in newborn rats. J Neurochem 46:1417-1428.

Knudsen G. M., Pettigrew K., Patlak C. S., Hertz M. M., Paulson O. B. (1990a) Kinetic analysis of blood-brain barrier transport in man: Quantitative evaluation of tracer backflux and capillary heterogeneity. Microvasc Res 39:28-49.

Knudsen G. M., Pettigrew K., Patlak C. S., Hertz M. M., Paulson O. B. (1990b) Asymmetrical transport of amino acids across the blood-brain barrier in man. J Cereb Blood Flow Metabol 10:698-703.

Kuwabara H., Evans A. C., Gjedde A. (1990) Michaelis-Menten constraints improved cerebral glucose metabolism and regional lumped constant measurements with [^{18}F]Fluorodeoxyglucose. J Cereb Blood Flow Metab 10:180-189.

Phelps M. E., Huang S.-C., Selin C., Sokoloff L., Kuhl D. E. (1979) Tomographic measurement of local cerebral glucose metabolic rate in humans with (F-18)2-fluoro-2-deoxy-D-glucose: validation of method. Ann Neurol 6:371-388.

Redies C., Gjedde A. (1989) Double-label and conventional deoxyglucose methods: A practical guide for the user. Cerebrovasc. Brain Metabol Rev 1:319-367.

Reivich M., Kuhl D., Wolf A., Greenberg J., Phelps M., Ido T., Casella V., Fowler J., Hoffman E., Alavi A., Som P., Sokoloff L. (1979) The [^{18}F]fluorodeoxyglucose method for the measurement of local cerebral glucose utilization in man. Circ Res 44:127-137.

Schroeder T., Holstein P., Lassen N. A., Engell H. C. (1986) Measurement of cerebral blood flow by intravenous Xenon-133 technique and a mobile system. Neurol Res 8:237-242.

Sokoloff L., Reivich M., Kennedy C., DesRosiers M. H., Patlak C. S., Pettigrew K. D., Sakurada D., Shinohara M. (1977) The [^{14}C]deoxyglucose method for the measurement of local cerebral glucose utilization: theory, procedure, and normal values in the conscious and anesthetized albino rat. J Neurochem 8:897-916.

NUCLEAR MAGNETIC RESONANCE STUDIES OF HUMAN BRAIN IN VIVO: ANATOMY, FUNCTION, AND METABOLISM

J. Frahm

Biomedizinische NMR
Max-Planck-Institut für biophysikalische Chemie
Göttingen, Germany

INTRODUCTION

In the past decade considerable progress in the development of in vivo nuclear magnetic resonance (NMR) has reached a point where NMR becomes an indispensable and unique tool for truly noninvasive studies of anatomy, function, and metabolism. It allows us to examine structural, physiologic, and biochemical aspects of life processes within the intact body, and to expand our investigations from tissue extracts and excised organs to animals and humans without potential disturbances by tissue preparation or anesthesia.

At present most researchers are best familiar with the applications of two- and three-dimensional NMR imaging (MRI) in medical diagnosis. State-of-the-art MRI techniques offer unsurpassed capabilities to visualize anatomy at the highest spatial resolution and to detect pathologic alterations with excellent soft tissue contrasts (e.g., see Stark and Bradley 1991). In addition, and beyond anatomical imaging, more recent developments in spatially resolved NMR provide access to functional imaging (joint movements, cardiac performance, blood flow, tissue perfusion) and allow complementary insights into in vivo metabolism (skeletal muscle, central nervous system). It is the purpose of this contribution to present some of the new possibilities in functional MRI and metabolic NMR spectroscopy (MRS). Rather than being comprehensive, selective examples will illustrate technical approaches and preliminary human applications. All examinations were carried out at a magnetic field strength of 2.0 Tesla using a whole-body MRI/MRS system (Siemens Magnetom).

FUNCTIONAL NMR IMAGING OF BRAIN PERFUSION

MRI studies of tissue perfusion are based on the administration of a paramagnetic contrast agent inducing regional susceptibility differences during the first-pass transit through the tissue. In the initial phase after intravenous bolus injection the effect is due to a heterogeneous distribution of the paramagnetic complexes in the vascular spaces. Thus, regional differences result in the generation of local magnetic field gradients that cause signal losses in gradient-echo MRI sequences with long echo times. The time course of dynamic contrast changes may be best followed by serial fast scan NMR images that match the aforementioned experimental conditions.

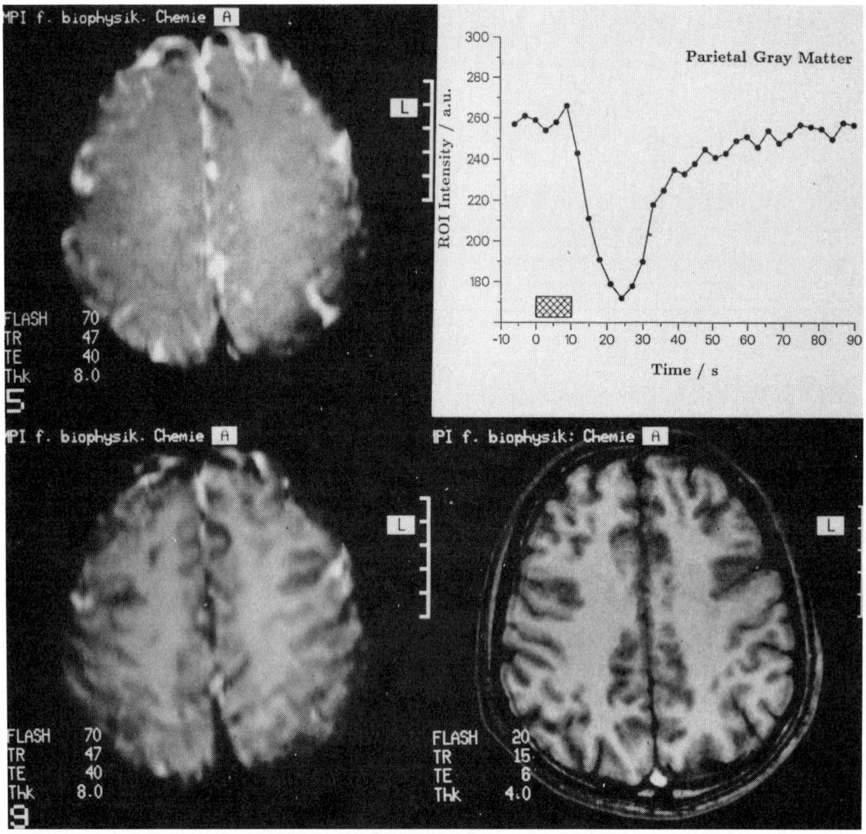

FIGURE 1. Functional MRI of cerebral perfusion in the normal human brain using transient susceptibility-contrasts before (upper left, image #5), during (lower left, image #9), and after the first-pass transit of a bolus of 0.1 mmol/kg Gd-DTPA through the tissue. Preferential perfusion of gray matter is demonstrated by comparison with a T1-weighted NMR image depicting differences in tissue T1 relaxation times rather than functional properties (lower right). Quantitative evaluations of cerebral blood volume and flow may be performed using region-of-interest intensities extracted from the entire series of sequential images (upper right).

Transient susceptibility contrasts after bolus injection of Gd-DTPA have been first described after the development of low flip angle, gradient-echo FLASH MRI (Frahm et al, 1986). Such techniques allowed continuous imaging with measuring times of individual images as short as one second. Initial applications of the concept dealt with in vivo perfusion studies of rabbit kidney and brain (Matthaei et al. 1985, Haase et al. 1986). Figure 1 shows data from a recent study of human brain perfusion in a healthy young volunteer under resting conditions (eyes closed, ear plugs). Dynamic scanning of contrast changes after a 10 s bolus of 0.1 mmol/kg Gd-DTPA was accomplished by means of a susceptibility-sensitized refocused FLASH sequence (TR/TE/α=47/40/70degree) with a matrix of 64x128 data points and a time resolution of three seconds. The resulting 128x128 images covered a field-of-view of 200 mm with a slice thickness of 8 mm yielding a spatial resolution of 1.5x1.5x8 mm^3 and a signal-to-noise ratio (SNR) of 60:1.

The upper left image (#5) in Figure 1 corresponds to a pre-contrast acquisition during bolus injection, while the lower left image (#9) characterizes the maximum contrast 12 seconds later. Contrast is directly associated with differences in regional cerebral blood volume (CBV) and clearly delineates the expected preferential perfusion of gray matter relative to white matter. This functional observation becomes obvious by a comparison to a pre-contrast T1-weighted RF-spoiled 3D FLASH image (TR/TE/α=15/6/20degree, 4 mm partition) of the same section (lower right of Fig. 1). In this case the gray/white matter contrast is entirely due to differences in the tissue T1 relaxation times. The upper right graph in Fig. 1 depicts a quantitative evaluation of the dynamic intensity changes with time in parietal gray matter. The curve originates from region-of-interest values extracted from the full set of images acquired before, during, and after contrast agent administration. The maximum intensity difference represents a 35% decrease of the gray matter NMR signal. Evaluations of absolute CBV or cerebral blood flow (CBF) images become possible with knowledge of the arterial input function.

Even better time resolution and/or more slices for perfusion studies of human brain may be obtained in conjunction with echo-planar imaging (e.g., Belliveau et al. 1990, Rosen et al. 1990). Corresponding studies have been carried out to calculate CBV maps of normal physiologic activity, e.g. demonstrating the increase of capillary blood flow in the primary visual cortex during photic stimulation (Belliveau et al. 1991). Thus, it may be concluded that functional MRI techniques bear considerable potential for future contributions to the understanding of both normal and pathologic brain function. Furthermore, advanced NMR modalities provide access to related anatomical and angiographic aspects, and offer insights into regional brain metabolism and its interrelation to function on an individual basis.

NMR SPECTROSCOPY OF CEREBRAL METABOLITES

In vivo NMR studies of metabolism exploit the original strength of spectroscopic applications in analytical chemistry to discriminate between nuclei in different chemical environments. For a given substance and nucleus the resulting pattern of resonance frequencies or "chemical shifts", i.e. the NMR spectrum, generally allows an identification of the molecular structure and a determination of the concentration. In tissues of unknown composition the "fingerprint" spectra of potential candidates serve to retrieve substances from the complex tissue spectrum. Further discriminative power originates from the use of multiple biologically relevant nuclei such as protons (hydrogen atoms) and phosphorus. Table 1 summarizes the NMR properties of major nuclei suitable for in vivo applications together with their primary metabolic targets.

With the exception of protons, most nuclei suffer from a low sensitivity. For in vivo applications this drawback translates into large volumes-of-interest (VOI) and/or long measuring times. For example, acquisition times of 10-20 minutes should not be exceeded for human studies. Some of the interesting nuclei such as carbon-12 and oxygen-16 do not have a nuclear magnetic moment. In these cases, the overall sensitivity not only depends on the NMR sensitivity of proper isotopes and the biological abundance of respective molecules, but also on the natural abundance of the isotope and/or the administration, biodistribution, and metabolism of isotope-enriched substrates such as carbon-13 enriched glucose or oxygen-17 enriched air. Of course, where feasible, the use of magnetically labeled compounds offers unique possibilities for the investigation of metabolic pathways, fluxes, and compartmentation under in vivo conditions.

Table 1. Major nuclei suitable for in vivo NMR applications.

Nucleus	γ MHz/Tesla	Relative Sensitivity	Metabolism
Proton	42.6	1	Amino Acids, Glycolysis, Neurotransmitters, Fatty Acids
Fluorine	40.1	0.834	Drugs
Phosphorus	17.2	0.066	High Energy Phosphates
Carbon-13	10.7	0.016	Fatty Acids, Substrate Fluxes
Deuterium	6.5	0.010	Substrate Fluxes
Oxygen-17	5.8	0.029	Oxidation Water

In general, meaningful MRS studies of an intact organism must be performed in conjunction with spatial localization. First attempts relied on the use of surface radiofrequency (RF) coils to study phosphate metabolites and bioenergetic aspects in skeletal muscle (Ackerman et al. 1980). Such techniques were extended to investigations of cerebral metabolism (Prichard and Shulman 1986) but often required surgical interventions to properly place the RF antenna with respect to the brain of an experimental animal. Most current examinations of animals and humans take advantage of two complementary MRI-based approaches that accomplish a true three-dimensional spatial discrimination: (i) single voxel techniques that restrict the excitation of the NMR signal to an image-controlled VOI, and chemical shift imaging (CSI) techniques that cover the spectral information of a cross-sectional plane in a low-resolution three-dimensional image with two spatial dimensions and one spectral (chemical shift) dimension.

This work entirely focusses on applications of the single voxel stimulated-echo (STEAM) sequence for obtaining fully localized phosphorus and proton NMR spectra of human brain. The method applies three slice-selective RF pulses as known from MRI. With the use of magnetic field gradients in orthogonal directions, the sequence excites nuclei in three perpendicular planes. Since the stimulated echo signal from a series of three RF pulses only originates from those nuclei that have experienced all pulses, a selected VOI is simply defined by the spatial intersection of the three planes. Its size and position may be directly determined from NMR images. Details of the STEAM technique have been described elsewhere (Frahm et al. 1987, Frahm et al. 1990).

Localized Phosphorus MRS

Figure 2 shows a localized phosphorus NMR spectrum from a 125 ml VOI in the occipital brain of a young adult (Merboldt et al. 1990). The spectrum was obtained within a measuring time of 13 minutes and exhibits resonances from the α-, β-, and γ-phosphates of adenosine triphosphate (ATP), phosphocreatine (PCr), and inorganic (ortho-) phosphates (Pi). The chemical shift difference between the latter two is a linear measure of the intracellular pH yielding 7.02 ± 0.02 in a number of volunteers. In the spectrum shown in Fig. 2 the chemical shift resolution was improved by broadband proton decoupling during the phosphorus acquisition. This proton irradiation removes proton-phosphorus spin-spin splittings from the ATP-resonances as well as from the phosphomonoester and phosphodiester spectral ranges. Accordingly, resonances from glycerophosphocholine (GPC), glycerophosphoethanolamine (GPE), phosphoethanolamine (PE), and of dinucleotides such as nictonamide adenine dinucleotide (NAD) become identifiable. Differences in the residual linewidths of the α-ATP versus the β- and γ-ATP resonances indicate Mg^{2+}-ATP binding to occur between the latter two. From the respective chemical shift difference the free Mg^{2+} concentration was estimated to 680 μM in this case.

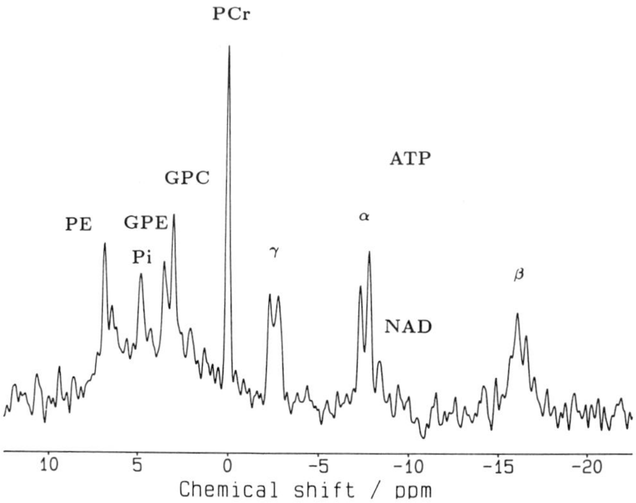

FIGURE 2. Localized proton-decoupled phosphorus NMR spectrum from a 50x50x50 mm^3 volume-of-interest (VOI) in the occipital brain of a young healthy volunteer (STEAM, TR/TE/TM=3000/3/5 ms, 256 accumulations, 2 Hz line broadening). Resonance assignments are due to the α-, β-, and γ-phosphates of adenosine triphosphate (ATP), nicotinamide adenine dinucleotides (NAD), phosphocreatine (PCr), glycerophosphocholine (GPC), glycerophosphoethanolamine (GPE), inorganic (ortho-) phosphates (Pi), and phosphoethanolamine (PE). Chemical shifts are given in parts per million (ppm) and referenced to 0 ppm for PCr. Reprinted with permission from (Merboldt et al. 1990).

While single voxel phosphorus techniques may be most suitable for quantitative evaluations of relaxation times and metabolite concentrations, the low sensitivity of the nucleus renders CSI strategies an attractive alternative for qualitative mapping of cerebral phosphate distributions and their response to function and pathology. In general, however, human applications of phosphorus MRS must be considered most promising for studies of "global" processes and generalized neurodegenerative diseases rather than for investigations of focal events and small lesions.

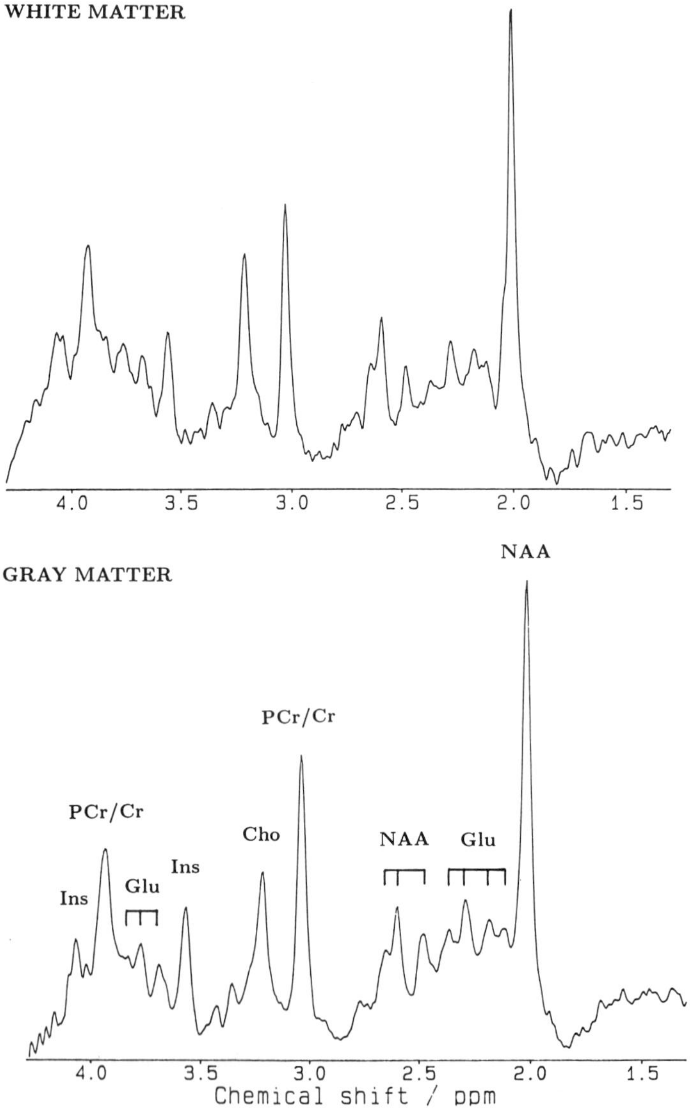

FIGURE 3. Localized proton NMR spectra from a 20x30x20 mm^3 VOI in right parietal white matter (upper trace) and from a 16x30x30 mm^3 VOI in midsagittal parietal gray matter (lower trace) of a young healthy volunteer (STEAM, TR/TE/TM=6000/20/30 ms, 64 accumulations, 0.5 Hz line broadening). Resonance assignments are due to N-acetylaspartate (NAA), glutamate (Glu), phosphocreatine and creatine (PCr/Cr), choline-containing compounds (Cho), and myo-inositol (Ins). Chemical shifts are given in parts per million (ppm) and referenced to 2.01 ppm for the CH_3 group of NAA.

Localized Proton MRS

Figure 3 shows representative proton NMR spectra of gray and white matter in a young healthy adult. In contrast to phosphorus applications the proton STEAM sequence has to be preceeded by water suppression. The corresponding sequence part minimizes the signal from tissue water protons that is at least a factor of 10,000 more intense than that of metabolites.

The identification of resonances involved high-resolution high-field NMR spectra of tissue extracts from experimental animals and human biopsy and autopsy material as well as investigations of model solutions under exactly the same conditions as used in vivo (Michaelis et al. 1991a). Major resonances are assigned to N-acetylaspartate (NAA), glutamate (Glu), phosphocreatine and creatine (PCr/Cr), choline-containing compounds (Cho), and myo-inositol (Ins). Due to its exclusive neuronal localization NAA may serve as a marker for vital neuronal and axonal parenchyma. On the other hand, myo-inositol predominantly occurs in glia cells. Since both inositols and cholines participate as precursors in the build-up of lipid membranes their common increase in a number of brain pathologies often indicates proliferative processes and/or gliosis. Similar observations are made in infants during the first four months of age. The identification of additional substances include the dipeptide N-acetylaspartylglutamate in adult white matter (Frahm et al. 1991a). It appears as a 2.045 ppm shoulder of the NAA resonance at 2.01 ppm (top spectrum in Fig. 3) that becomes resolved in the absence of line broadening. More recently, the singlet resonance at 3.35 ppm has been identified as scyllo-inositol (Michaelis et al., 1992). Resonances from glucose, e.g. at 3.43 ppm, provide insights into glycolysis and allow an assessment of the energy metabolism when combined with information available from creatines and lactate. In fact, lactate, glutamine, alanine, mobile lipids, and aromatic resonances are frequently elevated under pathologic conditions, e.g. in brain tumors (Bruhn et al. 1989, Frahm et al. 1991b).

Table 2. Absolute concentrations (mM) of metabolites in adult human brain in vivo as obtained from fully relaxed (TR=6000 ms), short-echo time (TE=20 ms), localized proton NMR spectra (STEAM) of young healthy adults (n=32, aged 18-28 years). The data are taken from (Michaelis et al. 1991b).

Metabolite	White Matter	Gray Matter	Cerebellum
N-Acetylaspartate	8.3*	10.7	10.5
Glutamate	6.6	9.0	7.9
Phosphocreatine + Creatine	5.4	6.6	8.6
Choline-containing Compounds	1.8	1.2	2.6
myo-Inositol	4.1	4.9	6.3

*Corrected for a 20% contribution from N-acetylaspartylglutamate.

One of the attractive properties of NMR is the direct and linear proportionality of the NMR signal, i.e. the resonance signal area, to the concentration of the respective nuclei. In order to facilitate the evaluation of metabolite concentrations, T1 saturation of signal areas was avoided by the acquisition of fully relaxed spectra (TR=6000), while T2 losses were minimized at short echo times (TE=20 ms). For absolute quantification of metabolites the in vivo NMR data were referenced to external standards of known composition measured under matched experimental conditions (Michaelis et al. 1991b). First results for adult human brain are summarized in Table 2. While most of the in vivo concentrations agree with biochemical expectations, significant discrepancies are seen for creatines in white and gray matter. The total creatine values given are well below the in vitro value of 10 mM reported for perchloric acid extracts of brain tissue. Since phosphocreatine concentrations of 5 mM have been found in phosphorus NMR spectra of human brain in vivo, it would be most likely to assume a limited NMR visibility of creatine. However, further studies are required to clarify this point and to assess potential deviations of the creatine kinase equilibrium. It should also be noted that the concentrations in Table 2 reveal significant regional differences. Further attempts aim at a "metabolic atlas" of the normal human brain. In addition, care must be taken to account for the age dependence of metabolite patterns particularly during early brain maturation.

Since noninvasive studies rely on the use of relatively large RF coils circumscribing the entire head, the minimum accessible VOI will be of the order of 1 mL (10x10x10 mm^3), assuming reasonable measuring times and adequate spectral SNR. Obviously, such macroscopic dimensions result in spectra comprising signals from both neural and glial structures as well as from vascular compartments and/or cerebrospinal fluid. Considerable improvements in spatial resolution will only be possible by sacrificing the non-invasiveness of the NMR approach, e.g. by means of miniaturized RF coils attached to catheters. Since the SNR is mainly determined by the relative dimensions of the used RF antenna and the desired VOI, the spectral quality remains unchanged if both are scaled in proportion. Despite these complications state-of-the-art MRS techniques offer fascinating means for investigations of brain pathologies. Selected studies of neurochemical and neurophysiologic aspects of normal human brain will be discussed below.

BRAIN FUNCTION AND METABOLISM

A correlation of function and metabolism requires the introduction of a stimulus to elucidate dynamic aspects associated with alterations of the parameter under investigation. NMR offers a number of different approaches to meet this requirement: (i) the use of RF labels in magnetization transfer experiments, e.g. studies of the creatine kinase kinetics, (ii) the administration of exogenous substances, e.g. contrast agents, drugs, or isotopically enriched substrates, and (iii) variations in the physiologic activity of the subjects.

Cerebral Uptake of Ethanol

A new class of human NMR studies are sequential acquisitions and follow-up

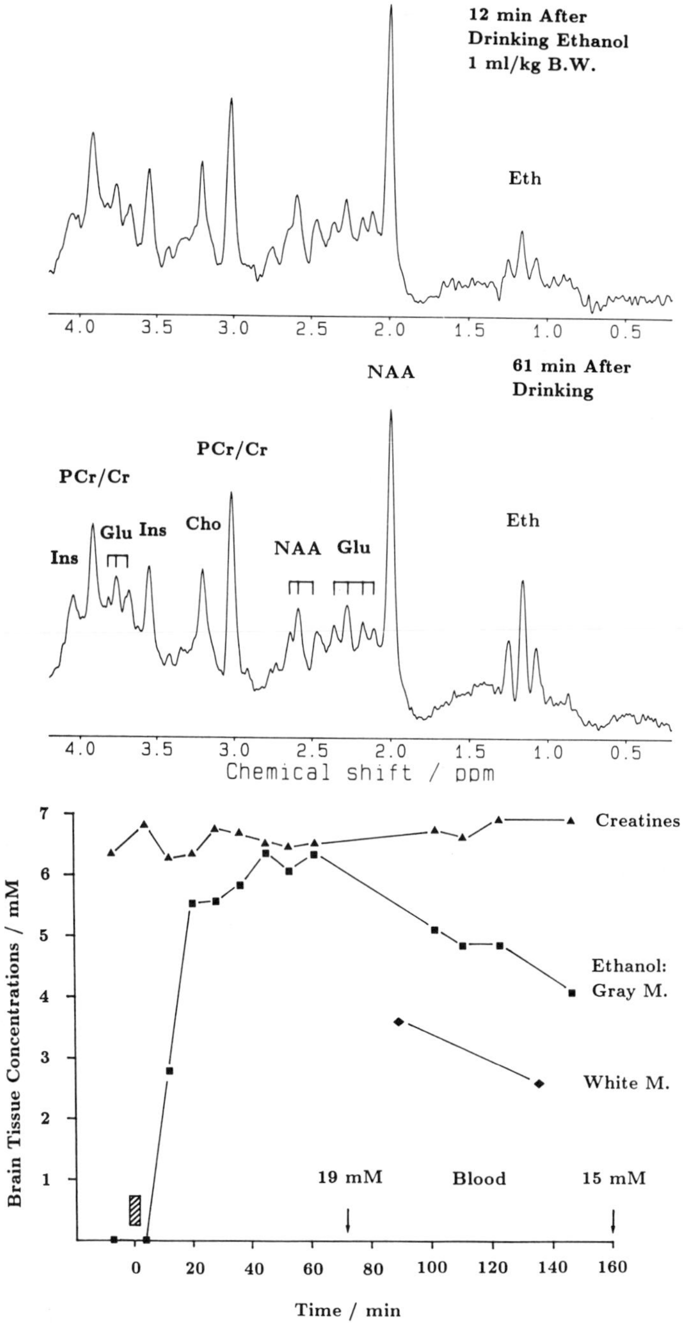

FIGURE 4. Cerebral uptake of ethanol in a young healthy volunteer after drinking alcohol. Localized proton NMR spectra (TR=3000 ms, 128 accumulations, other parameters as in Fig. 3) from a 16x30x30 mm^3 VOI in midsagittal parietal gray matter 12 minutes (upper trace) and 61 minutes (middle trace) after drinking (1 ml/kg ethanol, 5 minutes). The graph (bottom part) shows the time course of brain ethanol concentrations in gray and white matter relative to creatine (reference) and in comparison to blood plasma values.

studies of the cerebral uptake of drugs. Applications include investigations of uptake kinetics, regional distributions, drug metabolism, effects on the neurometabolism, and effects during treatment of a disease. A first illustrative example is the uptake of ethanol after oral consumption of alcohol as shown in Fig. 4.

Only 12 minutes after drinking, proton NMR spectra from parietal gray matter revealed the triplet signal from the CH_2-coupled CH_3-resonance of ethanol at 1.20 ppm. This signal appeared simultaneously with the occurrence of "classical" symptoms such as alterations of mood and speech. The lower part of Fig. 4 shows the time course of the gray matter ethanol concentration quantified relative to creatines. No alterations were observed for other metabolite pools. The ethanol concentration rapidly increased to about 6 mM after drinking 1 ml/kg body weight. Maximum tissue concentrations were reached after 40-75 minutes for different volunteers. This period was followed by a very slow decrease of tissue concentrations (approx 1.5 mM/hour).

Gray and white matter spectra from the same volunteer and the same study revealed regional variations in the ethanol concentration with a gray/white matter ratio of 1.6. More important, there was a remarkable difference between gray matter concentrations and blood ethanol values. About 70 minutes after drinking the tissue/plasma ratio was only 0.3 with a further decrease to 0.2 after 160 minutes. This behavior indicates that ethanol is more rapidly cleared from brain tissue than from blood. However, brain ethanol levels as low as 20-30% of blood ethanol are not yet understood and remain the subject of further investigations.

Glucose Homeostasis

Another important question is concerned with the homeostasis and transport of glucose under both normoglycemic and (transient) hyperglycemic conditions. Preliminary results on normal volunteers were obtained following standard glucose infusion protocols. The upper trace in Fig. 5 shows a spectrum from parietal gray matter of a young healthy volunteer with a plasma glucose concentration of 4.6 mM after overnight fasting. To improve the visibility of the glucose resonance at 3.43 ppm the same spectrum is expanded in the second trace of Fig. 5. After 23 minutes of 10 mg/kg/min glucose infusion, the resulting gray matter spectrum in the third trace in Fig. 5 revealed a significant increase in the cerebral glucose signal. The bottom part of Fig. 5 shows the full time course of the gray matter glucose concentration (squares, left scale) quantified relative to creatines for this particular volunteer. The data are compared to capillary plasma glucose levels (circles, right scale) determined in between the 6.5 minute spectral acquisitions.

The temporal profiles of plasma and brain glucose concentrations do not indicate a delay in the increase or decrease of brain levels after changing plasma concentrations. This is even better demonstrated for higher infusion rates (20 mg/kg/min, not shown) where spectral changes are less affected by SNR problems. The absolute concentrations of plasma and tissue glucose, however, differ by about a factor of 5-7 leading to brain/plasma ratios of 0.15-0.2 in the awake human brain (gray matter). These numbers are of relevance for the understanding of glucose transport alterations in long-term diabetics where remarkable differences in the plasma/tissue ratios were found in different patients despite similar plasma levels (Bruhn et al. 1991).

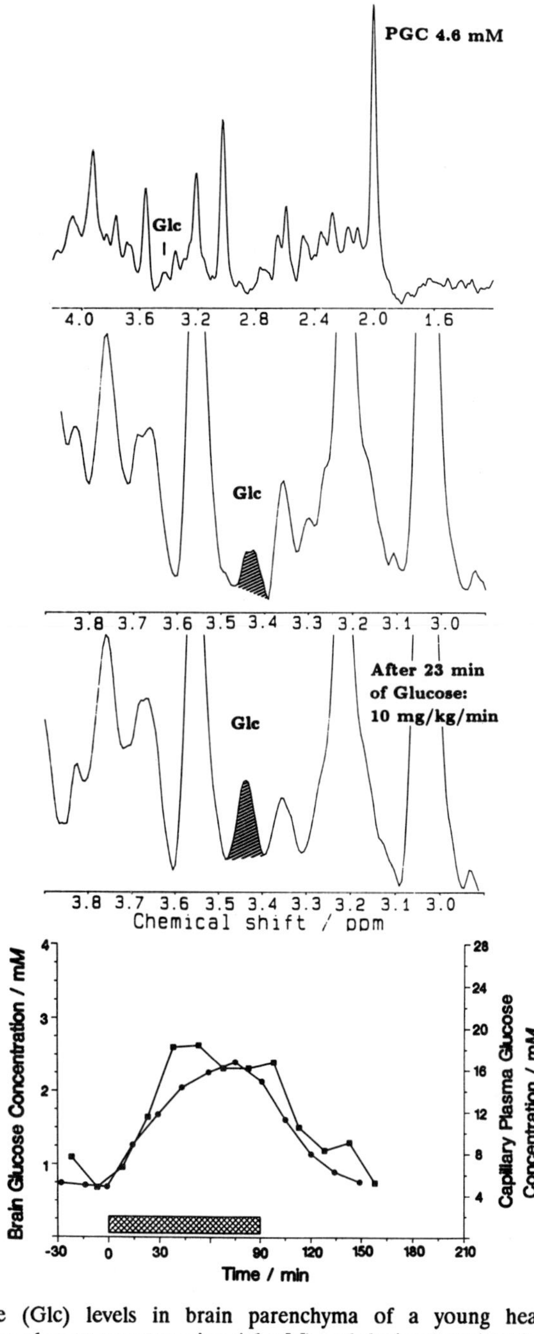

FIGURE 5. Glucose (Glc) levels in brain parenchyma of a young healthy volunteer under normoglycemia (plasma glucose concentration 4.6 mM) and during transient hyperglycemia following intraveneous infusion of glucose (10 mg/kg/min) for 90 minutes. Localized proton NMR spectra (parameters as in Fig. 4) from a 20x30x30 mm^3 VOI in midsagittal parietal gray matter before (upper two traces) and 23 minutes after start of the glucose infusion (third trace). The graph (bottom part) shows the time course of the gray matter glucose concentration relative to creatines (squares, left scale) and in comparison to capillary plasma values (circles, right scale).

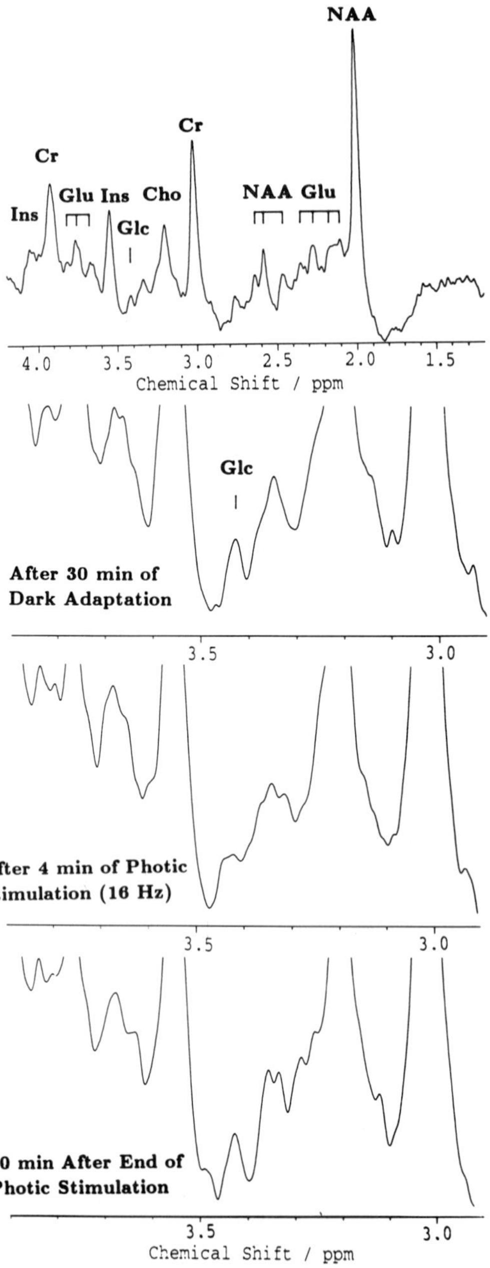

FIGURE 6: Decrease of the glucose (Glc) level in the primary visual cortex of a young healthy volunteer during photic stimulation. Localized proton NMR spectra (surface coil reception, other parameters as in Fig. 4) from a 20x20x20 mm^3 VOI in the occipital lobe (predominantly gray matter) after at least 30 minutes of rest and dark adaptation (upper two traces), during the initial phase of photic stimulation (third trace), and 10 minutes after the end of stimulation in the dark state (bottom trace). Reprinted with permission from (Merboldt et al. 1992).

Lactate and Glucose during Photic Stimulation

Until recently, the coupling of physiologic activity to glucose utilization, blood flow, and oxygen extraction in humans has been the exclusive subject of studies using positron emission tomography (PET). In vivo NMR now allows us to gathercomplementary information, e.g. on the concentrations of lactate and glucose in the human visual cortex during rest and photic stimulation.

The upper trace in Fig. 6 shows a proton NMR spectrum from an 8 ml VOI in the occipital lobe of a young healthy volunteer during rest and after at least 30 minutes of dark adaptation (ear plugs, lights off, eyes closed). Under these conditions basal lactate levels showed considerable interindividual differences ranging from below detectability (< 0.3 mM) to about 1 mM. In contrast to previous NMR reports (Prichard et al. 1991, Sappey-Marinier et al. 1990), no consistent alterations were observed during photic stimulation (16 Hz, binocular photic stimulation, 4x5 matrices, red light). On the other hand, local brain glucose levels were significantly reduced (approx 50%) during the entire period of photic stimulation and recovered to resting levels (0.8 ± 0.3 mM) within 10 minutes after the end of stimulation in the dark state (Merboldt et al. 1992).

The decrease of the steady-state glucose level in the primary visual cortex during photic stimulation is best understood as the establishment of a new equilibrium due to enhanced delivery (blood flow) and enhanced consumption as demonstrated by PET (e.g., see Fox et al. 1988). In addition, the present observations prove the accumulation of lactate to be effectively decoupled from enhanced glucose consumption. Thus, if increased production of lactate does not lead to its accumulation in brain parenchyma, then glucose consumption must be followed by a rapid efflux of lactate from the tissue in the activated state (Ackerman and Lear 1989, Lear and Kasliwal 1991).

CONCLUDING REMARKS

The development of in vivo NMR provides unique noninvasive tools for studies of function and metabolism in man. Applications range from muscle bioenergetics to neurophysiologic activation. Even though still in its infancy, NMR is expected to result in significant contributions to the "Decade of the Brain".

ACKNOWLEDGMENTS

Financial support by the Bundesminister für Forschung und Technologie (BMFT) of the Federal Republic of Germany (Grant 01 VF 8606/6) is gratefully acknowledged.

REFERENCES

Ackermann J.J.H., Grove T.H., Wong G.G., Gadian D.G., Radda G.K. (1980) Mapping of metabolites in whole animals by 31P NMR using surface coils. Nature 283:167-170.

Ackermann R.F. and Lear J.L. (1989) Glycolysis-induced discordance between glucose metabolic rates measured with radiolabeled fluorodeoxyglucose and glucose. J. Cereb. Blood Flow Metab. 9:774-785.

Belliveau J.W., Rosen B.R., Kantor H.L., et al. (1990) Functional cerebral imaging by susceptibility-contrast NMR. Magn. Reson. Med. 14:538-546.
Belliveau J.W., Kennedy D.N., McKinstry R.C., et al. (1991) Functional mapping of the human visual cortex by magnetic resonance imaging. Science 254:716-719.
Bruhn H., Frahm J., Gyngell M.L., et al. (1989) Noninvasive differentiation of tumors using localized H-1 MR spectroscopy in vivo: Initial experience in patients with cerebral tumors. Radiology 172:541-548.
Bruhn H., Michaelis T., Merboldt K.D., Hänicke W., Gyngell M.L., Frahm J. (1991) Monitoring cerebral glucose in diabetics by proton MRS. Lancet 337:745-746.
Fox P.T., Raichle M.E., Mintun M.A., Dence C. (1988) Nonoxidative glucose consumption during focal physiologic neural activity. Science 241:462-464.
Frahm J., Haase A., Matthaei D. (1986) Rapid NMR imaging of dynamic processes using the FLASH technique. Magn. Reson. Med. 3:321-327.
Frahm J., Merboldt K.D., Hänicke W. (1987) Localized proton spectroscopy using stimulated echoes. J. Magn. Reson. 72:502-508.
Frahm J., Michaelis T., Merboldt K.D., Bruhn H., Gyngell M.L., Hänicke W. (1990) Improvements in localized proton NMR spectroscopy of human brain. Water suppression, short echo times, and 1 ml resolution. J. Magn. Reson. 90:453-464.
Frahm J., Michaelis T., Merboldt K.D., Hänicke W., Gyngell M.L., Bruhn H. (1991a) On the N-acetyl methyl resonance in localized 1H-NMR spectra of human brain in vivo. NMR Biomed. 4:201-204.
Frahm J., Bruhn H., Merboldt K.D., Hänicke W., Mursch K., Markakis E. (1991b) Localized proton NMR spectroscopy of brain tumors. Methodologic improvements using short-echo time STEAM sequences. J. Comput. Assist. Tomogr. 15:915-922.
Haase A., Matthaei D., Hänicke W., Frahm J. (1986) Dynamic digital subtraction imaging using fast low-angle shot movie sequences. Radiology 160:537-541.
Lear J.L. and Kasliwal R.K. (1991) Autoradiographic measurement of cerebral lactate transport rate constants in normal and activated conditions. J. Cereb. Blood Flow Metab. 11:576-580.
Matthaei D., Frahm J., Haase A., Hänicke W. (1985) Regional physiological functions depicted by sequences of rapid magnetic resonance images. Lancet ii:893.
Merboldt K.D., Chien D., Hänicke W., Gyngell M.L., Bruhn H., Frahm J. (1990)Localized 31P-NMR spectroscopy of the adult human brain in vivo using stimulated echo (STEAM) sequences. J. Magn. Reson. 89:343-361.
Merboldt K.D., Bruhn H., Hänicke W., Michaelis T., Frahm J. (1992) Decrease of glucose in the human visual cortex during photic stimulation. Magn. Reson. Med. 25:187-194.
Michaelis, T., Merboldt K.D., Hänicke W., Gyngell M.L., Bruhn H., Frahm J. (1991a) On the identification of cerebral metabolites in localized 1H-NMR spectra of human brain in vivo. NMR Biomed. 4:90-98.
Michaelis T, Bruhn H., Gyngell M.L., Hänicke W., Merboldt K.D., Frahm J. (1991b) Quantification of cerebral metabolites in man. Results using short-echo time localized proton MRS. Book of Abstracts, Society of Magnetic Resonance in Medicine, Tenth Annual Meeting, San Francisco, USA, p. 387.
Michaelis T., Helms G., Merboldt K.D., Hänicke W., Bruhn H., Frahm J. (1992) Identification of scylloinositol in proton NMR spectra of human brain in vivo. NMR Biomed. (in press)
Prichard J.W., Shulman R.G. (1986) NMR spectroscopy of brain metabolism in vivo. Ann Rev Neurosci 9:61-85.

Prichard J.W., Rothman D.L., Novotny E., et al. (1991) Lactate rise detected by 1H NMR in human visual cortex during physiologic stimulation. Proc. Natl. Acad. Sci. USA 88:5829-5831.
Rosen B.R., Belliveau J.W., Vevea J.M., Brady T.J. (1990) Perfusion imaging with NMR contrast agents. Magn. Reson. Med. 14:249-265.
Sappey-Marinier D., Calabrese G., Hugg J., Deicken R., Fein G., Weiner M. (1990) Increased lactate in human visual cortex during photic stimulation. Book of Abstracts, Society of Magnetic Resonance in Medicine, Ninth Annual Meeting, New York, p. 106.
Stark D.D. and Bradley W.G. (1991), Eds., Magnetic resonance imaging, 2nd Edition, Mosby, St. Louis.

AUTORADIOGRAPHIC AND BIOCHEMICAL IMAGING IN CEREBRAL ISCHEMIA

G. Mies

Max-Planck-Institute for Neurological Research
Department of Experimental Neurology
Cologne, Germany

INTRODUCTION

The central nervous system consists of innumerable anatomical subunits, each of which participate in one or several functional activities during the normal state of the brain. The maintenance of cerebral activity is an extremely energy demanding process, the energy requirements being satisfied predominantly by oxidative metabolism of glucose. Sufficient perfusion levels, therefore, are essential for adequate substrate delivery. In order to study interrelationships between regional functional activity, perfusion and metabolism in more detail, two prerequisites must be fulfilled: first, techniques must allow the simultaneous measurement of hemodynamic and metabolic parameters, and second, these methods have to provide a resolution sufficient to identify brain regions of interest.

In pathological conditions, the functional relationships between the various brain areas may be disrupted, the severity of the disturbances possibly being reflected by regional hemodynamic and metabolic abnormalities. More conclusive investigations on the pathophysiology of ischemic processes aimed to the understanding of this mutual interrelationship mandate the evaluation of both blood flow and metabolic activity. Cerebral ischemia, for example, is by definition the reduction of blood flow below a critical perfusion level which is inadequate of sufficient substrate delivery for the maintenance of energy metabolism. In consequence, ischemic and non-ischemic brain regions have to be differentiated to allow the precise localization of affected brain areas and to elucidate ongoing pathophysiological processes in more detail.

With the establishment of autoradiographic techniques, it has become possible to measure quantitatively local cerebral blood flow (Landau et al. 1955; Freygang and Sokoloff 1958; Reivich et al. 1969; Sakurada et al. 1978), cerebral glucose utilization (Sokoloff et al. 1977) or protein synthesis (Smith et al. 1984) in individual animals, allowing the determination of regional cerebral hemodynamics or metabolism. With the introduction of quantitative double or triple tracer autoradiography (Lear et al. 1981; Mies et al. 1981; 1983; 1986), the above mentioned autoradiographic methods in fact

Optical Imaging of Brain Function and Metabolism, Edited by
U. Dirnagl *et al.*, Plenum Press, New York, 1993

can be carried out in the same experiment. It is therefore possible to perform measurements of local cerebral blood flow and brain metabolism during the same physiological or pathophysiological state in a single subject. In addition, these autoradiographic techniques can be combined with bioluminescence methods. These allow the regional determination of biochemical substrates such as ATP, glucose, and lactate. Also fluoroscopic methods can be used for the assessment of regional tissue pH or redox state.

In the following the potential of autoradiographic and biochemical imaging techniques will be demonstrated with respect to their suitability for the investigation of regional pathophysiological processes occurring in focal cerebral ischemia.

METHODS

Autoradiographic techniques

Measurement of cerebral blood flow.

The autoradiographic method for the determination of regional cerebral blood flow is based on a technique first described by Landau et al. (1955), who used ^{131}I-iodotrifluoromethane. Labeled iodoantipyrine is used today as a more convenient tracer. The algorithm for the calculation of local cerebral blood flow has been given by Sakurada et al. (1978):

$$C_i^*(T) = \lambda K \int_0^T C_a^*(t)\, e^{-K(T-t)}\, dt.$$

Regional cerebral blood flow f (= Kλ) can be calculated from the time course of the arterial tracer radioactivity in blood $C_a^*(t)$ during intravenous infusion of labeled iodoantipyrine and the local tissue radioactivity $C_i^*(T)$ at the end of the iodoantipyrine infusion. λ represents the blood-tissue partition coefficient of iodoantipyrine which equals 0.8.

Measurement of cerebral glucose consumption

The energy demands of the central nervous system are covered almost exclusively by oxidative metabolism of glucose. Regional cerebral glucose metabolism can be measured autoradiographically using ^{14}C-labeled deoxyglucose (^{14}C-DG) with a method described by Sokoloff et al. (1977). This technique is based on the following principle: DG is transported bidirectionally between blood and tissue by the same carrier that transports glucose across the blood-brain barrier. In tissue, DG is phosphorylated to deoxyglucose-6-phosphate (DGP) by hexokinase, the first enzyme in the glycolytic sequence. In contrast to glucose being metabolized to CO_2 and water, DGP is a poor substrate to other glycolytic enzymes. Consequently, labeled DGP remains essentially trapped in brain tissue. During systemic circulation of ^{14}C-DG from time t = 0 to t = T, the time course of arterial plasma ^{14}C-radioactivity $C_p^*(t)$ is determined by scintillation counting and arterial plasma glucose content is measured enzymatically. Since the accumulated amount of the tissue metabolite pool (DGP) can be derived from the difference between ^{14}C-tissue radioactivity $C_i^*(T)$ assessed autoradiographically and the free amount of tissue DG estimated kinetically, the regional metabolic rate for glucose per unit mass of tissue (R_i) is calculated using the following equation (Sokoloff et al. 1977):

$$R_i = \frac{C_i^*(T) - K_1^* \int_0^T C_p^*(t)\, e^{-(k_2^*+k_3^*)(T-t)}\, dt}{LC\left(\int_0^T [\, C_p^*(t) \,/\, C_p \,]\, dt - \int_0^T [\, C_p^*(t) \,/\, C_p \,]\, e^{-(k_2^*+k_3^*)(T-t)}\, dt \right)}$$

where k_1^* and k_2^* are the rate constants for the transport of DG from plasma to tissue and vice versa, respectively, and k_3^* is the rate constant for phosphorylation of DG to DGP. LC is the so-called lumped constant, i.e. a correction factor for the differences in the kinetic behavior of DG and glucose.

Local cerebral protein synthesis

The same principle described for the deoxyglucose method can be applied to the measurement of cerebral protein synthesis. When ^{14}C-leucine (^{14}C-leu) is applied intravenously, the rate of leucine incorporation into cerebral proteins (R_i) can be calculated by a similar equation (Smith et al. 1984). Due to the rapid clearance of plasma ^{14}C-leucine radioactivity $C_p^*(t)$ and the possibility of autoradiographic quantitation of the regional tissue protein radioactivity $C^*_{prot}(T)$, the original equation can be simplified to:

$$R_i = \frac{C^*_{prot}(T)}{\lambda \int_0^T [\, C_p^*(t) \,/\, C_p \,]\, dt}$$

where λ is a constant reflecting the steady state ratio of the specific activity of leucyl-bound tRNA, the actual precursor, and that of arterial plasma leucine. When tissue protein radioactivity is normalized by the respective total tissue radioactivity, the ratio $C^*_{prot}(T)/C_i^*(T)$ can be used as a measure of the cerebral protein synthesizing capacity.

Quantitative single tracer autoradiography

The quantitative determination of regional tissue radioactivity is based on the relationship between radionuclide-induced film blackening and the related amount of the isotope in the tissue. For this purpose several animals are each given graded doses of an isotope covering a wide range. Each brain is removed and is divided in half. One hemisphere is weighed, homogenized and the suspension of brain tissue assayed for its isotope concentration. The other hemisphere is frozen, sectioned at 20 μm and dried on a hot plate. Cryostat sections from several brains are then exposed transiently to X-ray film together with ^{14}C-methylacrylate standards. From the plot of the measured ^{14}C-concentrations in the brain hemispheres and their corresponding optical densities, a calibration curve can be constructed. This permits assigning a brain equivalent concentration of a radionuclide to each plastic standard by comparing its optical density with the calibration curve. Other radioactive cryostat sections are then exposed to X-ray film together with these calibrated plastic standards. Regional optical densities are measured in the autoradiogram which are converted to tissue radioactivity by using the relationship between extinction and isotope brain equivalent concentration of the plastic standard.

Quantitative double tracer autoradiography

The above mentioned autoradiographic techniques can be combined by labeling the compounds used to assess given parameters with radionuclides which differ in their half-lives and emitted energy spectra. Local tissue isotope radioactivity of more than one radionuclide can be differentiated as follows:

Half-life principle. Two radiocompounds are administered systemically and one isotope is short-lived while the other isotope possesses a long half-life, i.e. decay of the latter isotope is negligible during the procedure of quantitative autoradiography. Cerebral blood flow, for example, can be measured with ^{131}I-iodoantipyrine and cerebral glucose utilization with ^{14}C-deoxyglucose. ^{131}I-iodoantipyrine is given systemically in such high dose that a 1 day exposure of a cryostat section to X-ray film results in an autoradiogram exclusively due to ^{131}I-iodine. Under these conditions, cross-contamination by ^{14}C-carbon is negligible. The half-life of ^{131}I-iodine is 8 days and thus ^{131}I-iodine is decayed almost completely after 10 weeks. The same brain section is then reexposed to X-ray film for 14 days yielding an autoradiogram only from disintegrations of ^{14}C-carbon. From the same brain section, therefore, it is possible to quantify the regional tissue radioactivity necessary for the calculation of cerebral blood flow or glucose utilization.

Principle of photographic differentiation. A different strategy of autoradiographic isotope separation can be employed when two long-lived isotopes with different energy spectra are present in the same brain section. Conventional X-rays films are covered by a protection surface which is sufficient to absorb low-energy particle from the disintegrations of radionuclides such as ^{3}H-tritium but not the more energy-rich ones from ^{14}C-carbon. The so-called ^{3}H-sensitive films lack a protection layer and exposure of brain sections containing both isotopes to such photographic material will yield autoradiograms in which film blackening stems from both radionuclides. The ^{3}H-autoradiograms are then obtained by digital subtraction of the optical density induced by ^{14}C-radioactivity from total film blackening.

Principle of chemical or physical isotope removal. In order to improve autoradiographic differentiation between two long-lived isotopes of different maximal energy emission spectra, removal of the high-energy emitting radionuclide can be used to facilitate the autoradiographic detection of the remaining low-energy emitting isotope. A most convenient approach without interfering with tissue composition is the removal of one ^{14}C-labeled compound by sublimation, e.g. ^{14}C-DMO used for the measurement of brain tissue pH (Hakim and Arrieta, 1986) whereas the second one is non-volatile. Removal of one labeled tracer compound may also be achieved by incubation of the brain section in a dissolvent which reacts chemically with the labeled substrate, e.g. dimethylpropane for the removal of labeled iodoantipyrine (Diemer and Rosenorn, 1981) or allowing one tracer compound to be removed by passive diffusion, e.g. chloroform treatment of brain sections for removal of the diffusible iodoantipyrine (Furlow et al., 1983; Ginsberg et al., 1986). The effect of section incubation on possible loss of the other labeled compound has to be verified, however. When labeled amino acids are used for the measurement of cerebral protein synthesis, trichloroacetic acid or formaldehyde treatment of tissue sections is of particular interest. These chemicals precipitate all proteins whereas free labeled amino acids and metabolites are removed from the brain section. In such an instance the accumulated metabolite radioactivity is obtained by direct measurement rather than by estimation with the operational equation (see above).

Bioluminescence imaging

The bioluminescence methods are designed to provide the regional distribution of biochemical substrates such as ATP, glucose and lactate from the intact brain section. The principle of bioluminescence imaging is the following: cryostat sections from the in situ frozen brain are covered with a slice of frozen enzyme solution containing the reactants necessary for a substrate-specific reaction and for an enzymatically coupled photon-producing system. When the tissue/enzyme layer is thawed at room temperature, the resulting bioluminescence can be recorded on photographic material in a dark room. With this technique it has become possible to image the regional distribution of ATP (Kogure and Alonso, 1978), glucose (Paschen et al. 1981) and lactate (Paschen 1985). Since the various bioluminescence reactions are stoichiometric, film blackening correlates with the substrate concentrations in tissue (Paschen et al. 1985). Thus quantification of biochemical substrates at local levels is possible.

Fluoroscopic imaging

pH-dependent fluorescence.

Regional tissue pH (the composite of extracellular fluid pH and intracellular pH) can be measured quantitatively with the umbelliferone method described by Csiba et al. (1983). For this purpose, cryostat sections from the in-situ frozen brain are placed on umbelliferone-soaked paper at -20°C. Upon thawing of the section and from pH-standards, the resultant pH-related fluorescence of umbelliferone is photographed at 450 nm after the slice irradiation with UV light at 340 nm (pH-dependent fluorescence) and 370 nm (pH-independent fluorescence).

NADH fluorescence

Increased NADH fluorescence is a sensitive indicator of oxidative disturbances of the mitochondrial respiratory chain. The cut surface of the frozen brain is illuminated with UV-light and the NADH fluorescence is recorded photographically at 450 nm (Welsh and Rieder, 1978). Although the NADH signal is quenched by hemoglobin, it indicates a disturbance of the tissue redox state.

Experimental models

Regional ischemia by extracranial artery occlusion

Following occlusion of one common carotid artery, unilateral graded cerebral ischemia can be produced in the Mongolian gerbil. This is due to an incomplete circle of Willis because the arteria communicans posterior is absent. Collateral blood supply from the opposite hemisphere is little because of the small diameter of the anterior communicating artery.

Regional ischemia by intracranial artery occlusion

A method used for the experimental induction of regional cerebral ischemia is the occlusion of the middle cerebral artery in the cat or the rat (Sundt and Waltz, 1966; Tamura et al. 1981). This is most commonly achieved via a retro-orbital approach which does not involve retraction of parts of brain. In the cat, however, the size of ischemic region depends on the variability of the collateral blood supply, i.e. pial Heubner's anastomoses.

Tracer application

^{14}C-deoxyglucose (15 μCi/100 g body weight) was given as an intravenous bolus to measure cerebral glucose utilization. Blood samples were frequently collected for the determination of the time course of changes in specific activity of deoxyglucose. Thirty minutes later, ^{131}I-iodoantipyrine (gerbils) or ^{131}I-isopropyliodoamphetamine (cats) were injected for the measurement of cerebral blood flow. Brains were frozen in situ by immersing the head in liquid nitrogen or by funnel freezing.

Tissue preparation and regional evaluation

Brains were removed from the skull at -25°C and 20 μm thick coronal tissue sections were prepared in a cryostat at -20°C and were used alternately for quantitative autoradiography, substrate bioluminescence and pH fluorescence as described above. In addition NADH fluorescence was photographed from the coronally cut brain surface using a 450 nm emission filter.

Optical densities of all images were assessed using a microdensitometer which was driven by a laboratory computer. The pixels of digitized autoradiograms were quantitated as cerebral blood flow (ml/100g/min) and glucose utilization (μmol/100g/min) based on respective tracer tissue radioactivities and the time course of blood or plasma specific activity. In bioluminescence images, ischemic alterations in regional substrate contents were expressed as percentage of the opposite control level. Unspecific fluorescence in pH-dependent umbelliferone images and pH standards was corrected to obtain a pH-calibrated image.

APPLICATION OF METHODS

Focal ischemia of gerbil brain

Figure 1 illustrates the regional variability in ischemic cerebral blood flow rates and parallel changes in glucose consumption, glucose and ATP content as well as NADH fluorescence. In one animal, focal cerebral ischemia was confined to the hippocampal region whereas in the two others almost the whole right hemisphere was severely ischemic. These qualitative findings suggest already that in this experimental stroke model, a standardized ischemic territory cannot be produced. Regional techniques are necessary for the localization and for the interpretation of pathobiochemical processes within the ischemic lesion. Even in the ischemic territory, differences in the regional glucose consumption existed. In areas of mild ischemia (Figure 1A, 1B) a conspicuous increase in glucose utilization was detectable, and ATP and glucose levels were reduced. This suggests that high metabolic rates for glucose were insufficient to satisfy the energy requirements of brain tissue. The most likely cause is limited oxygen but not glucose availability. This interpretation is also supported by the beginning of the increase in NADH fluorescence. In areas of severe ischemia (Figure 1C), however, glycolysis ceased due to the ischemia-caused oxygen and glucose deficiency.

This battery of regional techniques is suited to examine the relationship between local cerebral blood flow and energy state failure, anaerobic stimulation of glycolysis, and protein synthesis inhibition in the border zone between normal to ischemic brain areas. As shown in Figure 2 (top), ATP levels, i.e., the energy state, remained within a physiological range at blood flow rates varying from 20 - 90 ml/100g/ min. The begin-

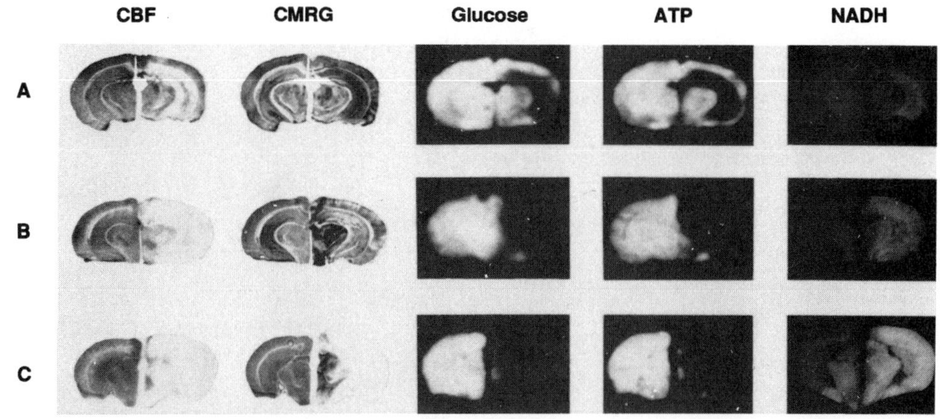

FIGURE 1. Autoradiographic, bioluminescent and fluoroscopic imaging in animals subjected to focal cerebral ischemia by unilateral occlusion of a common carotid artery. Autoradiograms of cerebral blood flow and glucose utilization are obtained by double tracer autoradiography from the same brain section. The regional distribution of glucose and ATP content is determined from cryostat sections by substrate specific bioluminescence. The cerebral redox state of the brain is assessed by photographic recording of NADH fluorescence. A: Animal without neurological symptoms. A moderate reduction of blood flow and increase of glucose consumption occurred in hippocampus leading to a more pronounced decrease of tissue glucose content than of tissue ATP level. B: Animal with mild neurological symptoms. Moderate ischemia affected almost the entire hemisphere which induced stimulation of glycolysis and a drastic reduction in glucose and ATP. The redox state of brain, however, was only minimally affected. C: Animal with neurological deficits. Severe ischemia induced a breakdown of metabolism, substrate depletion and severe disturbance of cerebral redox state.

ning of the decline in ATP content was observed at 20 ml/100g/min. A further decline in blood flow led to a sharp decline in the ATP level and at values below 10 ml/100g/min tissue ATP content was depleted. Similar ischemic thresholds were noticed for the disturbance of electrolyte homeostasis which resulted in tissue potassium loss (Mies et al. 1984). Figure 2 (bottom) illustrates that an elevation in glucose consumption occurred at blood flow values much higher than the disturbance of energy metabolism, namely at 35 ml/100g/min. Maximal metabolic rates for glucose were reached in brain regions perfused at blood flow rates ranging from 20 - 25 ml/100g/min. Since the local energy state was preserved at perfusion values as low as 20 ml/100g/min, the energy yield of presumably anaerobic glycolysis apparently was sufficient to satisfy the energy demand of brain tissue. When blood flow decreased below 20 ml/100g/min, cerebral rates of glucose utilization declined and ceased at 10 ml/100g/min in parallel with energy metabolism. Cerebral protein synthesis in the peri-infarct border zone was affected at much higher perfusion levels than stimulation of glycolysis (Xie et al. 1989; Mies et al. 1991). The beginning reduction of protein synthesis was noticed at blood flow values as high as 80 ml/100g/min and protein synthesis was inhibited at a flow rate below 40 ml/100g/min. The failure of protein synthesis in the infarct surrounding, however, was not associated with the ischemic breakdown of cerebral energy metabolism which occurred at a much lower flow rate, i.e. at 10 ml/100g/min in gerbil and at 18 ml/100g /min in rat brain.

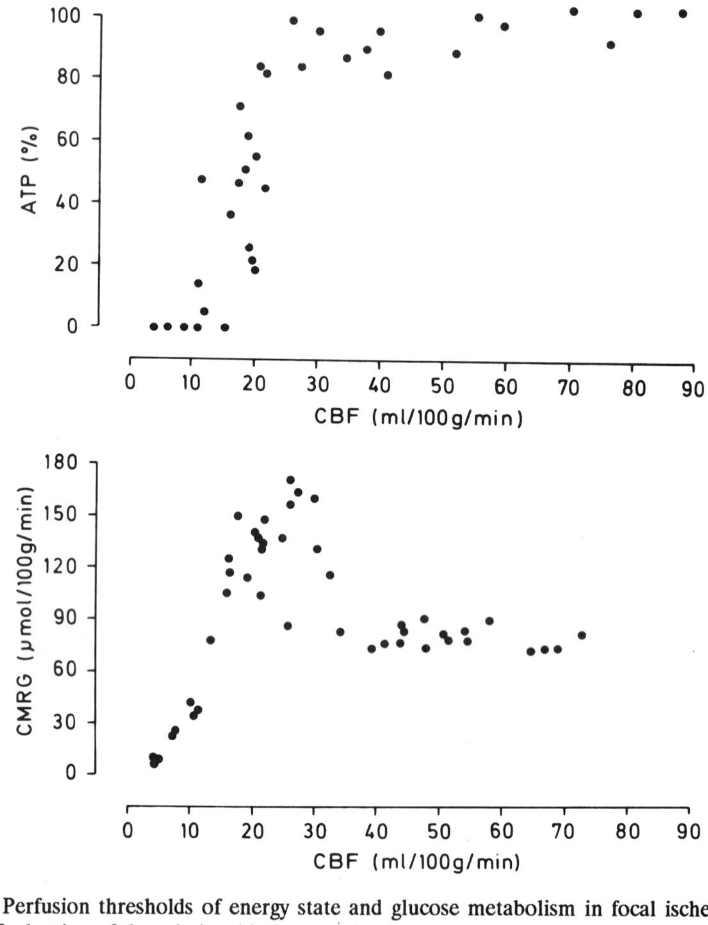

FIGURE 2. Perfusion thresholds of energy state and glucose metabolism in focal ischemia of gerbil brain. Top: Evaluation of the relationship between local cerebral blood flow and ATP level revealed an ischemic threshold of energy state failure of approximately 10 ml/100g/min. Bottom: A steady increase in glucose utilization was noticed at a flow rate of approximately 35 ml/100g/min which suggests the activation of anaerobic glycolysis. Below 20 ml/100g/min, glucose consumption declined and ceased at 10 ml/100g/min.

The findings obtained in the experimental brain stroke model of rodents demonstrate that defined thresholds of blood flow for the manifestation of cerebral metabolic disturbances exist in the border zone of the infarct. The ischemic threshold is reached at perfusion levels between 10 - 20 ml/100g/min and results in the failure of cerebral energy metabolism due to deficits in glucose and oxygen delivery. If the local flow situation is not reversed within the resuscitation time window of brain tissue, an infarct will develop. Disturbances of cerebral glucose metabolism and impairment of brain protein synthesis were identified at much higher and distinctly different perfusion values. At such oligemic blood flow levels, however, the deteriorated glucose and amino acid metabolism may only be tolerated transiently. Tissue perfused above the ischemic threshold and at oligemic blood flow values may be defined as zone around the infarct in which tissue still is structurally intact but metabolically endangered. From the hemodynamic point of view this peri-infarct border zone would be accessible to pharmacological interventions to delay the onset of irreversible tissue injury until a more sufficient perfusion can be restored.

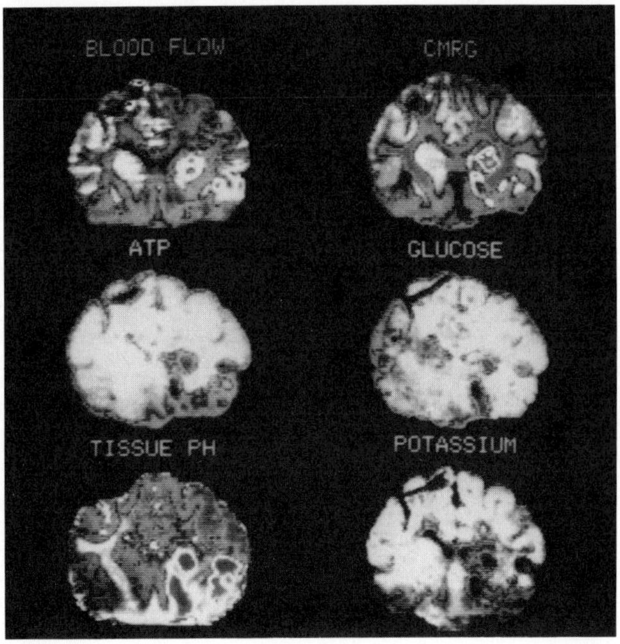

FIGURE 3. Multiparametric imaging in focal ischemia of cat brain. In addition to double tracer autoradiography of blood flow and glucose use, and bioluminescence of regional ATP and glucose levels, the fluoroscopic measurement of tissue pH and the histochemical visualization of tissue potassium content were applied to neighboring tissue section. Regional techniques allow the differentiation of the pathophysiological sequel leading to ischemic brain damage: reduction in local perfusion, alteration of glycolytic activity and subsequent changes in tissue pH, impairment of glucose availability and energy state, and depolarization of cell membranes as indicated by tissue potassium depletion.

Focal ischemia in cat brain

In Figure 3, images of cerebral blood flow, glucose utilization, glucose and ATP content, tissue pH and potassium distribution are shown from the same animal following permanent occlusion of the right middle cerebral artery (MCA) in cat for two hours. An enormous variability of the relationship between local perfusion and glucose metabolism was observed within the basolateral part of the right MCA territory. In the oval and ring-shaped hyperemic areas or in foci with mild or severe ischemia, glucose utilization was either enhanced, reduced or had ceased. Images of regional ATP, glucose and potassium distribution revealed a circumscribed depletion in three spherical areas whereas in the others ATP and potassium levels were reduced but glucose

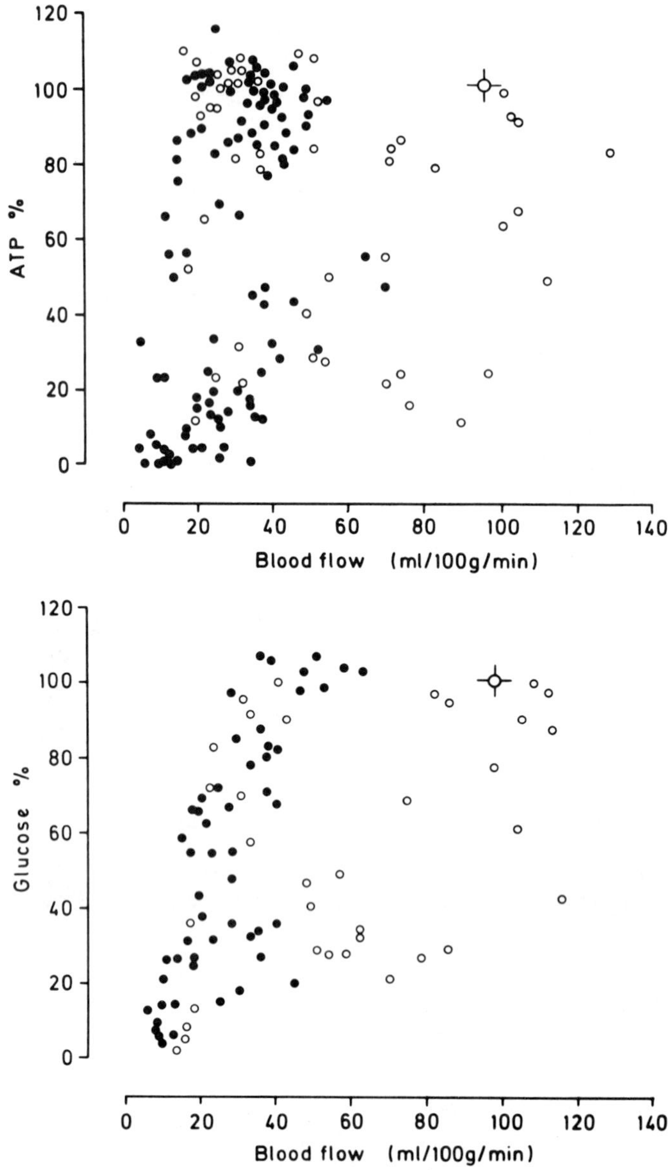

FIGURE 4. Relationship between local cerebral blood flow and ATP levels or tissue glucose content in focal ischemia of cat brain. Flow-ATP- and flow-tissue glucose data sets were grouped into those from the hyperemic (○) and oligemic (●) rim around the ischemic lesions. Top: Note the variability of the data pairs from perfusion-ATP content evaluation which most likely results from the local reperfusion of ischemic areas. Despite the scatter an ischemic threshold of energy state failure of 10 - 15 ml/100g/min is evident. Bottom: A decline of tissue glucose content was observed at a flow rate of 40 ml/100g/min indicating an oligemic threshold of limited tissue glucose availability. In both instances much higher flow rates are necessary to restore control levels of ATP and tissue glucose content.

content remained within the physiological range. Local dissociations also existed between regional tissue energy state and perfusion as well as glucose metabolism: ATP was depleted but blood flow in the same area was hyperemic or glucose utilization was enhanced at low ATP levels. As apparent from Figure 3, the pH image comprehended larger confluent brain areas in which pH values ranged from 6.0 - 6.5 thus delineating brain tissue with severe metabolic disturbances (see Hossmann et al. 1985). Following the occlusion of MCA, reductions in blood flow must have been at least transiently of such magnitude as to limit oxygen availability and subsequently to induce anaerobic stimulation of glycolysis and tissue acidosis. In brain tissue with acidotic pH-values, however, flow rates were ischemic, oligemic and hyperemic. Hyperemia within the MCA territory most likely is due to pH-induced vasodilatation (Kuschinsky et al. 1972) and/or spontaneous recirculation (Heiss et al. 1976).

The regional evaluation of the relationship between blood flow and energy state, or glucose content revealed a most variable interdependency between the various parameters. In contrast to the defined ischemic and oligemic perfusion thresholds in the peri-infarct rim of gerbil or rat brain, an enormous scatter in the flow-energy state or the flow-tissue glucose content data was observed following MCA-occlusion in the cat (Figure 4: ○ hyperemic and • oligemic infarct transition zone). Within the MCA territory, nevertheless, oligemic areas can be conjoined to a cluster which suggests an ischemic threshold of energy state failure at flow rates below 10 - 15 ml/100g/min. Including data sets from the hyperemic infarct rim for interpretation (Figure 4, top: ○) it becomes apparent that very high flow rates have to be reached to restore ATP levels close to control levels. An oligemic threshold of the decline of glucose content in peri-infarct areas was observed at a flow value of about 40 ml/100g/min (Figure 4, bottom). Again, areas were present in the hyperemic border zone of the infarct (Figure 4 bottom: ○), in which a flow rate of approximately 100 ml/100g/min seems to be necessary to restore normal tissue glucose levels.

These results indicate that in higher species the pathophysiology of focal ischemia is increasingly complex and much more difficult to assess. The most likely explanation for the lack of a consistent threshold relationship between regional perfusion and energy state, or tissue glucose content is recirculation of previously ischemic areas. This situation seems to be similar to that observed in humans where spontaneous recirculation is a common finding (Olsen, 1986). Therefore, the measurement of regional energy state may be more relevant for predicting the outcome of clinical stroke than the assessment of cerebral blood flow and glucose metabolism.

In summary, current techniques for the regional determination of hemodynamic and metabolic parameters offer the possibility to investigate their interrelationship and allow a more comprehensive evaluation of the pathophysiology and pathobiochemistry of the evolving ischemic lesion. The perspectives of these techniques for the further understanding of the pathophysiology of focal brain ischemia, in consequence, are most promising for the development of future therapeutic strategies.

REFERENCES

Csiba L., Paschen W., Hossmann K.-A. (1983) A topographic quantitative method for measuring brain tissue pH under physiological and pathophysiological conditions. Brain Res 289:334-337.

Diemer N.H., Rosenorn J. (1981) Determination of local cerebral blood flow and glucose metabolism or transfer by means of a double autoradiographic method. J Cereb Blood Flow Metab 1 (Suppl. 1):S72-S73.

Freygang W.H., Sokoloff L. (1958) Quantitative measurement of regional circulation in the central nervous system by use of radioactive gas. Adv Biol Med Phys 6:263-279.

Furlow T.W., Martin R.M., Harrison L.E. (1983) Simultaneous measurement of local glucose utilization and blood flow in the rat brain: an autoradiographic method using two tracers labeled with carbon-14. J Cereb Blood Flow Metab 3:62-66.

Ginsberg M.D., Smith D.W., Wachtel M.S., Gonzalez-Carvajal M., Busto R. (1986) Simultaneous determination of local cerebral glucose utilization and blood flow by carbon-14 double-label autoradiography: Method of procedure and validation studies in the rat. J Cereb Blood Flow Metab 6: 273-285.

Hakim A.M., Arrieta M. (1986) Cerebral acidosis in focal ischemia: I. A method for the simultaneous measurement of local cerebral pH with cerebral glucose utilization or cerebral blood flow in the rat. J Cereb Blood Flow Metab 6:667-675.

Heiss W.-D., Hayakawa T., Waltz A.G. (1976) Cortical neuronal function during ischemia. Effects of occlusion of one middle cerebral artery on single-unit activity in cats. Arch Neurol 33:813-820.

Hossmann K.A., Mies G., Paschen W., Csiba L., Bodsch W., Rapin J.R., Le Poncin-Lafitte M., Takahashi K. (1985) Multiparametric imaging of blood flow and metabolism after cerebral artery occlusion in cats. J Cereb Blood Flow Metab 5:97-107.

Kogure K., Alonso O.F. (1978) A pictorial representation of endogenous brain ATP by a bioluminescent method. Brain Res 154:273-284.

Kuschinsky W., Wahl M., Bosse O., Thurau K. (1972) Perivascular potassium and pH as determinants of local pial arterial diameter in cats. A microapplication study. Circ Res 31:240-247.

Landau W.M., Freygang W.H., Rowland R.P., Sokoloff L., Kety S.S. (1955) The local circulation of the living brain: values in the unanesthetized and anesthetized cat. Trans Amer Neurol Ass 80:125-129.

Lear J.L., Jones S.C., Greenberg J.H., Fedora T.J., Reivich M. (1981) Use of 123I and 14C in a double radionuclide autoradiographic technique for simultaneous measurement of LCBF and LCMRgl. Theory and method. Stroke 12:589-597.

Mies G., Niebuhr I., Hossmann K.-A. (1981) Simultaneous measurement of blood flow and glucose glucose metabolism by autoradiographic techniques. Stroke 12:581-588.

Mies G., Bodsch W., Paschen W., Hossmann K.-A. (1983) Experimental application of triple-labeled quantitative autoradiography for the measurement of cerebral blood flow, glucose metabolism and protein biosynthesis. In: Methods in Positron Emission Tomography. Phelps M.E., Heiss W.-D. (Eds). Springer Verlag, Berlin, Heidelberg, pp. 19-28.

Mies G., Kloiber O., Drewes L.R., Hossmann K.-A. (1984) Cerebral blood flow and regional potassium distribution during focal ischemia of gerbil brain. Ann Neurol 16:232-237.

Mies G., Bodsch W., Paschen W., Hossmann K.A. (1986) Triple-tracer autoradiography of cerebral blood flow, glucose utilization, and protein synthesis in rat brain. J Cereb Blood Flow Metab 6:59-70.

Mies G., Ishimaru S., Xie Y., Seo K., Hossmann K.-A. (1991) Ischemic thresholds of cerebral protein synthesis and energy state following middle cerebral artery occlusion in rat. J Cereb Blood Flow Metab 11:753-761.

Olsen T.S. (1986) Regional cerebral blood flow after occlusion of the middle cerebral artery. Acta Neurol. Scand. 73:321-337.

Paschen W., Niebuhr I., Hossmann K.-A. (1981) A bioluminescent method for the demonstration of regional glucose distribution in brain slices. J Neurochem 36:513-517.

Paschen W. (1985) Regional quantitative determination of lactate in brain sections. A bioluminescent approach. J Cereb Blood Flow Metab 5:609-612.

Paschen W., Mies G., Kloiber O., Hossmann K.-A. (1985) Regional quantitative determination of brain glucose in tissue sections: a bioluminescent approach. J. Cereb. Blood Flow Metab. 5:465-468.

Reivich M., Jehle J., Sokoloff L., Kety S.S. (1969) Measurement of regional cerebral blood flow with antipyrine-14C in awake cats. J Appl Physiol 27:296-300.

Sakurada O., Kennedy C., Jehle J., Brown J.D., Carbin G.L., Sokoloff L. (1978) Measurement of local cerebral blood flow with iodo(14C)-antipyrine. Am J Physiol 234:H59-H66.

Smith C.B., Crane A.M., Kadekaro M., Agranoff D.W., Sokoloff L. (1984) Stimulation of protein synthesis and glucose utilization in the hypoglossal nucleus induced by axotomy. J Neurosci 4:2489-2496.

Sokoloff L., Reivich M., Kennedy C., Des Rosiers M.H., Patlak C.S., Pettigrew K.D., Sakurada O., Shinohara M. (1977) The (14C)-deoxyglucose method for the measurement of local cerebral glucose utilization: Theory, procedure and normal values in the conscious and anesthetized albino rat. J Neurochem 28:897-916.

Sundt T.J., Waltz A. (1966) Experimental cerebral infarction: retro-orbital, extradural approach for occluding the middle cerebral artery. Mayo Clin Proc 41:159-168.

Tamura A., Graham D.I., McCulloch J., Teasdale G.M. (1981) Focal cerebral ischaemia in the rat: 1. Description of technique and early neuropathological consequences following middle cerebral artery occlusion. J Cereb Blood Flow Metab 1:53-60.

Welsh F.A., Rieder W. (1978) Evaluation of in situ freezing of cat brain by NADH fluorescence. J Neurochem 31:299-309.

Xie Y., Mies G., Hossmann K.-A. (1989) Ischemic threshold of brain protein synthesis after unilateral carotid occlusion in gerbils. Stroke 20: 620-626.

CONTRIBUTORS

Dr. S.R. Arridge
Department of Computer Science, University College London, 11-20 Capper Street, London WC1E 6JA, England, UK

Dr. T. Bonhoeffer
MPI für Hirnforschung, Deutschordenstr.46, 6000 Frankfurt 71, Germany, Phone: (069) 6704-240, Fax/email: (069) 6704-433 / bonhoeffer@mpih-frankfurt.mpg.dbp.de

Dr. S.G. Borg
Neuroscience Research Group, University of Calgary, Calgary AB, Canada T2N 4N1, Phone: (403) 220 4594, Fax: (403) 283 4740

Dr. B. Chance
Institute for Biochemistry/Biophysics, University of Pennsylvania, Philadelphia, Pa 19104, USA, Phone: (215) 898 4342, Fax: (215) 898 1806

Dr. M. Cope
Department of Medical Physics and Bioengineering, University College London, 1st Floor Shropshire House, 11-20 Capper Street, London WC1E 6JA, England, UK, Phone: (071) 380 9700, Fax: (071) 380 9577

Dr. R.A. De Blasi
Istituto di Anestesiologia e Rianimazione, Universita di Roma, Roma, Italia

Dr. D.T. Delpy
Department of Medical Physics and Bioengineering, University College London, 1st Floor Shropshire House, 11-20 Capper Street, London WC1E 6JA, England, UK, Phone: (071) 380 9700, Fax: (071) 380 9577

Dr. U. Dirnagl
Dept. of Neurology, Neurovascular Research Laboratory, Marchioninistr.15, 8000 München 70, Germany, Phone: 089 70953145, Fax/email: 089 7004418 / ud@neuro.gh.uni-muenchen.dbp.de and Dept. of Neurology, Charité, Humboldt Universität, Schumannstr. 20/21, O-1040 Berlin, Germany, Phone: (030) 286 8318, Fax: (030) 286 2035

Dr. H.U. Dodt
MPI für Psychiatrie, Kraepelinstr.2, 8000 München 40, Germany, Phone: (089) 30622 344, Fax: (089) 30622 200

Dr. R.B. Dyer
Los Alamos National Laboratory, Los Alamos, NM 87545, USA, Phone: (505)-2550

Prof.Dr. H.R. Eggert
Klinik für Neurochirurgie,
Städtische Klinik, Mönchebergstr.41-43, 3500 Kassel, Germany, Phone: (0561) 980 2635

Prof.Dr. K.M. Einhäupl
Dept. of Neurology, Charité, Humboldt Universität, Schumannstr. 20/21, O-1040 Berlin, Germany, Phone: (030) 286 8318, Fax: (030) 286 2035

Dr. A. Eke
Experimental Research Department, Semmelweis Medical University, Ulloi ul 70/a, 1082 Budapest, Ungarn, Phone: (361) 134 3162, Fax: (316) 134 3162

Dr. C.E. Elwell
Department of Medical Physics and Bioengineering, University College London, 1st Floor Shropshire House, 11-20 Capper Street, London WC1E 6JA, England, UK, Phone: (071) 380 9700, Fax: (071) 380 9577

Dr. M. Essenpreis
Department of Medical Physics and Bioengineering, University College London, 1st Floor Shropshire House, 11-20 Capper Street, London WC1E 6JA, England, UK, Phone: (071) 380 9700, Fax: (071) 380 9577

Dr. P. Federico
Neuroscience Research Group, University of Calgary, Calgary AB, Canada T2N 4N1, Phone: (403) 220 4594, Fax: (403) 283 4740

Prof.Dr. M. Ferrari
Dipartimento die Scienze e Tecnologie, Biomedice e Biometria, Universita dell Aquila, Italia, Phone: (862) 433496, Fax: (862) 433433

Dr.F.Fink
Laser Labor Adlershof e.V., Rudower Chaussee 6, O-1199 Berlin, Germany

Dr. E.R. Flynn
Los Alamos National Laboratory, Los Alamos, NM 87545, USA, Phone: (505)-2550

Dr. J. Frahm
MPI f. biophys.Chemie, Postfach 2841, 3400 Göttingen, Germany, Phone: (0551) 201721, Fax: (0551) 201307

Dr. J.S. George
Los Alamos National Laboratory, Los Alamos, NM 87545, USA, Phone: (505)-2550

Dr. A.S. Goggin
Los Alamos National Laboratory, Los Alamos, NM 87545, USA, Phone: (505)-2550

Dr. A.M. Gorbach
National Institute of Health, Bldg.10, Room 5C101, Bethesda, Maryland 20892, USA, Phone: (301) 402 0100, Fax: (301) 496 1675

Dr. A. Grinvald
The Weizmann Institute of Science, Dept.of Neurobiology, 76100 Rehovot, Israel, Phone: 00972-8-342126, Fax/email: 00972-8-344129/ bitnet bngrinva@weizmann

Dr. R.L. Haberl
Dept. of Neurology, Neurovascular Research Laboratory, Marchioninistr.15, 8000 München 70, Germany, Phone: (089) 70953145, Fax: (089) 7004418

Dr. R.M. Harper
Department of Anatomy and Cell Biology and the Brain Research Institute
University of California at Los Angeles, Los Angeles, CA 90024-1763, USA, Phone: (310) 825 5303, Fax/email: (310) 206 5855 / IZZYT23@MVS.OAC.UCLA.EDU

Dr. Y. Itoh
Central Research Laboratory, Hitachi LTD, Kokubunjii, Japan

Dr. F. Kawaguchi
Central Research Laboratory, Hitachi LTD, Kokubunjii, Japan

Dr. G.M. Knudsen
Dept. of Neurology, Rigshospitalet, 9 Blegdamsvej, DK-2100 Copenhagen, Denmark, Phone: 35452380, Fax: 35360213

Dr. U. Ködel
Dept. of Neurology, Neurovascular Research Laboratory, Marchioninistr.15, 8000 München 70, Germany, Phone: (089) 70953145, Fax: (089) 7004418

Dr. F.E. LeBlanc
Neuroscience Research Group, University of Calgary , Calgary AB, Canada T2N 4N1, Phone: (403) 220 4594, Fax: (403) 283 4740

Dr. J.D. Lewine
Los Alamos National Laboratory, Los Alamos, NM 87545, USA, Phone: (505)-2550

Dr. G. Lewis
Somanetics Corp., 1653 East Marple Road, Troy, MI 48083-4208, USA, Phone: (313) 689-4294, Fax: (313) 689 4272

Dr. E. Lieke
Inst.f.Tierphysiologie, Neurobiologie, Königin Luise Str. 28/30, 1000 Berlin 33, Germany, Phone: (030) 8386537, Fax/email: (030) 8385455 / bitnet 9991(p)db0tuz01

Dr. B.A. MacVicar
Neuroscience Research Group, University of Calgary , Calgary AB, Canada T2N 4N1, Phone: (403) 220 4594, Fax: (403) 283 4740

Dr. A. Maki
Central Research Laboratory, Hitachi LTD, Kokubunjii, Japan

Dr. P. McCormick
2213 Cherry St., Toledo, OH 43608, USA, Phone: (419) 321 4492, Fax: (419) 321 3841

Dr. G. Mies
MPI für Neurologische Forschung, Gleueler Str 50, 5000 Köln 41, Germany, Phone: (0221) 4726 211, Fax: (0021) 4726 298

Dr. S. Mironov
Max Planck Institut für Psychiatrie, Am Klopferspitz 8a, 8030 Planegg-Martinsried, Germany, Phone: (089) 8578 3695, Fax: (089) 8578 2481

Dr. C. Nicholson
Department of Physiology and Biophysics, New York University Medical Center, 550 Fist Avenue, New York, NY 10016, USA, Phone: (212) 263 5421, Fax: (212) 689 9060

Dr. R. Paul
Laser Labor Adlershof e.V., Rudower Chaussee 6, O-1199 Berlin , Germany, Phone: (030) 4481672

Dr. A. Pfarrherr
Institut für Optik und Spektroskopie, Bornholmerstr. 6, O 1071 Berlin Ost , Germany, Phone: (030) 4481672

PD Dr. H.W. Pfister
Dept. of Neurology, Neurovascular Research Laboratory, Marchioninistr.15, 8000 München 70, Germany, Phone: (089) 70953145, Fax: (089) 7004418

Dr. G.R. Poe
Department of Anatomy and Cell Biology and the Brain Research Institute, University of California at Los Angeles, Los Angeles, CA 90024-1763, USA, Phone: (310) 825 5303, Fax/email: (310) 206 5855 / IZZYT23@MVS.OAC.UCLA.EDU

Dr. D.M. Rector
Department of Anatomy and Cell Biology and the Brain Research Institute, University of California at Los Angeles, Los Angeles, CA 90024-1763, USA, Phone: (310) 825 5303, Fax/email (310) 206 5855 / IZZYT23@MVS.OAC.UCLA.EDU

Dr. F. Safoue
Istituto di Anestesiologia e Rianimazione, Universita di Roma, Roma, Italia

Dipl.Phys. L. Schleinkofer
Hamamatsu Photonics Germany, Arzbergerstr. 10, 8036 Herrsching, Germany, Phone: (08152) 37555, Fax: (08152) 2658

Dr. N. Shinohara
Dept.of Neurology, Tokai University, Bohseidai, Isehara 259-11, Japan, Phone: (0463) 931121, Fax: (0463) 948764

Prof.Dr. Y. Shinohara
Dept.of Neurology, Tokai University, Bohseidai, Isehara 259-11, Japan, Phone: (0463) 931121, Fax: (0463) 948764

Dr. S. Takagi
Dept.of Neurology, Tokai University, Bohseidai, Isehara 259-11, Japan, Phone: (0463) 931121, Fax: (0463) 948764

Dr. L. Tao
Department of Physiology and Biophysics, New York University Medical Center, 550 Fist Avenue, New York, NY 10016, USA, Phone: (212) 263 5421, Fax: (212) 689 9060

Dipl.Biol. A. Them
Dept. of Neurology, Neurovascular Research Laboratory, Marchioninistr.15, 8000 München 70, Germany, Phone: (089) 70953145, Fax: (089) 7004418

Dr. P. van der Zee
Department of Medical Physics and Bioengineering, University College London, 1st Floor Shropshire House, 11-20 Capper Street, London WC1E 6JA, England, UK, Phone: (071) 380 9700, Fax: (071) 380 9577

Dr. A. Villringer
Dept. of Neurology, Neurovascular Research Laboratory, Marchioninistr.15, 8000 München 70, Germany, Phone: (089) 70953145, Fax: (089) 7004418

Dr. T.W.J. Watson
Neuroscience Research Group, University of Calgary , Calgary AB, Canada T2N 4N1, Phone: (403) 220 4594, Fax: (403) 283 4740

Dr. Q. Wei
Dipartimento di Fisica, Universita di Firenze, Firenze, Italia

Dr. G. Zaccanti, Dipartimento di Fisica, Universita di Firenze, Italia

Dr. D.F. Wilson
Dept.of Biochemistry and Biophysics, University of Pennsylvania, Philadelphia, Pennsylvania 19104, USA, Phone: (215) 898 6382 Fax: (215) 8984217

Dr. Y. Yamashita
Central Research Laboratory, Hitachi LTD, Kokubunjii, Japan

INDEX